The Geometry of Supermanifolds

Mathematics and Its Applications

Volume 71

The Geometry of Supermanifolds

by

Claudio Bartocci
Department of Mathematics,
University of Genoa,
Genoa, Italy

Ugo Bruzzo
Department of Mathematics,
University of Genoa,
Genoa, Italy

and

Daniel Hernández-Ruipérez
Department of Pure and Applied Mathematics,
University of Salamanca,
Salamanca, Spain

KLUWER ACADEMIC PUBLISHERS
DORDRECHT / BOSTON / LONDON

Library of Congress Cataloging-in-Publication Data

Bartocci, C. (Claudio), 1962-
The geometry of supermanifolds / by Claudio Bartocci, Ugo Bruzzo, and Daniel Hernández-Ruipérez.
p. cm. -- (Mathematics and its applications ; v. 71)
Includes bibliographical references and index.
ISBN 0-7923-1440-9 (acid-free paper)
1. Supermanifolds (Mathematics) I. Bruzzo, U. (Ugo) II. Hernández-Ruipérez, Daniel, 1954- III. Title. IV. Series: Mathematics and its applications (Kluwer Academic Publishers) ; 71.
QA614.44.B37 1991
514'.3--dc20 91-29109

ISBN 0-7923-1440-9

Published by Kluwer Academic Publishers,
P.O. Box 17, 3300 AA Dordrecht, The Netherlands.

Kluwer Academic Publishers incorporates
the publishing programmes of
D. Reidel, Martinus Nijhoff, Dr W. Junk and MTP Press.

Sold and distributed in the U.S.A. and Canada
by Kluwer Academic Publishers,
101 Philip Drive, Norwell, MA 02061, U.S.A.

In all other countries, sold and distributed
by Kluwer Academic Publishers Group,
P.O. Box 322, 3300 AH Dordrecht, The Netherlands.

Printed on acid-free paper

Printed in the Netherlands

To Professor J. B. Sancho Guimerá
on the occasion of his 65th birthday

SERIES EDITOR'S PREFACE

'Et moi, ..., si j'avait su comment en revenir, je n'y serais point allé.'

Jules Verne

The series is divergent; therefore we may be able to do something with it.

O. Heaviside

One service mathematics has rendered the human race. It has put common sense back where it belongs, on the topmost shelf next to the dusty canister labelled 'discarded non-sense'.

Eric T. Bell

Mathematics is a tool for thought. A highly necessary tool in a world where both feedback and non-linearities abound. Similarly, all kinds of parts of mathematics serve as tools for other parts and for other sciences.

Applying a simple rewriting rule to the quote on the right above one finds such statements as: 'One service topology has rendered mathematical physics ...'; 'One service logic has rendered computer science ...'; 'One service category theory has rendered mathematics ...'. All arguably true. And all statements obtainable this way form part of the raison d'être of this series.

This series, *Mathematics and Its Applications*, started in 1977. Now that over one hundred volumes have appeared it seems opportune to reexamine its scope. At the time I wrote

> "Growing specialization and diversification have brought a host of monographs and textbooks on increasingly specialized topics. However, the 'tree' of knowledge of mathematics and related fields does not grow only by putting forth new branches. It also happens, quite often in fact, that branches which were thought to be completely disparate are suddenly seen to be related. Further, the kind and level of sophistication of mathematics applied in various sciences has changed drastically in recent years: measure theory is used (non-trivially) in regional and theoretical economics; algebraic geometry interacts with physics; the Minkowsky lemma, coding theory and the structure of water meet one another in packing and covering theory; quantum fields, crystal defects and mathematical programming profit from homotopy theory; Lie algebras are relevant to filtering; and prediction and electrical engineering can use Stein spaces. And in addition to this there are such new emerging subdisciplines as 'experimental mathematics', 'CFD', 'completely integrable systems', 'chaos, synergetics and large-scale order', which are almost impossible to fit into the existing classification schemes. They draw upon widely different sections of mathematics."

By and large, all this still applies today. It is still true that at first sight mathematics seems rather fragmented and that to find, see, and exploit the deeper underlying interrelations more effort is needed and so are books that can help mathematicians and scientists do so. Accordingly MIA will continue to try to make such books available.

If anything, the description I gave in 1977 is now an understatement. To the examples of interaction areas one should add string theory where Riemann surfaces, algebraic geometry, modular functions, knots, quantum field theory, Kac-Moody algebras, monstrous moonshine (and more) all come together. And to the examples of things which can be usefully applied let me add the topic 'finite geometry'; a combination of words which sounds like it might not even exist, let alone be applicable. And yet it is being applied: to statistics via designs, to radar/sonar detection arrays (via finite projective planes), and to bus connections of VLSI chips (via difference sets). There seems to be no part of (so-called pure) mathematics that is not in immediate danger of being applied. And, accordingly, the applied mathematician needs to be aware of much more. Besides analysis and numerics, the traditional workhorses, he may need all kinds of combinatorics, algebra, probability, and so on.

In addition, the applied scientist needs to cope increasingly with the nonlinear world and the extra mathematical sophistication that this requires. For that is where the rewards are. Linear models are honest and a bit sad and depressing: proportional efforts and results. It is in the non-linear world that infinitesimal inputs may result in macroscopic outputs (or vice versa). To appreciate what I am hinting at: if electronics were linear we would have no fun with transistors and computers; we would have no TV; in fact you would not be reading these lines.

There is also no safety in ignoring such outlandish things as nonstandard analysis, superspace and anticommuting integration, p-adic and ultrametric space. All three have applications in both electrical engineering and physics. Once, complex numbers were equally outlandish, but they frequently proved the shortest path between 'real' results. Similarly, the first two topics named have already provided a number of 'wormhole' paths. There is no telling where all this is leading - fortunately.

Thus the original scope of the series, which for various (sound) reasons now comprises five subseries: white (Japan), yellow (China), red (USSR), blue (Eastern Europe), and green (everything else), still applies. It has been enlarged a bit to include books treating of the tools from one subdiscipline which are used in others. Thus the series still aims at books dealing with:

- a central concept which plays an important role in several different mathematical and/or scientific specialization areas;
- new applications of the results and ideas from one area of scientific endeavour into another;
- influences which the results, problems and concepts of one field of enquiry have, and have had, on the development of another.

Supermathematics involves both commuting variables, the stuff of 'ordinary, non-super' mathematics, and anticommuting variables. That is, in more physical terms, both bosonic and fermionic ones. Supersymmetry places bosonic and fermionic particles on the same footing and it seems at the moment impossible to formulate a good theoretical model that does not incorporate supersymmetry in an essential way. This means redeveloping all of analysis, geometry, and algebra in the super setting, to provide the mathematical framework for superphysics.

There is already a very substantial journal and preprint literature on supermathematics but little in systematic monograph form (with the notable exceptions of Bryce de Witt's book (1984), which is, however, very much from the physics point of view; the fundamental introduction (1987) by the pioneer F.A. Berezin based on his partially edited notes and published 7 years after his death; and the summer institute proceedings edited by Seifert, Clarke and Rosenblum (1984)). That means that there is ample room for, for instance, a good, systematic, self-contained volume on supermanifolds, which will, at the least, remove some of the confusion and controversy regarding the right definitions and points of view.

Here we have such a book by two authors who have contributed substantially to the development of the subject.

The shortest path between two truths in the real domain passes through the complex domain.

J. Hadamard

La physique ne nous donne pas seulement l'occasion de résoudre des problèmes ... elle nous fait pressentir la solution.

H. Poincaré

Never lend books, for no one ever returns them; the only books I have in my library are books that other folk have lent me.

Anatole France

The function of an expert is not to be more right than other people, but to be wrong for more sophisticated reasons.

David Butler

Amsterdam, August 1991

Michiel Hazewinkel

Table of Contents

Preface

This book is the outgrowth of a collaboration between the authors that dates back to 1986. Among the results of the authors that have been included, some are so far unpublished, whilst others have already appeared in various papers, even though they have undergone a complete reorganization and, in some cases, radical modifications. The authors' viewpoint about supermanifolds, and certain specific results presented here, have been influenced by fruitful discussions with several people, among whom we would expressly like to thank M. Batchelor, P. Bryant, R. Catenacci, G. Falqui, G. Landi, D. Leĭtes, A. López Almorox, G. Marmo, V. Pestov, J. Rabin, C. Reina, A. Rogers, and M. Rothstein. Special thanks are due to J. Muñoz Masqué; he did not participate in this job, but much of his work and insight is nevertheless present in it.

The research work this monograph is based upon was made possible by funding provided by the research project 'Metodi geometrici in relatività e teorie di campo' of the Italian Ministry for the Universities and Scientific and Technological Research (MURST); by 'Consiglio Nazionale delle Ricerche', also through its National Group for Mathematical Physics (GNFM); by 'Consejo Superior de Investigación Científica', and by the Spanish CICYT through the research project 'Geometría de las teorías gauge.'

This book has been typeset using the AMS-TeX macro package.

Genova and Salamanca
June 1991

Introduction

I must be cruel only to be kind.
This bad begins, and worse remains behind.
W. SHAKESPEARE

Supergeometry is usually employed in theoretical physics in a rather heuristic way, and, accordingly, most expositions of that subject are heavily oriented towards physical applications. By way of contrast, in this book we wish to unfold a consistent and systematic, if not exhaustive, investigation of the structure of geometric objects — called *supermanifolds* — which generalize differentiable manifolds by incorporating, in a sense, 'anticommuting variables.' Thus, we shall pay no attention to physical questions but will rather develop the theory from its very foundations, with special regard to global geometric aspects.

Let us, before delineating in greater detail the scope of our subject, start with a cursory historical survey.

Supersymmetry. The introduction of anticommuting variables dates back to the book by Berezin on second quantization [**Be**], where they were used to 'integrate over the fermions' by means of a formal device now called the Berezin integral. The paper by Berezin and G.I. Kats of 1970 [**BK**] is also noteworthy, where they introduced formal Lie groups with anticommuting parameters, studying their relationship with graded Lie algebras. However, a concrete and widespread interest in supergeometry began only with the appearance of supersymmetry in theoretical physics.

Before the discovery of supersymmetry, bose and fermi particles had to be treated on an unequal footing. Vector bosons could be considered as gauge particles, which mathematically means that the classical (non-quantum) field representing the particle is a connection on a principal bundle over space-time.

The group of vertical automorphisms of the principal bundle yields local (i.e. with parameters depending on the space-time position) symmetries of the field theory, which provide a clue to the renormalization of the quantum theory [**BcRS1,BcRS2**]. No such geometrical description was available for fermi particles, until Wess and Zumino [**WZ**] devised a field theory invariant under a symmetry which mixes bosons and fermions (actually, a year before Volkov and Akulov had already studied a field theory bearing a non-linear realization of the supersymmetry algebra [**VoA**]). That symmetry can be made local and this, oversimplifying the whole story, leads to supergravity, which can be regarded in a sense as a gauge theory with both bose and fermi gauge particles.

For the sake of simplicity, let us stick to the original Wess-Zumino model. One considers the four-dimensional Minkowski space-time, with pseudo-cartesian coordinates $\{x^i\}$, and over it two complex scalar fields A, F, together with a Dirac spinor field, ψ^α, $\alpha = 1\dots 4$. The Lagrangian of the model is (letting $\partial_i = \dfrac{\partial}{\partial x^i}$)

$$L = -\tfrac{1}{2}\partial_i A\, \partial^i A^* - \tfrac{1}{2} i\bar{\psi}\gamma^i\partial_i\psi + \tfrac{1}{2}FF^*,$$

where * denotes complex conjugation; L is invariant (up to first order in ε) under the transformations

$$\begin{aligned} A &\mapsto A + i\bar{\varepsilon}\psi \\ \psi &\mapsto \psi + \partial_i A\,\gamma^i\varepsilon + F\varepsilon \\ F &\mapsto F + i\bar{\varepsilon}\gamma^i\partial_i\psi \end{aligned} \tag{1}$$

provided that the parameters ε^α and the spinor components ψ^α anticommute among themselves:

$$\varepsilon^\alpha\varepsilon^\beta = -\varepsilon^\beta\varepsilon^\alpha, \qquad \varepsilon^\alpha\psi^\beta = -\psi^\beta\varepsilon^\alpha, \qquad \psi^\alpha\psi^\beta = -\psi^\beta\psi^\alpha.$$

The transformations described by (1), together with the usual space-time translations, constitute a $\mathbb{Z}_2$-graded Lie algebra called the *supersymmetry algebra*, whilst the fields A, ψ, F form a *supermultiplet* in that they carry a linear representation of that algebra.

This simple example shows that any classical (i.e. non-quantum) mathematical theory of a supersymmetric system of fields must involve some generalization of differential geometry where anticommuting objects can find a natural framework. A first step in this direction, albeit in a purely formal way, was taken by Salam and Strathdee [**SaS**], who introduced the concept of superspace, heuristically described as a space with a Euclidean topology, and parametrized by four real coordinates $\{x^i\}$ and four other coordinates $\{y^\alpha\}$ satisfying $x^i y^\alpha = y^\alpha x^i$, $y^\beta y^\alpha = -y^\alpha y^\beta$. A scalar field $\Phi(x,y)$ on superspace (superfield) can be developed in powers of the y's

$$\begin{aligned}\Phi(x,y) = \hat{\Phi}_0(x) + \sum_{1\leq\alpha\leq 4} y^\alpha\,\hat{\Phi}_\alpha(x) + \sum_{1\leq\alpha<\beta\leq 4} y^\alpha y^\beta\,\hat{\Phi}_{\alpha\beta}(x) \\ + \sum_{1\leq\alpha<\beta<\gamma\leq 4} y^\alpha y^\beta y^\gamma\,\hat{\Phi}_{\alpha\beta\gamma}(x) + y^1 y^2 y^3 y^4\,\hat{\Phi}_{1234}(x)\,; \quad (2)\end{aligned}$$

the coefficients of this expansion can be expressed in terms of the fields of the Wess-Zumino supermultiplet [**WsB**]. In this way one can reformulate the whole theory in terms of superfields, achieving considerable simplification (e.g. one can introduce Feynman supergraphs, one of which corresponds to several ordinary Feynman graphs).

A detailed report on the first developments of supersymmetry, together with a huge bibliography, can be found in [**FF**]. More recent accounts are [**Fre**] and [**GGRS**].

Supergeometry. The first attempt to provide a mathematically satisfactory framework for supergeometry was the Berezin-Leĭtes-Kostant theory [**BL,Kos**]. Briefly, one considers a smooth — say m-dimensional — manifold X, and enlarges its structure sheaf $\mathcal{C}_X$ (the sheaf of germs of smooth real functions on X) to a sheaf $\mathcal{A}$ of $\mathbb{Z}_2$-graded commutative $\mathbb{R}$-algebras, which is locally isomorphic with the sheaf of functions over X with values in the exterior algebra of $\mathbb{R}^n$. The pair $(X,\mathcal{A})$ is said to be an (m,n) dimensional graded manifold (supermanifold in the Russian and some Western literature). In one sense, one has left the set of points unchanged, whilst the structure sheaf has been extended (so to say, in physical language, one has enlarged the space of observables). Any differential geometric construction related to a manifold X can be formulated more or less straightforwardly in terms of the structure sheaf $\mathcal{C}_X$; then, replacing $\mathcal{C}_X$ by the sheaf $\mathcal{A}$, one can generalize ordinary differential geometry to the setting of graded manifolds. Thus, the tools used in graded manifold theory come mainly from algebraic geometry; indeed, it is quite natural to regard the pair $(X,\mathcal{A})$ as a ringed space; the fact that the geometry of the graded manifold can be constructed in terms of the sheaf $\mathcal{A}$ is then a standard feature of algebraic geometry. In terms of local coordinates $(x^1,\ldots,x^m,y^1,\ldots,y^n)$ on $(X,\mathcal{A})$, a section f of $\mathcal{A}$ can be developed in powers of the y's as the superfield Φ in

Eq. (2) (cf. Section III.1). The coefficients of this expansion, which are real functions on X, should represent the physical fields of bose (fermi) statistics if they multiply an even (odd) number of y's. However, in this way spinor fields are real-valued, and this is incompatible with supersymmetry (for a discussion of this issue, see [**DeS**]).

A different approach, which was initiated by DeWitt and Rogers [**DW,Rs1**], is more similar to differential geometry. The basic idea is to enlarge the space over which the manifold is modeled by replacing the real field by a larger set, containing both commuting and anti-commuting quantities. More precisely, one considers an exterior algebra B, which is naturally $\mathbb{Z}_2$-graded commutative, in the sense that

$$B = B_0 \oplus B_1, \qquad B_\alpha \cdot B_\beta \subset B_{\alpha+\beta} \quad \forall \alpha, \beta \in \mathbb{Z}_2,$$

$$ab = (-1)^{\alpha\beta}\, ba \qquad \text{if} \quad a \in B_\alpha,\ b \in B_\beta.$$

After introducing the space $B^{m,n} = B_0^m \times B_1^n$, one would define an (m,n) dimensional supermanifold as a topological space, together with an atlas of $B^{m,n}$-valued coordinate patches, whose transition functions fulfill a suitable 'smoothness' condition.

This 'smoothness' condition appears to be the crucial point of the issue. The choice by Rogers (the so-called G^∞ function) yields a structure sheaf whose sheaf of derivations (i.e. its tangent sheaf) is not locally free. This means that the local geometry of the supermanifold cannot be described by using local coordinates, which is decidedly undesirable in physical applications, not to say in mathematical developments, as well.

Also the so-called GH^∞ supermanifolds, introduced by Rogers [**Rs2**] to avoid the drawbacks of the G^∞ supermanifolds, have some unsatisfactory features. Even though the sheaf of derivations of a GH^∞ supermanifold is locally free, it turns out that in that category it is not possible to devise any sensible notion of 'graded tangent space', in such a way that the tangent spaces at the various points of a given manifold are all isomorphic, and are free modules over the algebra B.

This sketchy discussion shows that in order to be suitable for physical applications, and to represent a reasonable generalization of the category of differentiable manifolds, any notion of 'supermanifold' is subject to a certain number of mathematical requirements. One could think of turning the terms of the question upside-down, considering as axioms all the properties that a supermanifold should verify in order to give rise to a reasonable geometric theory. Along these guidelines, M. Rothstein [**Rt2**] has proposed a set of four axioms which for any choice of a graded-commutative algebra B determine a broad category of 'well-behaved' supermanifolds (cf. Section IV.7). Graded manifolds fit into this axiomatics when the special choice $B = \mathbf{R}$ is made; by contrast, both G^∞ and GH^∞ supermanifolds, with B a finite-dimensional exterior algebra, violate Rothstein's axioms, and this is the ultimate reason of their inadequacy. However, one should notice that for a particular class of infinite-dimensional algebras B Rogers' approach is consistent [**JP**].

G-supermanifolds. Despite all its problems, Rogers' approach nevertheless shows several desirable features, mainly due to the fact that the odd coordinates are incorporated in the geometric substratum; this should make the theory more suitable to non-trivial physical applications. Thus, even though deep modifications are required, it seems reasonable to preserve the basic philosophy of this approach.

On these grounds, in this monograph we wish to give a systematic exposition of the theory of *G-supermanifolds*. These objects, originally introduced in the papers [**BB1**] and [**BBH**], represent in a precise sense the geometric structures that are closest to Rogers' supermanifolds and satisfy Rothstein's axiomatics. The resulting theory is a generalization of standard differential geometry, but is basically more involved; roughly speaking, a G-supermanifold is a pair $(X, \mathcal{A})$, where X is a topological space and $\mathcal{A}$ is a sheaf of $\mathbf{Z}_2$-graded algebras, which in general is not a sheaf of functions. One requires that the ringed space $(X, \mathcal{A})$ be locally isomorphic with a suitably defined 'standard G-supermanifold.' This definition is similar to both the definition of Rogers' supermanifolds and that of graded manifolds, inasmuch as the study of G-supermanifolds necessitates tools from algebraic geometry (ringed spaces) as well as from classical differential geometry (atlases). Moreover, the structure sheaf of a G-supermanifold in general has non-trivial cohomology, contrary to the case of sheaves of smooth functions; in this sense, G-supermanifolds display an analogy with complex manifolds.

We shall pay special attention to the global properties of supermanifolds. This is motivated not only by their mathematical interest, but also by physical applications. Indeed, in recent years global differential geometry has played a central role in theoretical physics, and especially in field theory; in this connection let us mention the anomaly problem of gauge theories and Witten's topological field theory. Cohomological machinery has also been employed in connection with the anomalies of supersymmetric field theories, in the superspace formalism [**BoPT2,Buc**] or exploiting supermanifold techniques [**BruL,BBL**]. Strings provide another important example, in that local considerations are inadequate to furnish a complete description of the theory [**Ma1,Sin**].

Description of the contents. We have endeavoured to make the exposition self-contained and readable with a modest mathematical background, which should reduce to some basic algebra and differential geometry. In this sense, this work is also addressed to an audience of physicists , although — as the theory involves many mathematical subtleties — its style is quite different from that of the usual physical literature.

The first two Chapters contain foundational material which will be heavily used in the rest of the book. In particular, in Chapter I we describe some elements of the theory of $\mathbb{Z}_2$-graded rings, modules and algebras. Chapter II provides a self-contained introduction to sheaf theory, sheaf cohomology, and other cohomology theories (de Rham, Čech, Dolbeault), whose scope goes far beyond their use in supermanifold theory. It also embodies an exposition of ringed spaces in the graded setting and a characterization of differentiable manifolds as ringed spaces in terms of the spectra of their rings of smooth functions; this is motivated by the fact that a similar characterization holds in the case of graded manifolds (more precisely, the differentiable manifold underlying a graded manifold is the spectrum of the ring of global graded functions).

In Chapter III we provide a detailed account of different categories of supermanifolds and their interrelations; we start with graded manifolds, that are here included not only in view of their intrinsic significance, but also because their properties will be used in connection with other categories of supermanifolds. The classes of G^∞, GH^∞ and H^∞ functions are introduced, which allows us to define supermanifolds in the sense of Rogers; the discussion of their shortcomings leads us to introduce the notion of G-supermanifold.

The basic geometry of G-supermanifolds is developed in Chapter IV, where the notions of morphism, product, and bundle are defined in an essentially different way than in usual differential geometry, involving explicitly the graded Fréchet algebra structure of the rings of sections of the structure sheaves. We are then ready to unfold the exterior graded calculus on the graded cotangent bundle. Successively, we examine the subcategory of DeWitt supermanifolds, that exhibit rather simple geometric features and are interesting in view of their far-reaching physical applications. It turns out that DeWitt G-supermanifolds are in one-to-one correspondence with graded manifolds. Finally, in the last paragraph, we carry out a detailed analysis of Rothstein's axiomatics, showing that it is convenient to integrate the original four axioms with a further assumption on the completeness of the topology of the rings of sections. Whenever the ground algebra is a finite dimensional exterior algebra, this enlarged set of axioms characterizes G-supermanifolds uniquely.

Since the structure sheaf of a G-supermanifold in general is not acyclic, i.e. its Čech (or sheaf) cohomology may not be trivial, the cohomology defined via the complex of graded differential forms — called super de Rham cohomology — may be different from the de Rham cohomology of the associated smooth manifold. This situation is investigated in Chapter V, together with the graded Dolbeault cohomology of complex G-supermanifolds. Moreover, we prove that the structure sheaf of any DeWitt G-supermanifold is acyclic, so that its super and ordinary de Rham cohomologies do coincide. From this, we infer some further results on the geometric structure of DeWitt supermanifolds.

Chapter VI is devoted to the theory of vector bundles in the category of G-supermanifolds; namely, supervector bundles. After describing, for any given supervector bundle, a cohomological invariant whose vanishing is equivalent to the existence of connections on the bundle, we study superline bundles, defining in particular their obstruction class and Picard group. Subsequently, a theory of characteristic classes for complex supervector bundles is developed along the guidelines of Grothendieck's approach. This Chapter ends with a result about the representation of such characteristic classes in terms of curvature forms.

In Chapter VII the reader will find an outline of Lie theory for G-supermanifolds: as the set of points of a G-supermanifold does not embody all the information about it, the multiplication, identity and inverse morphisms of a G-Lie supergroup must be characterized as graded ringed space morphisms. Finally, we sketch the first rudiments of principal superfibre bundles and their associated bundles.

Chapter I

Elements of graded algebra

῝Ο γα μάν ἀριθμός ἔχει δύο μέν ἴδια εἴδεα,
περισσόν καί ἄρτιον, τρίτον δέ ἀπ' ἀμφοτέρων μιχθέντων,
ἀρτιοπέρισσον. ῾Εκατέρω δέ τῶ εἴδεος πολλαί μορφαί,
ἂν ἑκάστα ὡσαύτως μερίζεται.

(Number has two species, odd and even,
whilst the third is the even-odd, which is a mixture
of both. Many forms there are of both species, and
every thing on its own reveals them.)
PHILOLAOS

This introductory Chapter aims at establishing, together with the basic notation and terminology, some elementary results about $\mathbb{Z}_2$-graded algebra that we shall constantly use in the sequel. The topics covered include $\mathbb{Z}_2$-graded rings and modules, $\mathbb{Z}_2$-graded tensor algebra, Lie superalgebras, and matrices with entries in a $\mathbb{Z}_2$-graded commutative ring.

1. Graded algebraic structures

In general, given an arbitrary group G, one can introduce G-graded algebraic objects [**Bou**,**NVO**]. Since in order to develop a 'supergeometry' only $\mathbb{Z}_2$-graded structures are needed, we shall only consider here that particular case. We shall assume as a rule that

$$\text{graded} \equiv \mathbb{Z}_2\text{-graded}.$$

Definition 1.1. *A ring*[1] $(R,+,\cdot)$ *is said to be graded if* $(R,+)$ *has two subgroups* R_0 *and* R_1 *such that* $R = R_0 \oplus R_1$ *and* $R_\alpha \cdot R_\beta \subset R_{\alpha+\beta}$ for all $\alpha, \beta \in \mathbf{Z}_2$.

An element $a \in R$ is said to be *homogeneous* if either $a \in R_0$ or $a \in R_1$; on the set $h(R)$ of homogeneous elements an application $|\ |$ is defined which to each element assigns its degree,

$$| \ |: h(R) \to \mathbf{Z}_2$$
$$a \mapsto \alpha \Leftrightarrow a \in R_\alpha \,.$$

The elements of degree 0 (1) are called *even (odd)*.

Obviously, any ring R can be trivially graded: $R_0 = R$, $R_1 = \{0\}$. More generally, any algebraic object can be trivially graded in such a way. On the other hand, for each category of graded objects that we shall introduce, one can define a forgetful functor into the category of the corresponding non-graded objects.

EXAMPLE 1.1. Let R be a $\mathbf{Z}$-graded ring, namely, $R = \bigoplus_{p\in\mathbf{Z}} \hat{R}_p$ and $\hat{R}_p \cdot \hat{R}_q \subset \hat{R}_{p+q}$. Then R can be graded by taking R_0 as the sum of the even components and R_1 as the sum of the odd ones. ▲

For any graded ring R, a *graded commutator* $\langle\ ,\ \rangle : R \times R \to R$ is defined by letting

$$\langle a, b\rangle = ab - (-1)^{|a||b|} ba \quad \forall a, b \in h(R) \tag{1.1}$$

and then extending to all of $R \times R$ by additivity. The *centre* of R is defined as the set

$$C(R) \equiv \{a \in R \mid \langle a, b\rangle = 0 \ \ \forall b \in R\}\,,$$

i.e. $C(R)$ is the set of the elements of R which graded-commute with any other element.

A graded ring R is said to be *graded-commutative* if $\langle a, b\rangle = 0$ for all $a, b \in R$, that is, if $C(R) = R$. Therefore, in a graded-commutative ring, odd elements are *strictly nilpotent*, in the sense that $a \cdot a = 0$ for all $a \in R_1$. Any commutative ring is graded-commutative, if endowed with the trivial gradation.

Let R be a graded ring, and M a left (right) R-module.

Definition 1.2. *M is a left (right) graded R-module if it has two subgroups M_0 and M_1 such that $M = M_0 \oplus M_1$ and, for all $\alpha, \beta \in \mathbf{Z}_2$, one has $R_\alpha M_\beta \subset M_{\alpha+\beta}$ $(M_\alpha R_\beta \subset M_{\alpha+\beta})$.*

[1] In accordance with Bourbaki's terminology, any ring has an identity.

If R is a graded-commutative ring, any left graded R-module determines uniquely a right graded R-module, which, as a set, has the same elements as M, and is endowed with the following multiplication rule:

$$xa = (-1)^{|x||a|}ax\,; \tag{1.2}$$

vice versa, any right graded R-module determines uniquely a left graded R-module. Therefore, whenever R is graded-commutative, which we shall henceforth assume, we shall use the term 'graded R-module' without ambiguity.

Having fixed two graded R-modules M and N, we then say that a morphism $f: M \to N$ is R-linear on the right if $f(xa) = f(x)a$ for all $x \in M$ and $a \in R$. Unless otherwise stated, by 'linear' we mean 'linear on the right.' Moreover, we say that f has degree $|f| = \beta \in \mathbb{Z}_2$, if $f(M_\alpha) \subset N_{\alpha+\beta}$ for all $\alpha \in \mathbb{Z}_2$. The set $\mathrm{Hom}_R(M,N)$ of R-linear morphisms $M \to N$ (that will be denoted simply by $\mathrm{Hom}(M,N)$ whenever no ambiguity can arise) has a natural grading, with $f \in \mathrm{Hom}(M,N)_\alpha$ whenever $|f| = \alpha$. If R is graded-commutative, $\mathrm{Hom}(M,N)$ is a graded R-module, with the multiplication rule $(af)(x) = af(x)$.

In particular, we define the *graded R-dual M^** of the graded R-module M as $\mathrm{Hom}(M,R)$.

It is convenient to introduce the category ***R*-GMod**, whose objects are the graded R-modules and whose morphisms are the R-linear morphisms of degree 0. Therefore, by 'morphism of graded R-modules' we shall refer to a morphism in the category ***R*-GMod**.

Given a graded-commutative ring R, a *graded ideal I* of R is a graded submodule of R, i.e., a submodule of R such that the inclusion $I \to R$ is an even morphism. For instance, the subgroup $\mathfrak{N}_R = \{a \in R \mid a^q = 0 \text{ for some } q \in \mathbb{N}\}$ is a graded ideal of R, called the *ideal of nilpotents*.

One of the most basic results in commutative ring theory, namely the Nakayama lemma [**AtM**], can be generalized to the graded setting. Let us define the *radical* of a graded-commutative ring R as the graded ideal $\mathfrak{R}$ obtained by intersecting all maximal graded ideals of R. It is not difficult to show that $1-a$ is invertible whenever $a \in \mathfrak{R}$.

We state the graded Nakayama lemma together with two corollaries that will be required further on.

Proposition 1.1. (Graded Nakayama lemma) *Let R be a graded-commutative ring R, I a graded ideal contained in the radical $\mathfrak{R}$ of R, and M a graded* *finitely generated R-module.*

(1) *If $IM = M$, then $M = 0$.*

(2) *If N is a graded submodule of M, and $M = IM + N$, then $M = N$.*

(3) *If $x^1, \ldots, x^m$ are even elements and $y^1, \ldots, y^n$ are odd elements in M such that the images $(\bar{x}^1, \ldots, \bar{x}^m, \bar{y}^1, \ldots, \bar{y}^n)$ are generators of M/IM over R/I, then $(x^1, \ldots, x^m, y^1, \ldots, y^n)$ are generators of M over R.*

Proof. (1) Let us assume $M \neq 0$ and let $x^1, \ldots, x^m$ be even elements and $y^1, \ldots, y^n$ odd elements in M such that $(x^1, \ldots, x^m, y^1, \ldots, y^n)$ is a minimal set of generators for M. Then $m \neq 0$ or $n \neq 0$. If $m \neq 0$, one has $x^m \in IM$, so that $x^m = \sum_{i=1}^m a_i\, x^i + \sum_{\alpha=1}^n b_\alpha\, y^\alpha$ where the a's are even and the b's are odd elements in I. Then $(1 - a_m) \cdot x^m = \sum_{i=1}^{m-1} a_i\, x^i + \sum_{\alpha=1}^n b_\alpha\, y^\alpha$, and since $a_m \in \mathfrak{R}$, the element $1 - a_m$ is invertible, which means that $(x^1, \ldots, x^{m-1}, y^1, \ldots, y^n)$ are still generators for M, thus contradicting the minimality of $(x^1, \ldots, x^m, y^1, \ldots, y^n)$. The case $n \neq 0$ is similar.

(2) If $M = IM + N$, then $I(M/N) = (IM + N)/N = M/N$, so that by (1) the thesis follows.

(3) Let N be the graded submodule of M generated by the set $\{x^1, \ldots, x^m, y^1, \ldots, y^n\}$. It can then readily shown that $M = IM + N$, so that (2) yields the thesis. ■

Let us now turn our attention to the notion of *free* graded module.

Definition 1.3. *A graded R-module F is said to be free if it has a basis formed by homogeneous elements.*

A basis of F of finite cardinality is of *type* (m, n), if it is formed by m even elements $\{f_i^0 \in F_0 \mid i = 1 \ldots m\}$ and n odd elements $\{f_\alpha^1 \in F_1 \mid \alpha = 1 \ldots n\}$. One then has a canonical isomorphism

$$F \simeq \Big(\bigoplus_{i=1}^{m} R f_i^0\Big) \oplus \Big(\bigoplus_{\alpha=1}^{n} R f_\alpha^1\Big)\,.$$

For each pair of natural numbers m, n such that $m + n = p$, the R-module R^p can be regarded as a free graded R-module endowed with a basis of type (m, n), by letting

$$\begin{aligned} (R^{m+n})_0 &\equiv R^{m,n} = R_0^m \oplus R_1^n\,; \\ (R^{m+n})_1 &\equiv R^{\bar{m},\bar{n}} = R_0^n \oplus R_1^m\,. \end{aligned} \tag{1.3}$$

R^{m+n} equipped with this gradation will be denoted by $R^{m|n}$.

In ordinary module theory it is possible that a finitely generated free R-module F has bases of different cardinalities [**Bly**]. However, provided that a homomorphism $\rho: R \to k$ onto a commutative field k exists, one can prove that all bases of F are equipotent [**Bou**] . This result can be easily extended to the case of finitely generated free graded R-modules. In order to do this, we should notice that one can associate with any graded-commutative ring R a field k_R together with a surjective ring morphism $\sigma : R \to k_R$, which is usually called the *augmentation map*. Indeed, since R_0 is a commutative subring of R, it has at least one maximal ideal $\mathfrak{I}$, and the quotient $k_R = R_0/\mathfrak{I}$ is a field. σ is defined as the composition $R \to R_0 \to R_0/\mathfrak{I}$. If M is a graded R-module, we can associate with it a vector space V_M over k_R defined by considering k_R as an R-module by means of the augmentation map (i.e. $a \cdot x = \sigma(a)x \; \forall a \in R, \; x \in k_R$) and letting $V_M = M \otimes_R k_R$. A surjective map $\sigma : M \to V_M$ is defined as $\sigma(x) = x \otimes 1$ for all $x \in M$. V_M is a graded vector space, and it is trivial to verify that, if M has a basis of type (m, n), then $\dim(V_M)_0 = m$, $\dim(V_M)_1 = n$. This proves the following claim.

Proposition 1.2. *Let R be a graded-commutative ring. If F is a finitely generated, free graded R-module, then all bases of F are of the same type.* ∎

Under the hypotheses of Proposition 1.2, we can define the *rank* of a free graded, finitely generated R-module as the pair of natural numbers (m, n) which identifies the type of anyone of its bases.

The following example introduces a kind of graded commutative ring we shall be deeply concerned with in this book.

EXAMPLE 1.2. (Cf. [**Bou**]) Let R be a commutative ring, and M an R-module. The exterior algebra of M over R, denoted $\bigwedge_R M$, is a $\mathbb{Z}$-graded algebra, namely $\bigoplus_{p \in \mathbb{Z}} \bigwedge_R^p M$, and is alternating, i.e. $x^2 = 0$ for all $x \in \bigwedge_R^{2p+1} M$. If M is free and finitely generated, with a basis $\{e_i \,|\, i = 1 \ldots N\}$, then $\bigwedge_R M$ is a free finitely generated R-module, with a canonical basis (relative to the basis $\{e_i\}$) which can be described as follows. Let Ξ_N denote the set

$$\{\mu : \{1 \ldots r\} \to \{1 \ldots N\} \text{ strictly increasing } \big| \, 1 \leq r \leq N\} \cup \{\mu_0\},$$

where μ_0 is the empty sequence, and let

$$\beta_\mu = e_{\mu(1)} \wedge \cdots \wedge e_{\mu(r)} \quad \text{for} \quad \mu \neq \mu_0, \qquad \beta_{\mu_0} = 1.$$

Then $\{\beta_\mu | \mu \in \Xi_N\}$ is the canonical basis of $\bigwedge_R M$.

The cases $R = \mathbf{R}$ and $R = \mathbf{C}$ have a particular interest and deserve *ad hoc* notations:

$$\textstyle\bigwedge_{\mathbf{R}} \mathbf{R}^L \equiv B_L \quad ; \quad \bigwedge_{\mathbf{C}} \mathbf{C}^L \equiv C_L \, . \tag{1.4}$$

B_L is a vector space, with a canonical basis obtained from the canonical basis of $\mathbf{R}^L$ according to the above described procedure. If $\mathfrak{N}_L$ is the ideal of nilpotents of B_L, the vector space direct sum decomposition $B_L = \mathbf{R} \oplus \mathfrak{N}_L$ defines two projections

$$\sigma : B_L \to \mathbf{R} \quad ; \quad s : B_L \to \mathfrak{N}_L \, , \tag{1.5}$$

which are sometimes called *body* and *soul* maps. Obviously, the body map coincides with the augmentation map previously introduced. The exterior product in B_L will be denoted simply by juxtaposition. Analogous considerations and notations hold concerning C_L. ▲

Tensor products. Let us recall that we are considering a graded-commutative ring R. The *graded tensor product*[2] of two graded R-modules M, N is by definition the usual tensor product $M \otimes_R N$, obtained by regarding M as a right module, and N as a left module, equipped with the gradation

$$(M \otimes_R N)_\gamma = \bigoplus_{\alpha+\beta=\gamma} \left\{ \sum m_i \otimes n_j \mid m_i \in M_\alpha \, , \, n_j \in N_\beta \right\} .$$

Evidently, $M \otimes_R N$ has a natural structure of graded R-module:

$$a(x \otimes y) = ax \otimes y = (-1)^{|a|\,|x|} xa \otimes y = (-1)^{|a|\,|x|} x \otimes ay = (-1)^{|a|(|x|+|y|)} (x \otimes y) a. \tag{1.6}$$

The graded tensor product can be characterized as a 'universal object.' To this end, given graded R-modules M, N and Q, we introduce the set $\mathcal{L}(M,N;Q)_\alpha$ (with $\alpha \in \mathbf{Z}_2$) of the graded R-bilinear morphisms $f : M \times N \to Q$, homogeneous of degree α: if $f \in \mathcal{L}(M,N;Q)_\alpha$, then f is a morphism of degree α such that $f(xa, y) = f(x, ay) = (-1)^{|a|\,|y|} f(x,y)a$ for all $a \in R$. The set

$$\mathcal{L}(M,N;Q) \equiv \mathcal{L}(M,N;Q)_0 \oplus \mathcal{L}(M,N;Q)_1$$

is endowed with a structure of graded R-module by enforcing the multiplication rule $(fa)(x,y) = f(ax,y)$. In the same way, if $M_1, \ldots, M_n, Q$ are graded R-modules, one defines the graded R-module $\mathcal{L}(M_1, \ldots, M_n; Q)$ formed by the graded R-multilinear morphisms $M_1 \times \cdots \times M_n \to Q$.

[2] We only deal with tensor products of finite families of graded modules; a more general treatment can be found in [**Ma2**].

Proposition 1.3. *There are natural isomorphisms in the category* **R-GMod**

$$\mathcal{L}(M, N; Q) \simeq \mathrm{Hom}_R(M \otimes_R N, Q) \simeq \mathrm{Hom}_R(M, \mathrm{Hom}_R(N, Q)).$$

Proof. To prove the first isomorphism, let $\pi : \overline{M \times N} \to M \otimes_R N$ denote the canonical morphism in **R-GMod**, where $\overline{M \times N}$ is $M \times N$ equipped with the obvious gradation. As in the commutative case, it is easily verified that each $f \in \mathcal{L}(M, N; Q)$ determines uniquely an $\bar{f} \in \mathrm{Hom}_R(M \otimes_R N, Q)$ such that $f = \pi \circ \bar{f}$. The second isomorphism is established by the map

$$\lambda : \mathrm{Hom}_R(M \otimes_R N, Q) \to \mathrm{Hom}_R(M, \mathrm{Hom}_R(N, Q))$$
$$\lambda(g)(m)(n) = g(m \otimes n),$$

where $m \in M$, $n \in N$, and $g \in \mathrm{Hom}_R(M \otimes_R N, Q)$. ■

REMARK 1.1. Even though the isomorphisms of the previous Proposition are natural, the construction of $\mathcal{L}(M, N; Q)$ involves arbitrary conventions concerning the choice of signs. ▲

The graded tensor product enjoys properties analogous to those of the ordinary tensor product. For the sake of completeness, we state here the main ones.

Proposition 1.4. *Let M, M', M'' be graded R-modules; the following natural isomorphisms of graded R-modules hold:*

a) $M \otimes_R M' \simeq M' \otimes_R M$, *achieved by the morphism*

$$x \otimes x' \mapsto (-1)^{|x||x'|} x' \otimes x;$$

b) $(M \otimes_R M') \otimes_R M'' \simeq M \otimes_R (M' \otimes_R M'')$, *achieved by the morphism*

$$(x \otimes x') \otimes x'' \mapsto x \otimes (x' \otimes x'');$$

c) $R \otimes_R M \simeq M \simeq M \otimes_R R$. ■

If $f : M \to P$, $g : N \to Q$ are morphisms of graded modules over a graded ring R, the tensor product $f \otimes g : M \otimes_R N \to P \otimes_R Q$ is the morphism defined by the condition

$$(f \otimes g)(m \otimes n) = (-1)^{|g||m|} f(m) \otimes g(n). \tag{1.7}$$

In the following Section we shall develop a general theory of graded tensor calculus over a graded-commutative algebra.

2. Graded algebras and graded tensor calculus

Graded algebras. Even though there are, of course, classical examples of graded algebras (by which, in conformity with our conventions, we mean $\mathbb{Z}_2$-graded algebras), such as Clifford algebras, the interest in such structures exploded in the 70's, as a by-product of their use in supersymmetric quantum field theory. Nowadays a large body of literature is available concerning graded algebras, mainly over the real or complex numbers (usually called *superalgebras*), their representations, etc. Classical references are [**CNS,Ka1,Ka2,Sch**].

Here, as customary, we wish only to introduce the most common notions and basic results and fix the notation. Certain other properties of a particular class of graded algebras (the graded Lie algebras) will be described in Chapter VII while dealing with Lie supergroups.

Let R be a graded-commutative ring.

Definition 2.1. *A graded R-algebra P is a graded R-module endowed with a graded R-bilinear multiplication*

$$\begin{aligned} P \otimes P &\to P \\ x \otimes y &\mapsto x \cdot y\,. \end{aligned}$$

A graded R-algebra P is said to be graded-commutative if all graded commutators

$$\langle x, y\rangle = x \cdot y - (-1)^{|x||y|} y \cdot x\,,$$

defined on the analogy of Eq. (1.1), vanish.

Given a graded ring S (not necessarily graded-commutative), a morphism $\chi\colon R \to S$ defines a graded R-algebra structure over S if $\chi(R) \subset C(S)$.

EXAMPLE 2.1. The graded module B_L (C_L) introduced in Example (1.2), equipped with the exterior product, is a graded-commutative $\mathbf{R}$-algebra ($\mathbf{C}$-algebra). ▲

The *graded tensor product* $P \otimes_R Q$ of two graded R-algebras P and Q is defined as the tensor product of the underlying R-modules equipped with the multiplication naturally induced by those of P and Q:

$$(x_1 \otimes y_1) \cdot (x_2 \otimes y_2) = (-1)^{|y_1||x_2|}(x_1 \cdot x_2) \otimes (y_1 \cdot y_2).$$

Definition 2.2. *A graded Lie R-algebra (or Lie R-superalgebra) $\mathfrak{P}$ is a graded R-algebra, whose multiplication, called graded Lie bracket and denoted by* $[\ ,\ ]$, *satisfies the following identities:*

$$[x, y] = -(-1)^{|x||y|}[y, x]; \tag{2.1}$$

$$(-1)^{|x||z|}[x, [y, z]] + (-1)^{|y||x|}[y, [z, x]] + (-1)^{|z||y|}[z, [x, y]] = 0. \tag{2.2}$$

The conditions in Definition 2.2 are no more than the graded versions of the antisymmetry property and Jacobi identity holding in the case of an ordinary Lie algebra.

REMARK 2.1. Given a graded Lie algebra $\mathfrak{P}$, its even part $\mathfrak{P}_0$ is a Lie algebra over the ring R_0. ▲

EXAMPLE 2.2. (cf. [FröN]) Given a finite-dimensional real vector space V, let $\mathfrak{F}(V)$ be the algebra of V-valued exterior forms over V. One can define a graded Lie bracket so that $\mathfrak{F}(V)$ is made into a graded Lie $\mathbb{R}$-algebra, usually called the Frölicher-Nijenhuis algebra. This construction can be straightforwardly extended to the case of the algebra of vector forms over a differentiable manifold. ▲

An important class of graded Lie algebras can be constructed in terms of the notion of *graded derivation*. Let P be a graded-commutative R-algebra (we assume here, as in the remainder of this book, that P is associative with an identity).

Definition 2.3. *A homogeneous morphism $D \in \operatorname{End}_R P$ is a graded derivation of P (over R) if it fulfills the following condition (called the graded Leibniz rule)*

$$D(x \cdot y) = D(x) \cdot y + (-1)^{|x||D|} x \cdot D(y). \tag{2.3}$$

The graded R-submodule of $\operatorname{End}_R P$ generated by the graded derivations of P will be denoted by $\operatorname{Der}_R P$, or simply $\operatorname{Der} P$.

Proposition 2.1. Der P, *equipped with the graded Lie bracket*

$$[D_1, D_2] \equiv D_1 \circ D_2 - (-1)^{|D_1||D_2|} D_2 \circ D_1 , \tag{2.4}$$

is a graded Lie R*-algebra.* ■

By identifying R with the submodule $R \cdot 1 \subset P$, condition (2.3) implies that, for all $D \in \operatorname{Der} P$, $D(R) = 0$. We notice that $\operatorname{Der} P$ is a (left) graded P-module in a natural way, by letting $(xD)(y) = x \cdot D(y)$.

Definition 2.3 can be generalized to the case of derivations of P with values in a graded P-module M.

Definition 2.4. *A graded derivation of* P *over* R *with values in* M *is a homogeneous element* $D \in \operatorname{Hom}_R(P, M)$ *which fulfills a graded Leibniz rule formally identical with Eq. (2.3).*

The graded P-submodule of $\operatorname{Hom}_R(P, M)$ generated by the graded derivations of P with values in M will be denoted by $\operatorname{Der}_R(P, M)$.

Graded tensor calculus. We wish now to unfold in some detail a version of tensor calculus appropriate to the setting of graded-commutative algebras, which will be used in particular in Section IV.4 to develop a graded exterior differential calculus.

Throughout this Section, R will denote a graded-commutative algebra with unit and M, N two graded R-modules; all tensor products will be over R. We assume that R has characteristic 0.[3]

Following our conventions, $\operatorname{Hom}(M, N)$ will denote the set of right R-linear morphisms from M to N, with a left module structure given by $(ag)(m) = ag(m)$.

Proposition 2.2. *Let* M *and* N *be* R*-modules. There is a natural morphism of graded* R*-modules*

$$\phi : N \otimes M^* \rightarrow \operatorname{Hom}(M, N)$$

described by $\phi(n \otimes \omega)(m) = n\omega(m)$. *This induces a morphism*

$$\gamma : M^* \otimes N^* \rightarrow (M \otimes N)^*$$

[3]The *characteristic* of a graded ring R can be defined as follows. Let $\phi : \mathbf{Z} \rightarrow R_0$ be the unique ring morphism such that $1 \mapsto 1$. The kernel of ϕ is an ideal of $\mathbf{Z}$, and therefore is the set of multiples of an integer p, which is by definition the characteristic of R.

whose expression is

$$\gamma(\omega \otimes \eta)(m \otimes n) = (-1)^{|\eta|\,|m|}\,\omega(m)\,\eta(n)\,.$$

Both morphisms are bijective whenever M is free and finitely generated.

Proof. We only show explicitly the existence of γ. In fact,

$$M^* \otimes N^* \xrightarrow{\phi} \mathrm{Hom}(N,\, M^*) = \mathrm{Hom}(N,\, \mathrm{Hom}(M,\, R)) \simeq (N \otimes M)^* \simeq (M \otimes N)^*$$

by Proposition 1.3. ∎

By iterating γ one obtains a morphism, again denoted by γ,

$$\gamma : M_1^* \otimes \cdots \otimes M_n^* \to (M_1 \otimes \cdots \otimes M_n)^* \simeq \mathcal{L}(M_1, \ldots, M_n;\, R) \qquad (2.5)$$

explicitly given, for homogeneous $\omega^1, \ldots, \omega^n$ and $m_1, \ldots, m_n$, by:

$$\gamma(\omega^1 \otimes \cdots \otimes \omega^n)(m_1, \ldots, m_n) = (-1)^{\sum_{h<j} |\omega^j|\,|m_h|}\omega^1(m_1) \ldots \omega^n(m_n). \qquad (2.6)$$

Graded exterior algebra. Let M be a graded R-module, and let us denote by

$$T^p M = M \underbrace{\otimes \cdots \otimes}_{p} M$$

the p-th tensor power of M, graded as usual. We can consider as in the non-graded setting the *graded tensor algebra* of M,

$$\mathcal{T}(M) = \bigoplus_{p=0}^{\infty} T^p M, \qquad (2.7)$$

which is in a natural way a bigraded R-algebra (i.e., it has the usual $\mathbb{Z}$-gradation of the tensor algebra, together with the $\mathbb{Z}_2$-gradation it carries as a graded R-algebra).

The *graded exterior algebra* $\bigwedge_R M$ of M (in this section denoted simply by $\bigwedge M$) is defined as the quotient of $\mathcal{T}(M)$ by the graded ideal $\mathfrak{I}(M)$ generated by elements of the form $m_1 \otimes m_2 + (-1)^{|m_1|\,|m_2|}\, m_2 \otimes m_1$, with

m_1, m_2 homogeneous.[4] The product induced in $\bigwedge M$ by this quotient is denoted by $\wedge$ and is called the *(graded) wedge product*, as usual. If we let $\mathfrak{I}^p(M) = \mathfrak{I}(M) \cap T^pM$, since $\mathfrak{I}(M)$ is generated by homogeneous elements, we obtain $\mathfrak{I}(M) = \bigoplus_{p=0}^{\infty} \mathfrak{I}^p(M)$, and, therefore,

$$\bigwedge M = \bigoplus_{p=0}^{\infty} \bigwedge^p M$$

with $\bigwedge^p M = T^pM/\mathfrak{I}^p(M)$.

We wish to ascertain the relationship existing between the exterior algebra $\bigwedge M^*$ and the modules of alternating graded multilinear forms; this will be realized by a morphism analogous to (and indeed induced by) the morphism (2.5).

If $F_p \in \operatorname{Hom}(T^pM, R)$ and $F_q \in \operatorname{Hom}(T^qM, R)$ are homogeneous graded multilinear forms, $F_p \otimes F_q$ acts on a family of homogeneous elements according to the formula:

$$\begin{aligned}(F_p \otimes F_q)&(m_1, \dots, m_{p+q}) \\ &= (-1)^{|F_q|(|m_{p+1}|+\dots+|m_{p+q}|)} F_p(m_1, \dots, m_n) F_q(m_{p+1}, \dots, m_{p+q}).\end{aligned}$$

Let $\mathfrak{S}_p$ be the group of permutations of p objects. For any $\sigma \in \mathfrak{S}_p$, and any $F_p \in \operatorname{Hom}(T^pM, R)$, we write, for homogeneous elements $m_1, \dots, m_p \in M$,

$$F_p^\sigma(m_1, \dots, m_p) = (-1)^{\Delta_1(\sigma,m)} F_p(m_{\sigma(1)}, \dots, m_{\sigma(p)}),$$

where

$$\Delta_1(\sigma, m) = \sum_{1 \le i < j \le p} \sum_{\sigma(i) > \sigma(j)} |m_{\sigma(i)}| \, |m_{\sigma(j)}| . \tag{2.8}$$

Definition 2.5. *A graded multilinear form $F_p \in \operatorname{Hom}(T^pM, R)$ is said to be alternating if $F_p^\sigma = (-1)^{|\sigma|} F_p$ for every $\sigma \in \mathfrak{S}_p$, where $|\sigma|$ is the parity of the permutation σ.*

The set $\operatorname{Alt}(\underbrace{M \times \cdots \times M}_{p}; R) \equiv \operatorname{Alt}(M^p, R)$ of all alternating graded multilinear forms is a submodule of $\operatorname{Hom}(T^pM, R)$; one can introduce a projection

[4] In this discussion, 'homogeneous' refers, as usual, to the $\mathbb{Z}_2$-gradation.

morphism, which is no more than the graded anti-symmetrization:

$$\begin{aligned} A_p : \operatorname{Hom}(T^pM,\, R) &\to \operatorname{Alt}(M^p;\, R) \\ F_p &\mapsto A_p(F_p) = \frac{1}{p!} \sum_{\sigma\in\mathfrak{S}_p} (-1)^{|\sigma|} F_p^\sigma\,. \end{aligned}$$

Proposition 2.3. *The morphism A_p has the following properties:*

(1) $A_p(F_p) = F_p$ *for any alternating form* F_p*;*

(2) $A_{p+q}(F_q \otimes F_p) = (-1)^{pq+|F_p||F_q|} A_{p+q}(F_p \otimes F_q)$ *for homogeneous* F_p, F_q*;*

(3) $A_{p+q}(A_p(F_p) \otimes F_q) = A_{p+q}(F_p \otimes F_q)$.

∎

We now assume that M is a free and finitely generated module, so that we may identify $T^p(M^*)$ with $\operatorname{Hom}(T^pM,\, R)$. In this way, the morphism A_p yields the exact sequence of graded R-modules

$$0 \to \mathfrak{I}^p(M^*) \to T^pM^* \xrightarrow{A_p} \operatorname{Alt}(M^p;\, R) \to 0\,, \tag{2.9}$$

and therefore we obtain an isomorphism $\bigwedge^p M^* \simeq \operatorname{Alt}(M^p;\, R)$. Thus, for a free and finitely generated module M, the homogeneous elements in the graded exterior algebra $\bigwedge M^*$ can be interpreted as alternating graded multilinear forms on M (also simply called 'forms'). In particular, we may interpret the wedge product of two elements $\omega^p \in \bigwedge^p M^*$ and $\omega^q \in \bigwedge^q M^*$ as a graded multilinear form, which acts on homogeneous elements $m_1, \ldots, m_{p+q}$ according to:[5]

$$(\omega^p \wedge \omega^q)(m_1, \ldots, m_{p+q}) =$$
$$\tfrac{1}{(p+q)!} \sum_{\sigma\in\mathfrak{S}_{p+q}} (-1)^{|\sigma|+\Delta_2(\sigma,m,\omega^q)} \omega^p(m_{\sigma(1)}, \ldots, m_{\sigma(p)}) \omega^q(m_{\sigma(p+1)}, \ldots, m_{\sigma(p+q)})$$

where, in terms of the symbol $\Delta_1(\sigma, m)$ previously defined, we set

$$\Delta_2(\sigma, m, \omega^q) = \Delta_1(\sigma, m) + |\omega^q| \sum_{i=1}^{p} |m_{\sigma(i)}|\,. \tag{2.10}$$

[5] The numerical factors appearing in the following equation, as well as in other equations in this subsection, are determined by the choice of the projection $T^pM^* \to \operatorname{Alt}(M^p;\, R)$. Here we follow the conventions of [**KN**].

Definition 2.6. *The inner product of an element $m \in M$ and a p-form $\omega^p \in \bigwedge^p M^*$ is the $(p-1)$-form which, in the homogeneous case, is given by:*

$$(m\rfloor\omega^p)(m_1,\ldots,m_{p-1}) = p(-1)^{|m||\omega^p|}\omega^p(m,m_1,\ldots,m_{p-1}).$$

Since $ma\rfloor\omega^p = m\rfloor a\omega^p$, the inner product defines a morphism of graded R-modules

$$\rfloor : M \otimes \textstyle\bigwedge^p M^* \to \bigwedge^{p-1} M^*.$$

By means of a direct calculation, which requires a careful book-keeping of all signs and morphisms involved, one can prove the following important property of the inner product.

Proposition 2.4. *If m is homogeneous, the inner product $m\rfloor$ is a graded derivation of bidegree $(-1,|m|)$, that is:*

$$m\rfloor(\omega^p \wedge \omega^q) = (m\rfloor\omega^p)\wedge\omega^q + (-1)^{p+|m||\omega^p|}\omega^p \wedge (m\rfloor\omega^q)$$

for homogeneous ω^p. ∎

Graded symmetric algebra. Finally, we should like to offer a few words on the graded symmetric algebra of a graded module M. We consider a morphism $S : \mathcal{T}(M) \to \mathcal{T}(M)$, called the *(graded) symmetrizer*, defined on elements of $\mathcal{T}(M)$ of the type $m_1 \otimes \cdots \otimes m_p$, with m_i homogeneous, as follows:

$$S(m_1 \otimes \cdots \otimes m_p) = \frac{1}{p!}\sum_{\sigma\in\mathfrak{S}_p}(-1)^{\Delta_1(\sigma,m)} m_{\sigma(1)} \otimes \cdots \otimes m_{\sigma(p)}.$$

Let $S(M)$ be the quotient of $\mathcal{T}(M)$ by the ideal $\mathfrak{J}(M)$ generated by elements of the form $m_1 \otimes m_2 - (-1)^{|m_1||m_2|} m_2 \otimes m_1$, with m_1 and m_2 homogeneous. Equipped with the product induced by the quotient (which we denote by $\odot$), $S(M)$ is a graded-commutative R-algebra called the *(graded) symmetric algebra* of M.

Since the ideal $\mathfrak{J}(M)$ is homogeneous with respect to the $\mathbb{Z}$-gradation, i.e. it verifies $\mathfrak{J}(M) = \oplus_{p=0}^{\infty}\mathfrak{J}(M) \cap T^p M$, one arrives at the decomposition

$$S(M) = \bigoplus_{p=0}^{\infty} S^p(M).$$

Each module $S^p(M)$ can be injected into T^pM simply by letting $[t] \mapsto S(t)$, where $[t]$ is the equivalence class of t. One obviously has $t_1 \odot t_2 = S(t_1 \otimes t_2)$ for all $t_1, t_2 \in S(M)$.

3. Matrices

Given a graded-commutative ring R, an R-module morphism $R^{m|n} \to R^{p|q}$ can be regarded, relative to the canonical bases of $R^{m|n}$ and $R^{p|q}$, as a $(p+q) \times (m+n)$ matrix with entries in R,

$$X = \begin{pmatrix} X_1 & X_2 \\ X_3 & X_4 \end{pmatrix}, \tag{3.1}$$

which acts on column vectors in $R^{m|n}$ from the left. The set $M_R[(p+q)\times(m+n)]$ of such matrices can be graded so as to be naturally isomorphic to the graded R-module $\mathrm{Hom}_R(R^{m|n}, R^{p|q})$, by decreeing that:

X is even if X_1 and X_4 have even entries, while X_2 and X_3 have odd entries;

X is odd if X_1 and X_4 have odd entries, while X_2 and X_3 have even entries.

The set of matrices of the form (3.1), equipped with this gradation, will be denoted by $M_R[p|q; m|n]$. The set of square matrices $M_R[m|n]$ (which are obtained by letting $p = m$, $q = n$) is a graded R-algebra.

The usual notions of trace and determinant of a matrix can be extended to the matrices in $M_R[m|n]$, thus obtaining the concepts of *graded trace* and *Berezinian* (also called *supertrace* and *superdeterminant*, respectively). For any matrix $X \in M_R[p|q; m|n]$, regarded as a morphism $X : R^{m|n} \to R^{p|q}$, we define the *graded transpose* of X — denoted by X^{gt} — as the matrix corresponding to the morphism $X^* : (R^{p|q})^* \to (R^{m|n})^*$ dual to X. With reference to Eq. (3.1), one obtains the following relations, where the superscript t denotes the usual matrix transposition:

$$\begin{pmatrix} X_1 & X_2 \\ X_3 & X_4 \end{pmatrix}^{gt} = \begin{cases} \begin{pmatrix} X_1^t & X_3^t \\ -X_2^t & X_4^t \end{pmatrix} & \text{if} \quad |X| = 0 \\ \begin{pmatrix} X_1^t & -X_3^t \\ X_2^t & X_4^t \end{pmatrix} & \text{if} \quad |X| = 1 \end{cases} \tag{3.2}$$

The graded transposition behaves naturally with respect to matrix multiplication:

$$(XY)^{gt} = (-1)^{|X||Y|} Y^{gt} X^{gt} .$$

In view of the isomorphism (1.8), a matrix $X \in M_R[m|n]$ singles out an element $\sum_i a_i^* \otimes a^i \in (R^{m|n})^* \otimes_R R^{m|n}$. The *graded trace* of X is the element $\operatorname{Str} X = \sum_i a_i^*(a^i) \in R$. Alternatively, one can give a direct characterization by letting, for all homogeneous $X \in M_R[m|n]$,

$$\operatorname{Str} X = \operatorname{Tr} X_1 - (-1)^{|X|} \operatorname{Tr} X_4 \tag{3.3}$$

where Tr designates the usual trace operation. The graded trace determines an R-module morphism $\operatorname{Str} : M_R[m|n] \to R$, which is natural with respect to graded transposition and matrix multiplication:

$$\begin{aligned} \operatorname{Str}(X^{gt}) &= \operatorname{Str} X \\ \operatorname{Str}(XY) &= (-1)^{|X||Y|} \operatorname{Str}(YX) . \end{aligned} \tag{3.4}$$

Let us notice that, by denoting by $I_{m|n}$ the identity matrix, one has $\operatorname{Str} I_{m|n} = m - n$.

In order to extend the notion of determinant, one must consider the subgroup $GL_R[m|n]$ of the matrices in $M_R[m|n]$ corresponding to even invertible endomorphisms. $GL_R[m|n]$ is the natural extension of the notion of general linear group, so that it will be called the *general graded linear group.*

Proposition 3.1. *A matrix $X \in M_R[m|n]_0$ is in $GL_R[m|n]$ if and only if $X_1 \in GL_R[m|0]$ and $X_4 \in GL_R[0|n]$, i.e. X is invertible if and only if X_1 and X_4 are invertible as ordinary matrices with entries in R_0.*

Proof. The claim can be proved by extending to the present setting the arguments given in Section 1.7.2 of [Leĭ]. ∎

Definition 3.1. [BL,ANZ] *Let $X \in GL_R[m|n]$. The Berezinian of X is the element in $GL_R[1|0]$ given by*

$$\operatorname{Ber} X = \begin{pmatrix} X_1 & X_2 \\ X_3 & X_4 \end{pmatrix} = \det(X_1 - X_2 X_4^{-1} X_3)(\det X_4^{-1}) . \tag{3.5}$$

In one sense, the following two results qualify the Berezinian as a generalization of the determinant.

Proposition 3.2. *The mapping* $\mathrm{Ber}: GL_R[m|n] \to GL_R[1|0]$ *is a group morphism, that coincides with the determinant whenever* $n = 0$*:*

$$\mathrm{Ber}(XY) = \mathrm{Ber}\, X\, \mathrm{Ber}\, Y \qquad \forall X\,, Y \in GL_R[m|n]\,. \tag{3.6}$$

■

Proposition 3.3. $\mathrm{Ber}(X^{gt}) = \mathrm{Ber}\, X$ *for all* $X \in GL_R[m|n]$. ■

Further properties of matrices with entries in a graded commutative ring can be stated in the case where R is the exterior algebra B_L (or C_L). We consider only the case of B_L, the other being identical. As far as notation is concerned, we set

$$GL_L[m|n] \equiv GL_{B_L}[m|n] \quad ; \quad GL_L[m|n; \mathbf{C}] \equiv GL_{C_L}[m|n]\,. \tag{3.7}$$

Using the notations of Example 1.2, we introduce the $\mathbf{R}$-algebra morphism

$$\begin{gathered} \sigma: M_{B_L}[m|n] \to M_{\mathbf{R}}[m+n] \\ (\sigma(X))_{ij} = \sigma(X_{ij})\,, \end{gathered} \tag{3.8}$$

where $M_{\mathbf{R}}[m+n]$ is the algebra of real $(m+n) \times (m+n)$ matrices. Denoting by $GL[m+n]$ the general real linear group, Proposition (3.1) implies

Corollary 3.1. *A matrix in* $X \in M_{B_L}[m|n]_0$ *is invertible if and only if* $\sigma(X) \in GL[m+n]$.

Proof. The 'only if' part is trivial, since σ is a ring morphism. To show the converse, it suffices to prove that a matrix $Z \in M_{B_L}[p|0]_0$ is invertible as a matrix with entries in $(B_L)_0$ if $\sigma(Z)$ is invertible. In the case $p = 1$ this is a consequence of the fact that in B_L the morphism σ is the natural projection $(B_L)_0 \to (B_L)_0/(\mathfrak{N}_L)_0$. The result is easily extended to $p > 1$ by induction. ■

We equip B_L with an l^1 norm by letting

$$\|x\| = \sum_{\mu \in \Xi_L} |x^\mu| \qquad \text{if} \qquad x = \sum_{\mu \in \Xi_L} x^\mu\, \beta_\mu\,.$$

This norm is submultiplicative, thus allowing one to prove easily that the exponential map $\exp: B_L \to B_L$, defined by

$$\exp x = \sum_{k=0}^{\infty} \frac{x^k}{k!} \tag{3.9}$$

converges. This map can be extended to matrices in $M_{B_L}[m|n]$, and one proves:

Proposition 3.4. [RiS] *For all matrices* $X \in GL_L[m|n]$

$$\mathrm{Ber}(\exp(X)) = \exp(\mathrm{Str}\, X).$$

■

Chapter II
Sheaves and cohomology

Prendre possession de l'espace est le geste premier des vivants, des hommes et des bêtes, des plants et des nuages ...

LE CORBUSIER

Sheaves provide a powerful tool for studying graded manifolds and supermanifolds. Sheaf theoretic techniques convey the elegant flavour of algebraic and analytic geometry. However, the use of sheaves is not merely a matter of aesthetics, but is unavoidable in the graded setting. Indeed, graded manifolds are intrinsically defined in terms of sheaves, while, on the other hand, supermanifolds share, together with complex spaces, the feature that their structure sheaf may have non-trivial cohomology; this in turn has implications in bundle theory etc.

After providing the basic definitions related to the concept of sheaf, in the second Section we describe some simple notions of homological algebra, and introduce sheaf cohomology via the Godement canonical flabby resolution of a sheaf. Other cohomology theories (de Rham, Čech, Dolbeault) are then outlined. The last Section extends to the graded setting the notion of *ringed space*, and contains the definition of *spectrum* of a ring, together with the proof that the spectrum of the ring of smooth functions over a differentiable manifold can be identified with the manifold.

Albeit the exposition is self-contained as far as possible, it obviously fails to be exhaustive. For general, more comprehensive treatments, the reader may refer to the works by Godement [**Go**], Bredon [**Bre**], Tennison [**Ten**], Kashiwara and Schapira [**KaS**], Hartshorne [**Har**], and Griffiths and Harris [**GrH**]. Anyhow, in the sequel a proper reference will be given whenever a result not included here is used.

1. Presheaves and sheaves

Let X be a topological space.

Definition 1.1. *A presheaf of abelian groups on X is a rule $\mathcal{P}$ which assigns an abelian group $\mathcal{P}(U)$ to each open subset U of X and a morphism (called restriction map) $\varphi_{U,V}: \mathcal{P}(U) \to \mathcal{P}(V)$ to each pair $V \subset U$ of open subsets, so as to verify the following requirements:*

(1) *$\varphi_{U,U}$ is the identity map;*
(2) *if $W \subset V \subset U$ are open sets, then $\varphi_{U,W} = \varphi_{V,W} \circ \varphi_{U,V}$.*

The elements $s \in \mathcal{P}(U)$ are called *sections* of the presheaf $\mathcal{P}$ on U. If $s \in \mathcal{P}(U)$ is a section of $\mathcal{P}$ on U and $V \subset U$, we shall write $s_{|V}$ instead of $\varphi_{U,V}(s)$. The restriction $\mathcal{P}_{|U}$ of $\mathcal{P}$ to an open subset U is defined in the obvious way.

Presheaves of rings are defined in the same way, by requiring that the restriction maps are ring morphisms. If $\mathcal{R}$ is a presheaf of rings on X, a presheaf $\mathcal{M}$ of abelian groups on X is called a *presheaf of modules* over $\mathcal{R}$ (or an $\mathcal{R}$-module) if, for each open subset U, $\mathcal{M}(U)$ is an $\mathcal{R}(U)$-module and for each pair $V \subset U$ the restriction map $\varphi_{U,V}: \mathcal{M}(U) \to \mathcal{M}(V)$ is a morphism of $\mathcal{R}(U)$-modules (where $\mathcal{M}(V)$ is regarded as an $\mathcal{R}(U)$-module via the restriction morphism $\mathcal{R}(U) \to \mathcal{R}(V)$).

Presheaves of graded algebraic objects can be considered as well; in this case the restriction maps are *even* morphisms in the relevant category.

The definitions in this Section are stated for the case of presheaves of abelian groups, but analogous definitions and properties hold for presheaves of rings and modules, possibly graded.

Definition 1.2. *A morphism $f: \mathcal{P} \to \mathcal{Q}$ of presheaves over X is a family of morphisms of abelian groups $f_U: \mathcal{P}(U) \to \mathcal{Q}(U)$ for each open $U \subset X$, commuting* *with the restriction morphisms; i.e., the following diagram commutes:*

$$\begin{array}{ccc} \mathcal{P}(U) & \xrightarrow{f_U} & \mathcal{Q}(U) \\ \varphi_{U,V}\downarrow & & \downarrow\varphi_{U,V} \\ \mathcal{P}(V) & \xrightarrow{f_V} & \mathcal{Q}(V) \end{array}$$

Definition 1.3. *The stalk of a presheaf $\mathcal{P}$ at a point $x \in X$ is the abelian group*[1]

$$\mathcal{P}_x = \varinjlim_U \mathcal{P}(U)$$

where U ranges over all open neighbourhoods of x, directed by inclusion. If $x \in U$ and $s \in \mathcal{P}(U)$, the image s_x of s in $\mathcal{P}_x$ is called the germ of s at x.

Then, two elements $s \in \mathcal{P}(U)$, $s' \in \mathcal{P}(V)$, U, V being open neighbourhoods of x, define the same germ at x, i.e. $s_x = s'_x$, if and only if there exists an open neighbourhood $W \subset U \cap V$ of x such that s and s' coincide on W, $s_{|W} = s'_{|W}$.

Definition 1.4. *A sheaf on a topological space X is a presheaf $\mathcal{F}$ on X which fulfills the following axioms for any open subset U of X and any cover $\{U_i\}$ of U.*

S1) *If two sections $s \in \mathcal{F}(U)$, $\bar{s} \in \mathcal{F}(U)$ coincide when restricted to any U_i, $s_{|U_i} = \bar{s}_{|U_i}$, they are equal, $s = \bar{s}$.*

S2) *Given sections $s_i \in \mathcal{F}(U_i)$ which coincide on the intersections, $s_{i|U_i \cap U_j} = s_{j|U_i \cap U_j}$ for every i, j, there exists a section $s \in \mathcal{F}(U)$ whose restriction to each U_i equals s_i, i.e. $s_{|U_i} = s_i$.*

---

[1]The definition of direct limit, denoted $\varinjlim$, is as follows. Let I be a directed set; a directed system of abelian groups is a family $\{G_i\}_{i\in I}$ of abelian groups, such that for all $i < j$ there is a group morphism $f_{ij}: G_i \to G_j$, with $f_{ii} = id$ and $f_{ij} \circ f_{jk} = f_{ik}$. On the set $\mathfrak{G} = \coprod_{i\in I} G_i$, where $\coprod$ denotes disjoint union, we put the following equivalence relation: $g \sim h$, with $g \in G_i$ and $h \in G_j$, if there exists a $k \in I$ such that $f_{ik}(g) = f_{jk}(h)$. The direct limit $\mathfrak{l}$ of the system $\{G_i\}_{i\in I}$, denoted $\mathfrak{l} = \varinjlim_{i\in I} G_i$, is the quotient $\mathfrak{G}/\sim$. Heuristically, two elements in $\mathfrak{G}$ represent the same element in the direct limit if they are 'eventually equal.' From this definition one naturally obtains the existence of canonical morphisms $G_i \to \mathfrak{l}$. The remark following Definition 1.3 should make this notion clearer; for more detail, the reader may consult [HiS].

Thus, roughly speaking, sheaves are presheaves defined by local conditions. The *stalk of a sheaf* is defined as in the case of a presheaf, and, moreover, a morphism of sheaves is nothing but a morphism of presheaves. If $f : \mathcal{F} \to \mathcal{G}$ is a morphism of sheaves on X, for every $x \in X$ the morphism f induces a morphism between the stalks, $f_x : \mathcal{F}_x \to \mathcal{G}_x$, in the following way: since the stalk $\mathcal{F}_x$ is the direct limit of the groups $\mathcal{F}(U)$ over all open U containing x, any $g \in \mathcal{F}_x$ is of the form $g = s_x$ for some open $U \ni x$ and some $s \in \mathcal{F}(U)$; then set $f_x(g) = (f_U(s))_x$.

A sequence of morphisms of sheaves $0 \to \mathcal{F}' \to \mathcal{F} \to \mathcal{F}'' \to 0$ is *exact* if for every point $x \in X$, the sequence of morphisms between the stalks $0 \to \mathcal{F}'_x \to \mathcal{F}_x \to \mathcal{F}''_x \to 0$ is exact. If $0 \to \mathcal{F}' \to \mathcal{F} \to \mathcal{F}'' \to 0$ is an exact sequence of sheaves, for every open subset $U \subset X$ the sequence of groups $0 \to \mathcal{F}'(U) \to \mathcal{F}(U) \to \mathcal{F}''(U)$ is exact, but the last arrow may fail to be surjective. An instance of this situation is contained in Example 1.4 below.

A more sophisticated definition of a sheaf can be given as follows. Let $\mathcal{F}$ be a presheaf on X, and $\{U_i\}$ a cover of an open subset $U \subset X$. If we denote by U_{ij} the intersection $U_{ij} = U_i \cap U_j$, and by $\varphi_i : \mathcal{F}(U) \to \mathcal{F}(U_i)$ and $\varphi_{ij} : \mathcal{F}(U_i) \to \mathcal{F}(U_{ij})$ the restriction morphisms, there exist a morphism

$$\mathcal{F}(U) \xrightarrow{r} \prod_i \mathcal{F}(U_i)$$
$$s \mapsto (\varphi_i(s))$$

and morphisms

$$\prod_i \mathcal{F}(U_i) \xrightarrow{r'} \prod_{ij} \mathcal{F}(U_i \cap U_j) \qquad \prod_j \mathcal{F}(U_j) \xrightarrow{r''} \prod_{ij} \mathcal{F}(U_i \cap U_j)$$
$$(s_i) \mapsto (\varphi_{ij}(s_i)) \qquad (s_j) \mapsto (\varphi_{ji}(s_j))$$

Then, axioms S1) and S2) are equivalent to the exactness of the sequence

$$0 \to \mathcal{F}(U) \xrightarrow{r} \prod_i \mathcal{F}(U_i) \xrightarrow{r'-r''} \prod_{ij} \mathcal{F}(U_{ij}) . \tag{1.1}$$

EXAMPLES 1.1–4.

1.1. Let G be an abelian group. Defining $\mathcal{P}(U) \equiv G$ for every open subset U and taking the identity maps as restriction morphisms, we obtain a presheaf, called the *constant presheaf* G. All stalks G_x of the presheaf G are isomorphic with the group G.

1.2. Let $\mathcal{C}_X(U)$ be the ring of real-valued continuous functions on an open set U of X. Then $\mathcal{C}_X$ is a sheaf (with the obvious restriction morphisms), the sheaf of continuous functions on X. The stalk $\mathcal{C}_x \equiv (\mathcal{C}_X)_x$ at x is the ring of germs of continuous functions at x.

1.3. In the same way one can define the following sheaves:

The sheaf $\mathcal{C}_X^\infty$ of differentiable functions on a differentiable manifold X.

The sheaves Ω_X^p of differential p-forms, and all the sheaves of tensor fields on a differentiable manifold X.

The sheaf of sections of a vector bundle $E \to X$ on a differentiable manifold X.

The sheaf of holomorphic functions on a complex manifold and the sheaves of holomorphic p-forms on it.

The sheaves of forms of type (p, q) on a complex manifold X.

1.4. Let X be a differentiable manifold, and denote by $d\colon \Omega_X^\bullet \to \Omega_X^\bullet$ the exterior differential. We can define the presheaves $\mathcal{Z}_X^p$ of closed differential p-forms, and $\mathcal{B}_X^p$ of exact p-differential forms,

$$\mathcal{Z}_X^p(U) = \{\omega \in \Omega_X^p(U) \mid d\omega = 0\},$$

$$\mathcal{B}_X^p(U) = \{\omega \in \Omega_X^p(U) \mid \omega = d\tau \quad \text{for some} \quad \tau \in \Omega_X^{p-1}(U)\}.$$

$\mathcal{Z}_X^p$ is a sheaf, but $\mathcal{B}_X^p$ may fail to be one. In fact, if $X = \mathbf{R}^2$, the presheaf $\mathcal{B}_X^1$ of exact differential 1-forms does not fulfill the second sheaf axiom: consider the form $\omega = \dfrac{x dy - y dx}{x^2 + y^2}$ defined on the open subset $U = X - \{(0,0)\}$. Since ω is closed on U, there is an open cover $\{U_i\}$ of U by open subsets where ω is an exact form, $\omega_{|U_i} \in \mathcal{B}_X^1(U_i)$ (this is Poincaré's lemma). But ω is not an exact form on U because its integral along the unit circle is different from 0.

As promised earlier, we notice that, while the sequence of sheaf morphisms $0 \to \mathsf{R} \to \mathcal{C}_X^\infty \xrightarrow{d} \mathcal{Z}_X^1 \to 0$ is exact, the morphism $\mathcal{C}_X^\infty(U) \xrightarrow{d} \mathcal{Z}_X^1(U)$ is not surjective. ▲

Étalé space. We wish now to describe how, given a presheaf, one can naturally associate with it a sheaf having the same stalks. As a first step we consider the case of a constant presheaf G on a topological space X. We can define another presheaf G_X on X by putting $G_X(U) = \{$locally constant functions

$f: U \to G\}$,[2] where $G(U) = G$ is included as the constant functions. It is clear that $(G_X)_x = G_x = G$ at each point $x \in X$ and that G_X is a sheaf, called the *constant sheaf with stalk* G. Notice that the functions $f: U \to G$ are the sections of the projection $\pi: \coprod_{x\in X} G_x \to X$ and the locally constant functions correspond to those sections which locally coincide with the sections produced by the elements of G.

Now, let $\mathcal{P}$ be an arbitrary presheaf on X. Consider the disjoint union of the stalks $\underline{\mathcal{P}} = \coprod_{x\in X} \mathcal{P}_x$ and the natural projection $\pi: \underline{\mathcal{P}} \to X$. The sections $s \in \mathcal{P}(U)$ of the presheaf $\mathcal{P}$ on an open subset U produce sections $s: U \hookrightarrow \underline{\mathcal{P}}$ of π, defined by $s(x) = s_x$, and we can define a new presheaf $\mathcal{P}^\natural$ by taking $\mathcal{P}^\natural(U)$ as the group of those sections $\sigma: U \hookrightarrow \underline{\mathcal{P}}$ of π such that for every point $x \in U$ there is an open neighbourhood $V \subset U$ of x which satisfies $\sigma_{|V} = s$ for some $s \in \mathcal{P}(V)$.

That is, $\mathcal{P}^\natural$ is the presheaf of all sections that locally coincide with sections of $\mathcal{P}$. It can be described in another way by the following construction.

Definition 1.5. *The set $\underline{\mathcal{P}}$, endowed with the topology whose base of open subsets consists of the sets $s(U)$ for U open in X and $s \in \mathcal{P}(U)$, is called the étalé space of the presheaf $\mathcal{P}$.*

$\mathcal{P}^\natural(U)$ turns out to be the set of all continuous sections $\sigma: U \hookrightarrow \underline{\mathcal{P}}$ of π. The presheaf $\mathcal{P}^\natural$ is actually a sheaf, whose stalks can be identified with those of $\mathcal{P}$, that is, $\mathcal{P}^\natural_x = \mathcal{P}_x$ at each point $x \in X$. $\mathcal{P}^\natural$ is called the *sheaf associated with the presheaf* $\mathcal{P}$. There is a presheaf morphism $\mathcal{P} \to \mathcal{P}^\natural$ which is an isomorphism if and only if $\mathcal{P}$ is a sheaf.

If $\mathcal{F}$ is a sheaf, we can construct, in the same manner, its étalé space $\underline{\mathcal{F}}$. The sheaf of sections of $\underline{\mathcal{F}}$ coincides with $\mathcal{F}$.

Definition 1.6. *Given a sheaf $\mathcal{F}$ on a topological space X and a subset (not necessarily open) $S \subset X$, the sections of the sheaf $\mathcal{F}$ on S are the continuous sections $\sigma: S \hookrightarrow \underline{\mathcal{F}}$ of $\pi: \underline{\mathcal{F}} \to X$. The group of such sections is denoted by* $\Gamma(S, \mathcal{F})$.

Definition 1.7. *Let $\mathcal{F}$, $\mathcal{G}$ be sheaves on a topological space X.*[3]

(1) *The direct sum of $\mathcal{F}$ and $\mathcal{G}$ is the sheaf $\mathcal{F} \oplus \mathcal{G}$ given, for every open sub-*

[2] A function is locally constant on U if it is constant on any connected component of U.

[3] Since we are dealing with abelian groups, i.e. with $\mathbb{Z}$-modules, the Hom modules and tensor products are taken over $\mathbb{Z}$.

set $U \subset X$, by $(\mathcal{F} \oplus \mathcal{G})(U) = \mathcal{F}(U) \oplus \mathcal{G}(U)$ with the obvious restriction morphisms.

(2) *For any open set $U \subset X$, let us denote by $\mathrm{Hom}(\mathcal{F}_{|U}, \mathcal{G}_{|U})$ the space of morphisms between the restricted sheaves $\mathcal{F}_{|U}$ and $\mathcal{G}_{|U}$; this is an abelian group in a natural manner. The sheaf of homomorphisms is the sheaf $\mathcal{H}om(\mathcal{F}, \mathcal{G})$ given by $\mathcal{H}om(\mathcal{F}, \mathcal{G})(U) = \mathrm{Hom}(\mathcal{F}_{|U}, \mathcal{G}_{|U})$ with the natural restriction morphisms.*

(3) *The tensor product of $\mathcal{F}$ and $\mathcal{G}$ is the sheaf $\mathcal{F} \otimes \mathcal{G}$ associated with the presheaf $U \to \mathcal{F}(U) \otimes \mathcal{G}(U)$.*

It should be noticed that in general $\mathcal{H}om(\mathcal{F}, \mathcal{G})(U) \not\simeq \mathrm{Hom}(\mathcal{F}(U), \mathcal{G}(U))$ and $\mathcal{H}om(\mathcal{F}, \mathcal{G})_x \not\simeq \mathrm{Hom}(\mathcal{F}_x, \mathcal{G}_x)$.

Direct and inverse images of presheaves and sheaves. Here we study the behaviour of presheaves and sheaves under change of base space. Let $f: X \to Y$ be a continuous map.

Definition 1.8. *The direct image by f of a presheaf $\mathcal{P}$ on X is the presheaf $f_*\mathcal{P}$ on Y defined by $(f_*\mathcal{P})(V) = \mathcal{P}(f^{-1}(V))$ for every open subset $V \subset Y$. If $\mathcal{F}$ is a sheaf on X, then $f_*\mathcal{F}$ turns out to be a sheaf.*

Let $\mathcal{P}$ be a presheaf on Y.

Definition 1.9. *The inverse image of $\mathcal{P}$ by f is the presheaf on X defined by*

$$U \to \varinjlim_{U \subset f^{-1}(V)} \mathcal{P}(V).$$

The inverse image sheaf of a sheaf $\mathcal{F}$ on Y is the sheaf $f^{-1}\mathcal{F}$ associated with the inverse image presheaf of $\mathcal{F}$.

The stalk of the inverse image presheaf at a point $x \in X$ is isomorphic with $\mathcal{P}_{f(x)}$. It follows that if $0 \to \mathcal{F}' \to \mathcal{F} \to \mathcal{F}'' \to 0$ is an exact sequence of sheaves on Y, the induced sequence

$$0 \to f^{-1}\mathcal{F}' \to f^{-1}\mathcal{F} \to f^{-1}\mathcal{F}'' \to 0$$

of sheaves on X, is also exact (that is, the inverse image functor for sheaves of abelian groups is exact).

The étalé space $\underline{f^{-1}\mathcal{F}}$ of the inverse image sheaf is the fibred product[4] $Y \times_X \underline{\mathcal{F}}$. It follows easily that the inverse image of the constant sheaf G_X on X with stalk G is the constant sheaf G_Y with stalk G, $f^{-1}G_X = G_Y$.

Flabby sheaves. We analyze the problem of extending the sections of a sheaf over a open set. Let X be a topological space.

Definition 1.10. *A sheaf $\mathcal{F}$ on X is flabby if for every pair $V \subset U$ of open subsets of X the restriction map $\mathcal{F}(U) \to \mathcal{F}(V)$ is surjective.*

This is equivalent to the condition that every section $s \in \mathcal{F}(U)$ can be extended to a global section $\sigma \in \mathcal{F}(X)$.

Proposition 1.1. *Let $0 \to \mathcal{F}' \xrightarrow{i} \mathcal{F} \xrightarrow{p} \mathcal{F}'' \to 0$ be an exact sequence of sheaves on X. If $\mathcal{F}'$ is flabby, the sequence of groups*

$$0 \to \mathcal{F}'(U) \xrightarrow{i} \mathcal{F}(U) \xrightarrow{p} \mathcal{F}''(U) \to 0$$

is exact for every open subset $U \subset X$.

Proof. We have only to prove that p is an epimorphism. Let $s'' \in \mathcal{F}''(U)$. Consider the set of the pairs (Z, s_Z) where Z is an open subset of U and s_Z is a section on $\mathcal{F}$ on Z such that $p(s_Z) = s''_{|Z}$. This set is not empty by definition of exact sequence of sheaves; it is ordered by inclusion and restriction of sections and is inductive. Now, Zorn lemma asserts that it has a maximal element, say (V, s_V). If $V = U$ there is nothing more to prove. If not, let $x \in U - V$ and let $W \subset U$ be an open neighbourhood of x such that there exists a section $s \in \mathcal{F}(W)$ fulfilling $p(s) = s''_{|W}$. Then $p(s_{V|V\cap W} - s_{|V\cap W}) = 0$, and so, $s_{V|V\cap W} - s_{|V\cap W} = i(s'_{V\cap W})$ for some section $s'_{V\cap W} \in \mathcal{F}'(V \cap W)$. Since $\mathcal{F}'$ is flabby, there exists a section $s' \in \mathcal{F}'(U)$ whose restriction to $V \cap W$ is $s'_{V\cap W}$. Now, the sections of $\mathcal{F}$, $s_V - i(s'_{\ |V})$ on V, and s on W, coincide on $V \cap W$, thus defining a section $\tilde{s}_{V\cup W} \in \mathcal{F}(V \cup W)$ such that $p(\tilde{s}_{V\cup W}) = s''_{|V\cup W}$, which is absurd because of the maximality of (V, s_V). ∎

Corollary 1.1. *Given an exact sequence of sheaves $0 \to \mathcal{F}' \xrightarrow{i} \mathcal{F} \xrightarrow{p} \mathcal{F}'' \to 0$ on a topological space X, if $\mathcal{F}'$ and $\mathcal{F}$ are flabby, so is $\mathcal{F}''$.* ∎

Glueing of sheaves. Let X be a topological space, $\{U_i\}$ an open cover of X, and, for every index i, let $\mathcal{F}_i$ be a sheaf on U_i.

[4]For a definition of fibred product see e.g. [**Hus**].

Let us write $U_{ij} = U_i \cap U_j$ and let us assume that there are sheaf isomorphisms

$$\theta_{ij}: \mathcal{F}_{j|U_{ij}} \xrightarrow{\sim} \mathcal{F}_{i|U_{ij}}$$

fulfilling the *glueing condition* ([**GroD**], Ch.0, 3.3)

$$\theta'_{ik} = \theta'_{ij} \circ \theta'_{jk}, \tag{1.2}$$

for every triple (i, j, k), where primes denote restrictions to $U_{ijk} = U_i \cap U_j \cap U_k$.

Proposition 1.2. *There exists a sheaf $\mathcal{F}$ on X and, for every index i, sheaf isomorphisms*

$$\theta_i: \mathcal{F}_{|U_i} \xrightarrow{\sim} \mathcal{F}_i,$$

such that $\theta_{i|U_{ij}} = \theta_{ij} \circ \theta_{j|U_{ij}}$ for every pair of indices (i, j). The sheaf $\mathcal{F}$ and the isomorphisms θ_i are characterized up to isomorphisms.

Proof. Let $\mathfrak{B}$ be the family of the open subsets $U \subset X$ that are contained at least in a U_i. For every open subset U in $\mathfrak{B}$, let us choose one of the U_i's so that $U \subset U_i$, and let us denote by $\mathcal{F}(U)$ the group $\bigcup_i \mathcal{F}_i(U)$. If $V \subset U$, and we have chosen $V \subset U_j$, let us define $\varphi_{U,V}: \mathcal{F}(U) \to \mathcal{F}(V)$ by $\varphi_{U,V} = \theta_{ji} \circ \varphi^i_{U,V}$, where $\varphi^i_{U,V}: \mathcal{F}_i(U) \to \mathcal{F}_i(V)$ is the restriction morphism of $\mathcal{F}_i$. The transitivity property, $\varphi_{U,W} = \varphi_{U,V} \circ \varphi_{V,W}$ for $W \subset V \subset U$, follows from the glueing condition (1.2). We have thus constructed an object which behaves like a sheaf, although $\mathcal{F}(U)$ is only defined for the open subsets U in $\mathfrak{B}$. If V is an arbitrary open subset, and $\{V_i\}$ is the family of the open subsets in $\mathfrak{B}$ that are contained in V, we define $\mathcal{F}(V)$ so that the sequence

$$0 \to \mathcal{F}(V) \to \prod_i \mathcal{F}(V_i) \xrightarrow{r'-r''} \prod_{ij} \mathcal{F}(V_{ij}),$$

constructed as (1.1), is exact. If $V \in \mathfrak{B}$ this definition is coherent with the previous one by (1.1), and the restriction morphisms are defined in the obvious way. The proof is then completed straightforwardly. ■

Let us consider another family of sheaves $\mathcal{G}_i$ on the U_i's endowed with sheaf isomorphisms

$$\zeta_{ij}: \mathcal{G}_{j|U_{ij}} \xrightarrow{\sim} \mathcal{G}_{i|U_{ij}}$$

fulfilling the glueing condition (1.2), so that they define a sheaf $\mathcal{G}$ on X and sheaf isomorphisms

$$\zeta_i: \mathcal{G}_{|U_i} \xrightarrow{\sim} \mathcal{G}_i$$

verifying $\zeta_{i|U_{ij}} = \zeta_{ij} \circ \zeta_{j|U_{ij}}$, as above. One can easily prove the following fact.

Lemma 1.1. *Given sheaf morphisms $f_i: \mathcal{F}_i \to \mathcal{G}_i$ such that the diagram*

$$\begin{array}{ccc} \mathcal{F}_{j|U_{ij}} & \xrightarrow{f_j} & \mathcal{G}_{j|U_{ij}} \\ {\scriptstyle\theta_{ij}}\downarrow & & \downarrow{\scriptstyle\zeta_{ij}} \\ \mathcal{F}_{j|U_{ij}} & \xrightarrow{f_i} & \mathcal{G}_{i|U_{ij}} \end{array}$$

is commutative, there exists a sheaf morphism $f: \mathcal{F} \to \mathcal{G}$ such that $\zeta_i \circ f_{|U_i} = f_i \circ \theta_i$ for every i.

2. Sheaf cohomology

In this Section, for expository reasons, we shall state all definitions and results in the commutative non-graded case, even though they also hold, with obvious changes, in the graded setting, provided all morphisms are understood to be even.

Differential complexes. First of all, we have to introduce some basic cohomological tools at a purely algebraic level. Let R be a commutative ring, and M an R-module.

Definition 2.1. *A differential on M is a morphism $d: M \to M$ of R-modules such that $d^2 \equiv d \circ d = 0$. The pair (M, d) is called a differential module.*

The elements of the spaces $Z(M, d) \equiv \operatorname{Ker} d$ and $B(M, d) \equiv \operatorname{Im} d$ are called *cocycles* and *coboundaries* of (M, d) respectively. The condition $d^2 = 0$ implies that $B(M, d) \subset Z(M, d)$, and the R-module

$$H(M, d) = Z(M, d)/B(M, d)$$

is called the *cohomology group* of the differential module (M, d). We shall often write $Z(M)$, $B(M)$ and $H(M)$, omitting the differential d when there is no risk of confusion.

Let (M, d) and (M', d') be differential R-modules.

Definition 2.2. *A morphism of differential modules is a morphism $f\colon M \to M'$ of R-modules which commutes with the differentials, $f \circ d' = d \circ f$.*

A morphism of differential modules maps cocycles to cocycles and coboundaries to coboundaries, thus inducing a morphism $H(f)\colon H(M) \to H(M')$.

Proposition 2.1. *Let $0 \to M' \xrightarrow{i} M \xrightarrow{p} M'' \to 0$ be an exact sequence of differential R-modules. There exists a morphism $\delta\colon H(M'') \to H(M')$ (called connecting morphism) and an exact triangle of cohomology*

$$\begin{array}{ccc} H(M) & \xrightarrow{H(p)} & H(M'') \\ {\scriptstyle H(i)}\nwarrow & & \swarrow{\scriptstyle \delta} \\ & H(M') & \end{array} \qquad (2.1)$$

Proof. The construction of δ is as follows: let $\xi'' \in H(M'')$ and let m'' be a cocycle whose class is ξ''. If m is an element of M such that $p(m) = m''$, we have $p(d(m)) = d(m'') = 0$ and then $d(m) = i(m')$ for some $m' \in M'$ which is a cocycle. Now, the cocycle m' defines a cohomology class $\delta(\xi'')$ in $H(M')$, which is independent of the choices we have made, thus defining a morphism $\delta\colon H(M'') \to H(M')$. One proves by direct computation that the triangle is exact. ∎

The above results can be translated to the setting of complexes of R-modules.[5]

Definition 2.3. *A complex of R-modules is a differential R-module $(M^\bullet, d)$ which is $\mathbb{Z}$-graded, $M^\bullet = \bigoplus_{n\in\mathbb{Z}} M^n$, and whose differential fulfills $d(M^n) \subset M^{n+1}$ for every $n \in \mathbb{Z}$.*

We shall usually write a complex of R-modules in the more pictorial form

$$\dots \xrightarrow{d_{n-2}} M^{n-1} \xrightarrow{d_{n-1}} M^n \xrightarrow{d_n} M^{n+1} \xrightarrow{d_{n+1}} \dots$$

For a complex $M^\bullet$ the cocycle and coboundary modules and the cohomology group split as direct sums of terms $Z^n(M^\bullet) = \operatorname{Ker} d_n$, $B^n(M^\bullet) = \operatorname{Im} d_{n-1}$ and $H^n(M^\bullet) = Z^n(M^\bullet)/B^n(M^\bullet)$ respectively. The groups $H^n(M^\bullet)$ are called the *cohomology groups* of the complex $M^\bullet$.

[5] Complexes of modules are also called $\mathbb{Z}$-graded differential modules, but we prefer to avoid this terminology, which could lead to confusion.

Definition 2.4. *A morphism of complexes of R-modules $f\colon N^\bullet \to M^\bullet$ is a collection of morphisms $\{f_n\colon N^n \to M^n \mid n \in \mathbf{Z}\}$, such that the following diagram commutes:*

$$\begin{array}{ccc} N^n & \xrightarrow{f_n} & M^n \\ {\scriptstyle d}\downarrow & & \downarrow{\scriptstyle d} \\ N^{n+1} & \xrightarrow[f_{n+1}]{} & M^{n+1} \end{array} .$$

For complexes, Proposition 2.1 takes the following form:

Proposition 2.2. *Let $0 \to N^\bullet \xrightarrow{i} M^\bullet \xrightarrow{p} P^\bullet \to 0$ be an exact sequence of complexes of R-modules. There exist connecting morphisms $\delta_n\colon H^n(P^\bullet) \to H^{n+1}(N^\bullet)$ and a long exact sequence of cohomology*

$$\dots \xrightarrow{\delta_{n-1}} H^n(N^\bullet) \xrightarrow{H(i)} H^n(M^\bullet) \xrightarrow{H(p)} H^n(P^\bullet) \xrightarrow{\delta_n}$$
$$\xrightarrow{\delta_n} H^{n+1}(N^\bullet) \xrightarrow{H(i)} H^{n+1}(M^\bullet) \xrightarrow{H(p)} H^{n+1}(P^\bullet) \xrightarrow{\delta_{n+1}} \dots$$

Proof. The connecting morphism $\delta\colon H^\bullet(P^\bullet) \to H^\bullet(N^\bullet)$ defined in Proposition 2.1 splits into morphisms $\delta_n\colon H^n(P^\bullet) \to H^{n+1}(N^\bullet)$ and the long exact sequence of the statement is obtained by developing the exact triangle of cohomology (2.1). ∎

Canonical flabby resolutions. We provide a direct definition of sheaf cohomology in terms of the so-called *Godement resolution* of a sheaf [Go], even though we shall see later on that sheaf cohomology can also be computed by means of other resolutions. Let $\mathcal{F}$ be a sheaf on a topological space X, and let $\pi\colon \underline{\mathcal{F}} \to X$ be the étalé space of $\mathcal{F}$. We can define a sheaf $C^0\mathcal{F}$ on X by putting:

$$C^0\mathcal{F}(U) \equiv \{\text{all sections } \sigma\colon U \hookrightarrow \underline{\mathcal{F}} \text{ of } \pi \text{ on } U \text{ such that } \sigma \circ \pi = \mathrm{Id}\} = \prod_{x\in U} \mathcal{F}_x$$

with the obvious restriction maps. There is a canonical immersion of sheaves $0 \to \mathcal{F} \xrightarrow{\epsilon} C^0\mathcal{F}$, since, for every open subset U, $\mathcal{F}(U)$ is the set of continuous sections $s\colon U \hookrightarrow \underline{\mathcal{F}}$ of π, while $C^0\mathcal{F}(U)$ is the set of all sections, continuous or not. This also shows that $C^0\mathcal{F}$, called the *sheaf of discontinuous sections* of $\mathcal{F}$, is flabby.

The quotient sheaf $\mathcal{F}_1$ given by $0 \to \mathcal{F} \xrightarrow{\epsilon} C^0\mathcal{F} \xrightarrow{p} \mathcal{F}_1 \to 0$ can be imbedded into $C^0\mathcal{F}_1$, giving rise to a new sequence $0 \to \mathcal{F}_1 \xrightarrow{\epsilon_1} C^0\mathcal{F}_1 \xrightarrow{p_1} \mathcal{F}_2 \to 0$; now $\mathcal{F}_2$ can be imbedded into $C^0\mathcal{F}_2$ obtaining a quotient sheaf $\mathcal{F}_3$, and so on. Then, letting $C^k\mathcal{F} = C^0\mathcal{F}_k$, we obtain a family of exact sequences of sheaves

$$0 \to \mathcal{F}_k \xrightarrow{\epsilon_k} C^k\mathcal{F} \xrightarrow{p_k} \mathcal{F}_{k+1} \to 0\,. \tag{2.2}$$

Definition 2.5. *The long exact sequence of sheaves*

$$0 \to \mathcal{F} \xrightarrow{\epsilon} C^\bullet\mathcal{F} \quad \text{i.e.} \quad 0 \to \mathcal{F} \xrightarrow{\epsilon} C^0\mathcal{F} \xrightarrow{d_0} C^1\mathcal{F} \xrightarrow{d_1} C^2\mathcal{F} \to \dots,$$

obtained from the exact sequences (2.2) by letting $d_k = \epsilon_{k+1} \circ p_k$, is called the canonical flabby resolution of the sheaf $\mathcal{F}$.

The global sections $C^\bullet\mathcal{F}(X)$ of $C^\bullet\mathcal{F}$ give rise to a complex of abelian groups

$$C^0\mathcal{F}(X) \xrightarrow{d_0} C^1\mathcal{F}(X) \xrightarrow{d_1} C^2\mathcal{F}(X) \xrightarrow{d_2} \dots$$

Definition 2.6. *The cohomology groups of X with values in the sheaf $\mathcal{F}$, or simply the cohomology groups of $\mathcal{F}$, are the cohomology groups $H^k(X, \mathcal{F})$ of the complex $C^\bullet\mathcal{F}(X)$, that is:*

$$H^k(X, \mathcal{F}) = H^k(C^\bullet\mathcal{F}(X)) = \operatorname{Ker} d_k / \operatorname{Im} d_{k-1}\,.$$

Notice that $H^0(X, \mathcal{F}) = \mathcal{F}(X)$ for every sheaf $\mathcal{F}$ on X.

The cohomology groups depend functorially on the sheaf in the sense that, given morphisms $f\colon \mathcal{F} \to \mathcal{F}'$ and $g\colon \mathcal{F}' \to \mathcal{F}''$, there exist morphisms $H^k(f)\colon H^k(X,\mathcal{F}) \to H^k(X,\mathcal{F}')$ and $H^k(g)\colon H^k(X,\mathcal{F}') \to H^k(X,\mathcal{F}'')$, naturally induced by the corresponding morphisms of complexes $f\colon C^\bullet(X,\mathcal{F}) \to C^\bullet(X,\mathcal{F}')$ and $g\colon C^\bullet(X,\mathcal{F}') \to C^\bullet(X,\mathcal{F}'')$, which satisfy $H^k(g) \circ H^k(f) = H^k(g \circ f)$.

Definition 2.7. *A sheaf $\mathcal{F}$ on X is acyclic if $H^k(X, \mathcal{F}) = 0$ for every $k > 0$.*

Lemma 2.1. *Flabby sheaves are acyclic.*

Proof. If $\mathcal{F}$ is a flabby sheaf, the sequence $0 \to \mathcal{F}(X) \to C^0\mathcal{F}(X) \to \mathcal{F}_1(X) \to 0$ obtained from $0 \to \mathcal{F} \to C^0\mathcal{F} \to \mathcal{F}_1 \to 0$ is exact (Proposition 1.1). Since $\mathcal{F}$ and $C^0\mathcal{F}$ are flabby, so is $\mathcal{F}_1$ by Corollary 1.1, and the sequence $0 \to \mathcal{F}_1(X) \to C^1\mathcal{F}(X) \to \mathcal{F}_2(X) \to 0$ is also exact. With this procedure we can show that the complex $C^\bullet\mathcal{F}(X)$ is exact, thus proving the claim. ∎

Proposition 2.3. *Let* $0 \to \mathcal{F}' \xrightarrow{i} \mathcal{F} \xrightarrow{p} \mathcal{F}'' \to 0$ *be an exact sequence of sheaves on a topological space* X*. There is a long exact sequence of cohomology groups*

$$\begin{aligned} 0 \to H^0(X, \mathcal{F}') &\xrightarrow{H(i)} H^0(X, \mathcal{F}) \xrightarrow{H(p)} H^0(X, \mathcal{F}'') \xrightarrow{\delta} H^1(X, \mathcal{F}') \xrightarrow{H(i)} \\ &\xrightarrow{H(i)} H^1(X, \mathcal{F}) \xrightarrow{H(p)} H^1(X, \mathcal{F}'') \xrightarrow{\delta} H^2(X, \mathcal{F}') \xrightarrow{H(i)} \dots \end{aligned} \tag{2.3}$$

Proof. The sequence $0 \to \mathcal{F}'_x \to \mathcal{F}_x \to \mathcal{F}''_x \to 0$ is exact for every point $x \in X$, and hence, the induced sequence $0 \to C^0\mathcal{F}'(X) \to C^0\mathcal{F}(X) \to C^0\mathcal{F}''(X) \to 0$ is exact. Iteration gives an exact sequence of complexes of groups $0 \to C^\bullet\mathcal{F}'(X) \to C^\bullet\mathcal{F}(X) \to C^\bullet\mathcal{F}''(X) \to 0$; by taking the corresponding long exact sequence of cohomology, as in Proposition 2.2, one can prove the claim. ∎

The exact sequence of cohomology depends functorially on the exact sequence of sheaves. This amounts to saying that given a commutative diagram of sheaves

$$\begin{array}{ccccccccc} 0 & \longrightarrow & \mathcal{F}' & \longrightarrow & \mathcal{F} & \longrightarrow & \mathcal{F}'' & \longrightarrow & 0 \\ & & \downarrow{\scriptstyle f} & & \downarrow{\scriptstyle g} & & \downarrow{\scriptstyle h} & & \\ 0 & \longrightarrow & \mathcal{G}' & \longrightarrow & \mathcal{G} & \longrightarrow & \mathcal{G}'' & \longrightarrow & 0 \end{array}$$

with exact rows, the associated commutative diagram of complexes

$$\begin{array}{ccccccccc} 0 & \longrightarrow & C^\bullet\mathcal{F}'(X) & \longrightarrow & C^\bullet\mathcal{F}(X) & \longrightarrow & C^\bullet\mathcal{F}''(X) & \longrightarrow & 0 \\ & & \downarrow{\scriptstyle f} & & \downarrow{\scriptstyle g} & & \downarrow{\scriptstyle h} & & \\ 0 & \longrightarrow & C^\bullet\mathcal{G}'(X) & \longrightarrow & C^\bullet\mathcal{G}(X) & \longrightarrow & C^\bullet\mathcal{G}''(X) & \longrightarrow & 0 \end{array}$$

induces in cohomology the following commutative diagram [Go]:

$$\begin{array}{ccccccccccc} \to & H^k(X,\mathcal{F}') & \to & H^k(X,\mathcal{F}) & \to & H^k(X,\mathcal{F}'') & \xrightarrow{\delta} & H^{k+1}(X,\mathcal{F}') & \to \\ & \downarrow{\scriptstyle H(f)} & & \downarrow{\scriptstyle H(g)} & & \downarrow{\scriptstyle H(h)} & & \downarrow{\scriptstyle H(f)} & \\ \to & H^k(X,\mathcal{G}') & \to & H^k(X,\mathcal{G}) & \to & H^k(X,\mathcal{G}'') & \xrightarrow{\delta} & H^{k+1}(X,\mathcal{G}') & \to \dots \end{array}$$

Other resolutions. We now show how the cohomology of a sheaf can be computed without resorting to the canonical flabby resolution.

Definition 2.8. *A resolution of a sheaf* $\mathcal{F}$ *on a topological space* X *is an exact sequence of sheaves* $0 \to \mathcal{F} \xrightarrow{\epsilon} \mathcal{R}^\bullet$, *i.e.* $0 \to \mathcal{F} \xrightarrow{\epsilon} \mathcal{R}^0 \to \mathcal{R}^1 \to \mathcal{R}^2 \to \dots$

A resolution is called *flabby* (*acyclic*, etc.) if all the sheaves $\mathcal{R}^k$ are flabby (acyclic, etc.). It turns out that the cohomology groups of a sheaf $\mathcal{F}$ can be calculated in terms of any acyclic resolution of $\mathcal{F}$.

Proposition 2.4. (Abstract de Rham theorem) *Given a resolution*

$$0 \to \mathcal{F} \xrightarrow{\epsilon} \mathcal{R}^0 \xrightarrow{d_0} \mathcal{R}^1 \xrightarrow{d_1} \dots$$

of a sheaf $\mathcal{F}$, *for every* $k \geq 0$ *there is a morphism*

$$H^k(\mathcal{R}^\bullet(X)) \to H^k(X, \mathcal{F}), \tag{2.4}$$

which for $k = 0$ *is an isomorphism. If* $H^k(X, \mathcal{R}^p) = 0$ *for* $0 \leq p \leq q-1$ *and* $1 \leq k \leq q$ *for a fixed integer* $q \geq 1$, *these morphisms are bijective for* $0 \leq k \leq q$. *In particular, if the resolution* $\mathcal{R}^\bullet$ *is acyclic, all morphisms (2.4) are bijective.*

Proof. The existence of the morphism for $k = 0$ is trivial. Let us define $\mathcal{Q}^k = \operatorname{Ker} d_k \colon \mathcal{R}^k \to \mathcal{R}^{k+1}$; then the sequence of sheaves

$$0 \to \mathcal{F} \xrightarrow{\epsilon} \mathcal{R}^0 \xrightarrow{d_0} \mathcal{Q}^1 \to 0 \tag{2.5}$$

is exact. The induced exact cohomology sequence contains the segments

$$0 \to H^0(X, \mathcal{F}) \to H^0(X, \mathcal{R}^0) \to H^0(X, \mathcal{Q}^1) \xrightarrow{\delta} H^1(X, \mathcal{F}) \to H^1(X, \mathcal{R}^0) \tag{2.6}$$

$$H^{k-1}(X, \mathcal{R}^0) \to H^{k-1}(X, \mathcal{Q}^1) \xrightarrow{\delta} H^k(X, \mathcal{F}) \to H^k(X, \mathcal{R}^0). \tag{2.7}$$

Sequence (2.6) provides a morphism

$$H^1(\mathcal{R}^\bullet(X)) = \frac{H^0(X, \mathcal{Q}^1)}{\operatorname{Im} H^0(X, \mathcal{R}^0)} \to H^1(X, \mathcal{F}),$$

which proves the first claim for $k = 1$. We now proceed by induction on k. To this end, we consider Eq. (2.7), and notice that $0 \to \mathcal{Q}^1 \xrightarrow{\epsilon} \mathcal{R}^{\bullet+1}$, where

$\mathcal{R}^{\bullet+1}$ is the complex $\mathcal{R}^1 \to \mathcal{R}^2 \to \mathcal{R}^3 \to \dots$, is a resolution. By the inductive hypothesis, we have morphisms $H^{k-1}(\mathcal{R}^{\bullet+1}(X)) \to H^{k-1}(X, \mathcal{Q}^1)$. The composition

$$H^k(\mathcal{R}^\bullet(X)) = H^{k-1}(\mathcal{R}^{\bullet+1}(X)) \to H^{k-1}(X, \mathcal{Q}^1) \xrightarrow{\delta} H^k(X, \mathcal{F})$$

provides the morphism we were looking for (the first equality making sense because $k > 1$).

We now assume that $H^k(X, \mathcal{R}^p) = 0$ for $0 \le p \le q-1$ and $1 \le k \le q$. Under this hypothesis, the exact sequence

$$0 \to \mathcal{Q}^k \to \mathcal{R}^k \xrightarrow{d_k} \mathcal{Q}^{k+1} \to 0$$

and sequence (2.5) yield

$$0 \to H^k(X, \mathcal{Q}^1) \to H^{k+1}(X, \mathcal{F}) \to 0, \qquad 1 \le k \le q-1,$$

$$0 \to H^0(X, \mathcal{Q}^k) \to H^0(X, \mathcal{R}^k) \to H^0(X, \mathcal{Q}^{k+1}) \to H^1(X, \mathcal{Q}^{k+1}) \to 0, \qquad 1 \le k \le q-1,$$

$$0 \to H^k(X, \mathcal{Q}^{p+1}) \to H^{k+1}(X, \mathcal{Q}^p) \to 0, \qquad 1 \le k \le q-1, \quad 1 \le p \le q-1.$$

These sequences entail that, for all $1 \le k \le q$,

$$\begin{aligned} H^k(X, \mathcal{F}) &\simeq H^{k-1}(X, \mathcal{Q}^1) \simeq H^{k-2}(X, \mathcal{Q}^2) \simeq \dots \\ &\simeq H^1(X, \mathcal{Q}^k) \simeq \frac{H^0(X, \mathcal{Q}^k)}{\operatorname{Im} H^0(X, \mathcal{R}^{k-1})} = H^k(\mathcal{R}^\bullet(X)). \end{aligned}$$

■

The morphisms (2.4) are natural, in the sense that, given a commutative diagram

$$\begin{array}{ccccccccc} 0 & \longrightarrow & \mathcal{F} & \longrightarrow & \mathcal{R}^0 & \longrightarrow & \mathcal{R}^1 & \longrightarrow & \dots \\ & & {\scriptstyle f}\downarrow & & \downarrow{\scriptstyle g} & & \downarrow{\scriptstyle g} & & \\ 0 & \longrightarrow & \mathcal{G} & \longrightarrow & \mathcal{S}^0 & \longrightarrow & \mathcal{S}^1 & \longrightarrow & \dots \end{array}$$

one obtains, for any $k \geq 0$, a commutative diagram

$$\begin{array}{ccc} H^k(\mathcal{R}^\bullet(X)) & \longrightarrow & H^k(X,\mathcal{F}) \\ {\scriptstyle H(g)}\downarrow & & \downarrow{\scriptstyle H(f)} \\ H^k(\mathcal{S}^\bullet(X)) & \longrightarrow & H^k(X,\mathcal{G}) \end{array}$$

Corollary 2.1. *Let $\mathcal{F}$, $\mathcal{G}$ be sheaves on a topological space X. There is a natural isomorphism*

$$H^k(X,\mathcal{F}\oplus\mathcal{G}) \simeq H^k(X,\mathcal{F})\oplus H^k(X,\mathcal{G}).$$

Proof. If $0 \to \mathcal{F} \to C^\bullet\mathcal{F}$ and $0 \to \mathcal{G} \to C^\bullet\mathcal{G}$ are the canonical flabby resolutions of $\mathcal{F}$ and $\mathcal{G}$, then the resolution $0 \to \mathcal{F}\oplus\mathcal{G} \to C^\bullet\mathcal{F}\oplus C^\bullet\mathcal{G}$ is flabby and therefore acyclic by Lemma 2.1. ■

Now we study the behaviour of sheaf cohomology under the direct image functor. Let $\mathcal{F}$ be a sheaf on a topological space X and let $f\colon X \to Y$ be a continuous map.

Lemma 2.2. *If there exists a flabby resolution $0 \to \mathcal{F} \xrightarrow{\epsilon} \mathcal{R}^\bullet$ of the sheaf $\mathcal{F}$ whose direct image $0 \to f_*\mathcal{F} \xrightarrow{\epsilon} f_*\mathcal{R}^\bullet$ is a resolution of $f_*\mathcal{F}$, then the cohomology groups of the sheaves $\mathcal{F}$, $f_*\mathcal{F}$ coincide:*

$$H^k(X,\mathcal{F}) \simeq H^k(Y,f_*\mathcal{F}), \qquad k \geq 0.$$

Proof. The direct image of a flabby sheaf is flabby, hence, the sequence $0 \to f_*\mathcal{F} \xrightarrow{\epsilon} f_*\mathcal{R}^\bullet$ is a flabby resolution of the sheaf $f_*\mathcal{F}$. The abstract de Rham theorem implies the thesis. ■

We state now a weak version of the Leray theorem [Go], which is anyhow sufficient to our purposes.

Proposition 2.5. *Let $f\colon X \to Y$ be a continuous map, and $\mathcal{F}$ a sheaf on X. If either:*

(1) *f is a closed immersion, or*
(2) *every point $y \in Y$ has a base of open neighbourhoods whose pre-images are acyclic for the sheaf $\mathcal{F}$,*

then the cohomology groups of the sheaves $\mathcal{F}$ and $f_*\mathcal{F}$ are isomorphic,

$$H^k(X, \mathcal{F}) \simeq H^k(Y, f_*\mathcal{F}), \qquad k \geq 0.$$

Proof. Let $0 \to \mathcal{F} \xrightarrow{\epsilon} \mathcal{R}^\bullet$ be a flabby resolution of a sheaf $\mathcal{F}$ on X. If f is a closed immersion, $f_*\mathcal{R}^\bullet$ is still a flabby resolution of $f_*\mathcal{F}$, which proves (1). For (2), one has to prove that $f_*\mathcal{R}^\bullet$ is a resolution of $f_*\mathcal{F}$; namely, that the sequence $0 \to (f_*\mathcal{F})_y \xrightarrow{\epsilon} (f_*\mathcal{R})^\bullet_y$ is exact for every point $y \in Y$, which is equivalent to the exactness of the sequences $0 \to f_*\mathcal{F}(V_i) \xrightarrow{\epsilon} f_*\mathcal{R}^\bullet(V_i)$, where $\{V_i\}$ is a basis of open neighbourhoods of y. Since the cohomology groups of these sequences are the groups $H^k(V_i, \mathcal{F})$, one concludes. ∎

The effect of inverse image in cohomology is described as follows. Let $\mathcal{G}$ be a sheaf on a topological space Y and let $f: X \to Y$ be a continuous map.

Proposition 2.6. *The map f induces morphisms of abelian groups*

$$f^\sharp: H^k(Y, \mathcal{G}) \to H^k(X, f^{-1}\mathcal{G}), \qquad k \geq 0,$$

called the inverse image in cohomology. In particular, taking $\mathcal{G}$ as a constant sheaf G_Y, one obtains morphisms

$$f^\sharp: H^k(Y, G_Y) \to H^k(X, G_X), \qquad k \geq 0.$$

Proof. Since the inverse image of sheaves preserves exact sequences, if $0 \to \mathcal{G} \to C^\bullet\mathcal{G}$ is the canonical flabby resolution of $\mathcal{G}$, then $0 \to f^{-1}\mathcal{G} \to f^{-1}(C^\bullet\mathcal{G})$ is a resolution of $f^{-1}\mathcal{G}$, so that, according to Proposition 2.4, there are morphisms $H^k(f^{-1}(C^\bullet\mathcal{G})(X)) \to H^k(X, f^{-1}\mathcal{G})$. Composing these with the natural morphisms $H^k(C^\bullet\mathcal{G}(Y)) \to H^k(f^{-1}(C^\bullet\mathcal{G})(X))$ one proves the claim. ∎

A particular class of acyclic resolutions which, in accordance with Proposition 2.4, can be used to compute sheaf cohomology, are the *injective resolutions* [**Gro2**]. A sheaf $\mathcal{F}$ is said to be injective if, for any exact sequence of sheaves $0 \to \mathcal{F}' \to \mathcal{F}''$ and any sheaf morphism $\mathcal{F} \to \mathcal{F}''$, there is a morphism $\mathcal{F} \to \mathcal{F}'$ such that the following diagram commutes

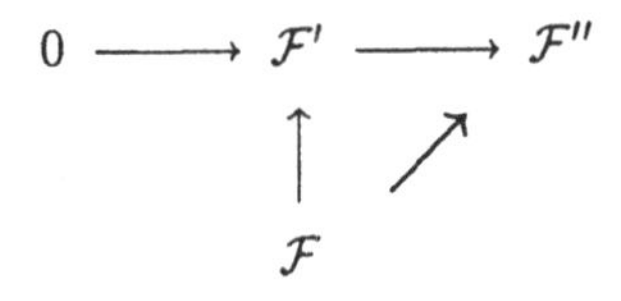

It can be shown [Har,Ten] that any sheaf has an injective resolution (one usually says that the category of sheaves of abelian groups has enough injectives). It is also possible to prove that injective sheaves are flabby, so that sheaf cohomology can be computed by means of injective resolutions.[6] This possibility will be exploited in Chapter V.

3. de Rham, Dolbeault, and Čech cohomologies

Soft and fine sheaves. Let us start by examining in deeper detail the problem of the prolongation of sections.

Definition 3.1. *A sheaf $\mathcal{F}$ is soft if every section of $\mathcal{F}$ on a closed subset $S \subset X$ is the restriction of some global section.*

Lemma 3.1. *Let X be a paracompact topological space, and S a closed subset of X. If $\mathcal{F}$ is a sheaf on X, any section $s \in \Gamma(S, \mathcal{F})$ can be extended to a section of $\mathcal{F}$ on an open neighbourhood W of S in X, that is:*

$$\Gamma(S, \mathcal{F}) = \varinjlim_{S \subset U} \mathcal{F}(U) \qquad (U \text{ open}).$$

Proof. Given a section $s \in \Gamma(S, \mathcal{F})$, there exists an open cover $\{U_i\}$ of S and sections $s_i \in \mathcal{F}(U_i)$ such that $s_{i|S\cap U_i} = s_{|S\cap U_i}$. Since X is paracompact, we can assume that the cover is locally finite, and, even more, that there exists another locally finite open cover $\{V_i\}$ of S such that the closure $\overline{V_i}$ of V_i is contained in U_i, $\overline{V_i} \subset U_i$. Let $W = \{x \in X \,|\, \text{if } x \in \overline{V_i} \cap \overline{V_j}, \text{ then } (s_i)_x = (s_j)_x\}$. The section $s \in \Gamma(S, \mathcal{F})$ can be extended to a section of $\mathcal{F}$ on W, so that it suffices to prove that W is a neighbourhood of S, which follows straightforwardly from a standard topological argument. ■

Corollary 3.1. *Every flabby sheaf on a paracompact space is soft.* ■

Lemma 3.2. *Let $0 \to \mathcal{F}' \xrightarrow{i} \mathcal{F} \xrightarrow{p} \mathcal{F}'' \to 0$ be an exact sequence of sheaves on a paracompact space X. If $\mathcal{F}'$ is soft, the sequence of groups*

$$0 \to \mathcal{F}'(U) \xrightarrow{i} \mathcal{F}(U) \xrightarrow{p} \mathcal{F}''(U) \to 0$$

[6] This means that the sheaf cohomology groups are the derived functors of the global section functor.

is exact for any open subset $U \subset X$.

Proof. The proof is similar to that of Proposition 1.1. ■

Corollary 3.2. *If $0 \to \mathcal{F}' \xrightarrow{i} \mathcal{F} \xrightarrow{p} \mathcal{F}'' \to 0$ is an exact sequence of sheaves on a paracompact space X, and $\mathcal{F}'$, $\mathcal{F}$ are soft, so is $\mathcal{F}''$.* ■

Proceeding as in Lemma 2.1, and applying the abstract de Rham theorem (Proposition 2.4), one proves the following result.

Corollary 3.3. *On a paracompact space X, any soft sheaf $\mathcal{F}$ is acyclic. Therefore, if $0 \to \mathcal{F} \xrightarrow{\epsilon} \mathcal{R}^\bullet$ is a soft resolution of $\mathcal{F}$, there are natural isomorphisms*

$$H^k(\mathcal{R}^\bullet(X)) \simeq H^k(X, \mathcal{F}), \qquad k \geq 0.$$

■

Definition 3.2. *A sheaf of rings $\mathcal{R}$ on a topological space X is fine if, for any locally finite oper cover $\mathfrak{U} = \{U_i\}_{i\in I}$ of X, there is a family $\{s_i\}_{i\in I}$ of global sections of $\mathcal{R}$ such that:*

(1) $\sum_{i\in I} s_i = 1$;
(2) *for every $i \in I$ there is a closed subset $S_i \subset U_i$ such that $(s_i)_x = 0$ whenever $x \notin S_i$.*

A sheaf $\mathcal{F}$ of abelian groups over X is fine if the sheaf of rings $\mathcal{H}om(\mathcal{F}, \mathcal{F})$ is fine.

The family $\{s_i\}$ is called a *partition of unity* subordinated to the cover $\mathfrak{U}$. For instance, the sheaf of continuous functions on a paracompact topological space as well as the sheaf of smooth functions on a differentiable manifold are fine, while the sheaves of complex or real analytic functions are not.

Lemma 3.3. *Any fine sheaf $\mathcal{F}$ on a paracompact space X is soft.*

Proof. Let $S \subset X$ be a closed subset, and $s \in \Gamma(S, \mathcal{F})$. Consider a cover $\{U_1, U_2\}$ of X, with $S \subset U_1$ and $U_2 = X - S$, and a subordinated partition of unity $\{h_1, h_2\}$ of $\mathcal{H}om(\mathcal{F}, \mathcal{F})$. By Lemma 3.1, we may take U_1 such that s can be extended to it. Now we define a global section s' of $\mathcal{F}$ which extends s by letting $s'_{|U_1} = h_1(s)$, $s'_{|X-U_1} = 0$. ■

In general, the converse holds true only for sheaves of rings.

Lemma 3.4. *Any soft sheaf of rings $\mathcal{R}$ on a paracompact space X is fine.*

Proof. [Go] Let $\{U_i\}_{i\in I}$ be a locally finite open cover of X, and $\{S_i\}_{i\in I}$ a closed cover such that $S_i \subset U_i$. Consider the set of pairs (J, F_J), where $J \subset I$, and $F_J = \{s_j \in \mathcal{R}(X)\}_{j\in J}$ is such that

$$\sum_{j\in J} s_j = 1 \qquad \text{on} \qquad S_J = \bigcup_{j\in J} S_j .$$

This set is ordered by inclusion of the sets J, is inductive and not empty, so that by Zorn lemma it has a maximal element, say $(\widehat{J}, F_{\widehat{J}})$. We have to prove that $\widehat{J} = I$. Indeed, if there is an $i \in I - \widehat{J}$, one can construct a global section s_i such that

$$s_{i|X-U_i} = 0, \qquad s_{i|S_{\widehat{J}}\cup S_i} = 1 - \sum_{j\in\widehat{J}} s_j .$$

Thus, s_i is defined on the closed set $S_{\widehat{J}} \cup S_i \cup (X - U_i)$ and can be extended to X, thereby contradicting the maximality of $(\widehat{J}, F_{\widehat{J}})$. ∎

Corollary 3.4. *If $\mathcal{R}$ is a fine sheaf of rings on a paracompact space X, then any $\mathcal{R}$-module is fine, and therefore soft.* ∎

For instance, if $\mathcal{F}$ is a fine sheaf of abelian groups on X, the sheaf $\mathcal{F} \otimes_{\mathbb{Z}} \mathcal{G}$ is fine for every sheaf $\mathcal{G}$ on X since it has a natural $\mathcal{H}om(\mathcal{F}, \mathcal{F})$-module structure.

de Rham and Dolbeault theorems. We possess now the tools for providing a simple proof of the classical de Rham theorem. Let us consider an n-dimensional differentiable manifold X, and let us denote by Ω^k the sheaf of differential k-forms on X.

Definition 3.3. *The de Rham cohomology groups of X are the cohomology groups $H^k_{DR}(X) = \operatorname{Ker} d_k / \operatorname{Im} d_{k-1}$ of the de Rham complex of differential forms*

$$\Omega^0(X) \xrightarrow{d_0} \Omega^1(X) \xrightarrow{d_1} \Omega^2(X) \xrightarrow{d_2} \dots \xrightarrow{d_{n-1}} \Omega^n(X)$$

where $d_k : \Omega^k(X) \to \Omega^{k+1}(X)$ is the exterior differential.

Proposition 3.1. (de Rham theorem) *There are natural isomorphisms*

$$H^k_{DR}(X) \simeq H^k(X, \mathbb{R}), \qquad k \geq 0.$$

Proof. The sequence $0 \longrightarrow \mathbf{R} \longrightarrow \Omega^0 \xrightarrow{d_0} \Omega^1 \xrightarrow{d_1} \dots$ is a fine resolution of the constant sheaf $\mathbf{R}$ (Poincaré lemma). By Corollary 3.3 one attains the thesis. ■

If $f: X \to Y$ is a smooth map of differentiable manifolds, the pullback f^* acting on differential forms commutes with the exterior differential operator, thus inducing a morphism $f^\sharp: H^\bullet_{DR}(Y) \to H^\bullet_{DR}(X)$ (which is no more than the morphism denoted $H(f^*)$ after Definition 2.2). In other words, $H^\bullet_{DR}(\cdot)$ is a contravariant functor from the category of smooth manifolds to the category of real vector spaces. Moreover, the morphisms $f^\sharp: H^k_{DR}(Y) \to H^k_{DR}(X)$ can be proved to coincide, via the de Rham theorem, with the inverse image morphisms $f^\sharp: H^k(Y, \mathbf{R}) \to H^k(X, \mathbf{R})$ defined in Proposition 2.6.

Another application of the abstract de Rham theorem is the Dolbeault theorem.

Definition 3.4. *The Dolbeault cohomology groups of a complex manifold X are the cohomology groups $H^{p,q}_{\bar\partial}(X) = \operatorname{Ker} \bar\partial_q / \operatorname{Im} \bar\partial_{q-1}$ of the Dolbeault complex*

$$\mathcal{O}^p(X) \longrightarrow \Omega^{p,0}(X) \xrightarrow{\bar\partial_0} \Omega^{p,1}(X) \xrightarrow{\bar\partial_1} \Omega^{p,2}(X) \xrightarrow{\bar\partial_2} \dots$$

where $\mathcal{O}^p$ is the sheaf of holomorphic p-forms, $\Omega^{p,q}$ the sheaf of differential forms of type (p,q), and $\bar\partial$ the Dolbeault (also called Cauchy-Riemann) operator (cf. e.g. [**Wel**] *or* [**GrH**]*).*

Proposition 3.2. *Let X be a complex manifold. There are natural isomorphisms*

$$H^{p,q}_{\bar\partial}(X) \simeq H^q(X, \mathcal{O}^p)$$

Proof. The sequence $0 \longrightarrow \mathcal{O}^p \longrightarrow \Omega^{p,0} \xrightarrow{\bar\partial_0} \Omega^{p,1} \xrightarrow{\bar\partial_1} \Omega^{p,2} \xrightarrow{\bar\partial_2} \dots$ is a soft resolution of the sheaf $\mathcal{O}^p$ of holomorphic p-forms (this is the Poincaré lemma for $\bar\partial$, also called the Dolbeault or Grothendieck lemma; cf. [**GrH**]). One once more concludes by Corollary 3.3. ■

Čech cohomology. Finally, we outline the foundations of still another cohomology theory, namely, the Čech cohomology of a sheaf. If the base space is paracompact, the Čech and sheaf cohomology coincide; in the next Chapters, we shall take advantage of this fact, and shall switch freely from one cohomology to the other, as best suits us.

Let $\mathfrak{U} = \{U_i\}_{i \in I}$ be a locally finite open cover of a topological space X, with I an ordered set; we let $U_{i_0,\dots,i_p} = U_{i_0} \cap \dots \cap U_{i_p}$. Let $\mathcal{F}$ be a sheaf on X.

Definition 3.5. *The group of p-cochains of $\mathcal{F}$ with respect to $\mathfrak{U}$ is*

$$\check{C}^p(\mathfrak{U}, \mathcal{F}) = \prod_{i_0<\cdots<i_p} \mathcal{F}(U_{i_0,\dots,i_p}).$$

A p-cochain α is described by its components $\alpha(i_0, \dots, i_p) \in \mathcal{F}(U_{i_0,\dots,i_p})$. One can define a map $d_p: \check{C}^p(\mathfrak{U}, \mathcal{F}) \to \check{C}^{p+1}(\mathfrak{U}, \mathcal{F})$ by

$$d_p(\alpha)(i_0, \dots, i_{p+1}) = \sum_{k=0}^{p+1} (-1)^k \alpha(i_0, \dots, \widehat{i_k}, \dots, i_{p+1})_{|U_{i_0,\dots,i_{p+1}}}$$

where, as usual, $\widehat{i_k}$ denotes the omission of i_k. One proves easily that $d_{p+1} \circ d_p = 0$, thus obtaining a complex of abelian groups $\check{C}^\bullet(\mathfrak{U}, \mathcal{F})$, i.e.

$$\check{C}^0(\mathfrak{U}, \mathcal{F}) \xrightarrow{d_0} \check{C}^1(\mathfrak{U}, \mathcal{F}) \xrightarrow{d_1} \check{C}^2(\mathfrak{U}, \mathcal{F}) \xrightarrow{d_2} \dots .$$

Definition 3.6. *The Čech cohomology groups of $\mathcal{F}$ with respect to $\mathfrak{U}$ are the cohomology groups $\check{H}^k(\mathfrak{U}, \mathcal{F}) = \operatorname{Ker} d_k / \operatorname{Im} d_{k-1}$ of the Čech complex of cochains $\check{C}^\bullet(\mathfrak{U}, \mathcal{F})$.*

The sheaf axioms entail the identifications $\check{H}^0(\mathfrak{U}, \mathcal{F}) = \mathcal{F}(X) = H^0(X, \mathcal{F})$.

Čech cohomology can be related to sheaf cohomology as follows [**Go,Har**]. Let us denote by $\mathcal{F}_{i_0,\dots,i_p}$ the sheaf $j_*(\mathcal{F}_{|U_{i_0,\dots,i_p}})$ where $j: U_{i_0,\dots,i_p} \hookrightarrow X$ is the natural immersion, and let us consider the sheaf of germs of p-cochains $\check{\mathcal{C}}^p(\mathfrak{U}, \mathcal{F}) = \prod_{i_0<\cdots<i_p} \mathcal{F}_{i_0,\dots,i_p}$, so that $\check{C}^p(\mathfrak{U}, \mathcal{F}) = \Gamma(X, \check{\mathcal{C}}^p(\mathfrak{U}, \mathcal{F}))$. By defining $d_p: \check{\mathcal{C}}^p(\mathfrak{U}, \mathcal{F}) \to \check{\mathcal{C}}^{p+1}(\mathfrak{U}, \mathcal{F})$ as above, one obtains a complex of sheaves $\check{\mathcal{C}}^\bullet(\mathfrak{U}, \mathcal{F})$.

Proposition 3.3. *The complex of sheaves $\check{\mathcal{C}}^\bullet(\mathfrak{U}, \mathcal{F})$ is a resolution of $\mathcal{F}$, which is called the Čech resolution of $\mathcal{F}$ with respect to the cover $\mathfrak{U}$. One then has natural morphisms*

$$\check{H}^k(\mathfrak{U}, \mathcal{F}) \to H^k(X, \mathcal{F}) \tag{3.1}$$

which are isomorphisms when the sheaves $\check{\mathcal{C}}^k(\mathfrak{U}, \mathcal{F})$ are acyclic.

Proof. The kernel of $d_0: \check{\mathcal{C}}^0(\mathfrak{U}, \mathcal{F}) \to \check{\mathcal{C}}^1(\mathfrak{U}, \mathcal{F})$ is clearly $\mathcal{F}$ (by the sheaf axioms). One then has to prove the exactness of $\check{\mathcal{C}}^\bullet(\mathfrak{U}, \mathcal{F})$ in degree higher than zero. To do so, if $\alpha \in \Gamma(V, \check{\mathcal{C}}^p(\mathfrak{U}, \mathcal{F}))$ is a cocycle on a neighbourhood V of a point x, one has to find a $(p-1)$-cochain $\beta \in \Gamma(W, \check{\mathcal{C}}^{p-1}(\mathfrak{U}, \mathcal{F}))$ defined on some open neighbourhood $W \subset V$ of x such that $d_{p-1}(\beta) = \alpha_{|W}$. As $x \in U_j$ for some index j, let us take $W = V \cap U_j$. Then, $k_p: \Gamma(W, \check{C}^p(\mathfrak{U}, \mathcal{F})) \to \Gamma(W, \check{C}^{p-1}(\mathfrak{U}, \mathcal{F}))$ defined by $k_p(\gamma)(i_0, \ldots, i_{p-1}) = \gamma(j, i_0, \ldots, i_{p-1})$ fulfills $(d_{p-1} \circ k_p + k_{p+1} \circ d_p)(\gamma) = \gamma$ for every p-cochain γ on W; that is, $\alpha_{|W} = d_{p-1}(k_p(\alpha_{|W}))$ as required. The second claim is a consequence of Proposition 2.4. ∎

The sheaves of germs of p-cochains of a sheaf $\mathcal{F}$ may fail to be acyclic. Thus, in general, Čech cohomology with respect to a given cover is not equivalent to sheaf cohomology, although if $\mathcal{F}$ is a flabby sheaf then all the sheaves $\check{C}^p(\mathfrak{U}, \mathcal{F})$ are flabby as well, so that the morphisms (3.1) are bijective. More generally, the following result, known as *Leray lemma*, holds.

Proposition 3.4. *If the cover $\mathfrak{U}$ is acyclic for the sheaf $\mathcal{F}$, that is, if $H^k(U_{i_0,\ldots,i_p}, \mathcal{F}) = 0$ for every $k \geq 1$ and every non-void intersection $U_{i_0,\ldots,i_p}$, one has isomorphisms*

$$\check{H}^k(\mathfrak{U}, \mathcal{F}) \simeq H^k(X, \mathcal{F}) \qquad k \geq 0. \tag{3.2}$$

Proof. Let us consider the exact sequence $0 \to \mathcal{F} \to C^0\mathcal{F} \to \mathcal{F}_1 \to 0$, with $C^0\mathcal{F}$ the sheaf of discontinuous sections of $\mathcal{F}$. Since $H^1(U_{i_0,\ldots,i_p}, \mathcal{F}) = 0$ for every $U_{i_0,\ldots,i_p}$, one obtains an exact sequence of complexes

$$0 \to \check{C}^\bullet(\mathfrak{U}, \mathcal{F}) \to \check{C}^\bullet(\mathfrak{U}, C^0\mathcal{F}) \to \check{C}^\bullet(\mathfrak{U}, \mathcal{F}_1) \to 0.$$

The corresponding exact sequence of cohomology yields

$$0 \to \check{H}^0(\mathfrak{U}, \mathcal{F}) \to \check{H}^0(\mathfrak{U}, C^0\mathcal{F}) \to \check{H}^0(\mathfrak{U}, \mathcal{F}_1) \to \check{H}^1(\mathfrak{U}, \mathcal{F}) \to 0$$

and

$$\check{H}^k(\mathfrak{U}, \mathcal{F}_1) \simeq \check{H}^{k+1}(\mathfrak{U}, \mathcal{F}) \qquad k \geq 1;$$

indeed $\check{H}^i(\mathfrak{U}, C^0\mathcal{F}) = 0$ for every $k \geq 1$, since $C^0\mathcal{F}$ is flabby. Comparing these results with the similar ones obtained from the exact sequence of sheaf cohomology induced by $0 \to \mathcal{F} \to C^0\mathcal{F} \to \mathcal{F}_1 \to 0$, one concludes. ∎

It is possible to define Čech cohomology groups depending only on the pair $(X, \mathcal{F})$, and not on a cover, by letting

$$\check{H}^k(X, \mathcal{F}) = \varinjlim_{\mathfrak{U}} \check{H}^k(\mathfrak{U}, \mathcal{F}).$$

The direct limit is taken over a cofinal[7] subset of the directed set of all covers of X (the order is of course the refinement of covers: a cover $\mathfrak{V} = \{V_j\}_{j \in J}$ is a refinement of $\mathfrak{U}$ if there is a map $f: I \to J$ such that $V_{f(i)} \subset U_i$ for every $i \in I$). The order must be fixed at the outset, since a cover may be regarded as a refinement of another in many ways. As two different cofinal families give rise to the same inductive limit [Go], the groups $\check{H}^k(X, \mathcal{F})$ are well defined.

The cohomology groups $\check{H}^k(X, \mathcal{F})$ can be equivalently described as the cohomology groups of the complex $\check{C}^\bullet(X, \mathcal{F}) = \varinjlim_{\mathfrak{U}} \check{C}^\bullet(\mathfrak{U}, \mathcal{F})$, since the cohomology groups of a direct limit of complexes of groups is equal to the direct limit of the corresponding cohomology groups of each complex. More generally, one can introduce the sheaf $\check{\mathcal{C}}^\bullet(X, \mathcal{F})$ associated with the presheaf

$$U \to \varinjlim_{\mathfrak{U}} \Gamma(U, \check{\mathcal{C}}^\bullet(\mathfrak{U}, \mathcal{F}))\,,$$

i.e. the direct limit sheaf of the system $\check{\mathcal{C}}^\bullet(\mathfrak{U}, \mathcal{F})$. The complex $\check{\mathcal{C}}^\bullet(X, \mathcal{F})$ is again a resolution of $\mathcal{F}$ [Go], called the *Čech resolution* of $\mathcal{F}$, so that the abstract de Rham theorem entails the existence of natural morphisms

$$\check{H}^k(X, \mathcal{F}) \to H^k(X, \mathcal{F}). \tag{3.3}$$

Proposition 3.5. *The morphisms (3.3) are bijective whenever X is paracompact.*

Proof. $\check{\mathcal{C}}^k(X, \mathcal{F})$ is a sheaf of $\check{\mathcal{C}}^0(X, \mathbb{Z})$-modules, where $\mathbb{Z}$ is the constant sheaf on X, whose stalks are the integers. Since X is paracompact, it is easy to demonstrate the isomorphism $\check{\mathcal{C}}^0(X, \mathbb{Z}) \simeq C^0\mathbb{Z}$, where $C^0\mathbb{Z}$ is the flabby sheaf of discontinous sections of $\mathbb{Z}$. By Corollary 3.1, Lemma 3.4, and finally Corollary 3.4, $\check{\mathcal{C}}^k(X, \mathcal{F})$ is soft, and hence acyclic. ∎

Propositions 3.1 and 3.5 imply the following form of the de Rham theorem.

Corollary 3.5. *If X is a differentiable manifold, there are natural isomorphisms*

$$H^k_{DR}(X) \simeq \check{H}^k(X, \mathbb{R}), \qquad k \geq 0\,.$$

∎

[7] A subset B of a directed set A is said to be *cofinal* in A if for every $a \in A$ there exists a $b \in B$ such that $a < b$.

4. Graded ringed spaces

In this Section we introduce some basic algebraic-geometric machinery necessary for the development of supergeometry in the following Chapters. We start by generalizing the notions of *ringed space* and *locally ringed space*, which belong to the realm of algebraic geometry (cf. [**GroD**]), to the graded setting. We shall spell out in some detail how ordinary real and complex differential geometry can be formulated within the framework of locally ringed spaces by means of the notion of *spectrum*.

A graded ring R is said to be *local* if it has a unique graded maximal ideal. If R and S are graded local rings, a ring morphism $f: R \to S$ is said to be *local* if it maps the maximal ideal of R into the maximal ideal of S.

Definition 4.1. *Let R be a graded-commutative ring. A graded ringed R-space is a pair $(X, \mathcal{A})$, where X is a topological space and $\mathcal{A}$ is a sheaf of graded-commutative R-algebras on X. If every stalk $\mathcal{A}_x$ is a local ring, $(X, \mathcal{A})$ is said to be a graded locally ringed space.*

Whenever all the graded objects involved in this definition are ordinary commutative objects endowed with the trivial gradation, the usual notion of (locally) ringed space is recovered.

If $(X, \mathcal{A})$ is a graded (locally) ringed space, and $V \subset X$ is an open subset, the pair $(V, \mathcal{A}_V)$, where $\mathcal{A}_V = \mathcal{A}_{|V}$, is also a graded (locally) ringed space. These graded (locally) ringed spaces will be called open subspaces of $(X, \mathcal{A})$.

Let $(X, \mathcal{A})$ and $(Y, \mathcal{B})$ be graded ringed R-spaces.

Definition 4.2. *A morphism of graded ringed R-spaces is a pair*

$$(f, \phi): (X, \mathcal{A}) \to (Y, \mathcal{B}),$$

where $f: X \to Y$ is a continuous map, and $\phi: \mathcal{B} \to f_*\mathcal{A}$ is an even morphism of sheaves of graded R-algebras. If $(X, \mathcal{A})$ and $(Y, \mathcal{B})$ are graded locally ringed spaces, a morphism $(f, \phi): (X, \mathcal{A}) \to (Y, \mathcal{B})$ as above is said to be a morphism of graded locally ringed spaces if the induced morphisms $\phi_y: \mathcal{B}_y \to \mathcal{A}_{f^{-1}(y)}$ are local for every point $y \in Y$.

The notion of morphism of graded (locally) ringed spaces includes that of isomorphism in the obvious way. More generally, two graded (locally) ringed R-spaces $(X, \mathcal{A})$ and $(Y, \mathcal{B})$ are said to be *locally isomorphic* if there exist open covers $\{U_i\}_{i\in I}$ of X and $\{V_i\}_{i\in I}$ of Y, together with a family of isomorphisms $(U_i, \mathcal{A}_{|U_i}) \xrightarrow{\sim} (V_i, \mathcal{B}_{|V_i})$.

Let $F = (f, \phi): (X, \mathcal{A}) \to (Y, \mathcal{B})$ be a morphism of graded ringed spaces, and $\mathcal{N}$ a sheaf of $\mathcal{B}$-modules.

Definition 4.3. *The inverse image of $\mathcal{N}$ by F is the sheaf of $\mathcal{A}$-modules given by*

$$F^*\mathcal{N} = \mathcal{A} \otimes_{f^{-1}\mathcal{B}} f^{-1}\mathcal{N}$$

where $\mathcal{A}$ is considered as a module over $f^{-1}\mathcal{B}$ by means of the sheaf morphism $f^{-1}\mathcal{B} \to \mathcal{A}$ induced by $\phi: \mathcal{B} \to f_*\mathcal{A}$.

Contrary to what happens in the case of the inverse image of sheaves of abelian groups, the inverse image of sheaves of modules may fail to be exact, i.e. in general it does not map exact sequences to exact sequences. In fact, it is exact whenever $\mathcal{A}$ is flat over $\mathcal{B}$ (cf. for instance [**Har**]).

Locally ringed spaces were introduced by Grothendieck to provide formal and unified foundations of algebraic geometry through the concept of scheme. Affine or projective varietes are among the simplest examples of locally ringed spaces. It is also possible to give a treatment of real and complex differential geometry in terms of locally ringed spaces. Thus, a differentiable manifold is a locally ringed $\mathbf{R}$-space $(X, \mathcal{C}^\infty_X)$ locally isomorphic with $(\mathbf{R}^n, \mathcal{C}^\infty_{\mathbf{R}^n})$, while a complex analytic manifold is a locally ringed $\mathbf{C}$-space $(X, \mathcal{O})$ locally isomorphic with $(\mathbf{C}^n, \mathcal{O}_{\mathbf{C}^n})$, with $\mathcal{O}_{\mathbf{C}^n}$ the sheaf of holomorphic functions on $\mathbf{C}^n$.

This characterization of differentiable manifolds (and, analogously, that of complex manifolds) agrees with the usual one because if U is an open set in X and $(f, \phi): (U, \mathcal{C}^\infty_{X\,|U}) \to (V, \mathcal{C}^\infty_{\mathbf{R}^n\,|V})$ is an isomorphism of locally ringed $\mathbf{R}$-spaces, then one necessarily has $\phi = f^*$. This fact is proved in terms of the notion of the *spectrum* of a ring (cf. Proposition 4.1).

The spectrum of a ring. We recall here some basic facts about the spectrum of a commutative, non-graded ring R [**AtM**]; the generalization to the graded setting is straightforward.

Definition 4.4. *The spectrum of R, denoted* Spec R, *is the set of all prime ideals of R.*

Spec R can be endowed with the so-called *Zariski topology*, which is generated by the basis of closed subsets

$$V(f) = \{\mathfrak{p} \in \operatorname{Spec} R \,|\, f \in \mathfrak{p}\},$$

where f is an element of R.

If $\mathfrak{I}$ is an ideal of R, the set $V(\mathfrak{I})$ of all prime ideals of R which contain $\mathfrak{I}$ is a closed subset of Spec R, and all closed subsets of Spec R can be written in this form. In particular, one has $V(\mathfrak{N}) = \operatorname{Spec} R$, where $\mathfrak{N}$ is the ideal of nilpotent elements.

A ring morphism $\phi: R \to S$ induces a map $\phi^*: \operatorname{Spec} S \to \operatorname{Spec} R$, defined as $\phi^*(\mathfrak{p}) = \phi^{-1}(\mathfrak{p})$, which is easily shown to be continuous with respect to the Zariski topology. If $\pi: R \to S = R/\mathfrak{I}$ is the quotient morphism with respect to an ideal $\mathfrak{I}$, then π^* is a homeomorphism from $\operatorname{Spec}(R/\mathfrak{I})$ onto the closed subset $V(\mathfrak{I})$,

$$\operatorname{Spec}(R/\mathfrak{I}) \simeq V(\mathfrak{I}). \tag{4.1}$$

In particular, one has a homeomorphism

$$\operatorname{Spec}(R/\mathfrak{N}) \simeq \operatorname{Spec} R. \tag{4.2}$$

Definition 4.5. *The maximal spectrum of a ring R is the subspace $\operatorname{Spec}_{\max} R \subset \operatorname{Spec} R$ of all maximal ideals of R, endowed with the Zariski topology.*

We focus now our attention on a commutative $\mathbf{R}$-algebra P; a ring morphism $\mathbf{R} \to P$ is defined by letting $x \mapsto 1 \cdot x$.

Definition 4.6. *The real spectrum of P is the subspace $\operatorname{Spec}_{\mathbf{R}} P \subset \operatorname{Spec}_{\max} P$ of all maximal ideals $\mathfrak{M}$ of P such that the composition of morphisms $\mathbf{R} \to P \to P/\mathfrak{M}$ is an isomorphism.*

It should be noticed that there is a one-to-one correspondence

$$\begin{aligned} \operatorname{Hom}_{\mathbf{R}\text{-alg}}(P, \mathbf{R}) &\to \operatorname{Spec}_{\mathbf{R}}(P) \\ \phi &\mapsto \operatorname{Ker} \phi \end{aligned}$$

which allows us to regard the real spectrum as the set of all $\mathbf{R}$-algebra morphisms from P into the field of real numbers.

The real spectrum of an algebra is functorial, since for every $\mathbf{R}$-algebra morphism $\psi: P' \to P$ the induced map $\psi^*: \operatorname{Spec}(P) \to \operatorname{Spec}(P')$ sends the real spectrum into the real spectrum, thus giving a continuous map

$$\psi^*\colon \operatorname{Spec}_{\mathbf{R}}(P) \to \operatorname{Spec}_{\mathbf{R}}(P'). \tag{4.3}$$

In terms of morphisms of $\mathbf{R}$-algebras, this map can be described by $\psi^*(\phi_{\mathfrak{M}}) = \phi_{\mathfrak{M}} \circ \psi$.

Any element of P induces a real function on the real spectrum $\operatorname{Spec}_{\mathbf{R}} P$. For every point $\mathfrak{M} \in \operatorname{Spec}_{\mathbf{R}} P$, let us denote by $\phi_{\mathfrak{M}}\colon P \to P/\mathfrak{M} \simeq \mathbf{R}$ the quotient morphism; then, for every $f \in P$, we have a function

$$\begin{aligned} f\colon \operatorname{Spec}_{\mathbf{R}} P &\to \mathbf{R} \\ \mathfrak{M} &\mapsto \phi_{\mathfrak{M}}(f). \end{aligned}$$

In general, this function is not continuous in the Zariski topology, but one can introduce in $\operatorname{Spec}_{\mathbf{R}} P$ another topology, called the Gel'fand topology, which is the coarsest topology which makes all such functions continuous. Under certain conditions [GlRC,Nai], that are for instance fulfilled by the ring of differentiable functions on a smooth manifold, the Zariski and Gel'fand topologies coincide.

In the case of a graded commutative ring $R = R_0 \oplus R_1$, one can define its spectrum $\operatorname{Spec} R$ as the set of its graded prime ideals, and the corresponding Zariski topology can be described by means of the closed subsets $V(f)$ for the homogeneous elements $f \in R$. The notion of maximal spectrum can be also given, and one thus sees that $\operatorname{Spec}_{\max} R = \operatorname{Spec}_{\max} R_0$. In a similar way, one can introduce the real spectrum $\operatorname{Spec}_{\mathbf{R}} P$ of a graded $\mathbf{R}$-algebra P; one has that:

$$\operatorname{Hom}_{\text{graded } \mathbf{R}\text{-alg}}(P, \mathbf{R}) \simeq \operatorname{Spec}_{\mathbf{R}} P = \operatorname{Spec}_{\mathbf{R}} P_0. \tag{4.4}$$

Differentiable manifolds as ringed spaces. Differentiable manifolds can be described algebraically, since they can be reconstructed as the spectra of their rings of differentiable functions.

Let X be a differentiable manifold, and let us take P as the ring $\mathcal{C}^\infty(X)$ of differentiable functions on X. For every point $x \in X$, there is an ideal $\mathfrak{M}_x$ described as $\mathfrak{M}_x = \{f \in \mathcal{C}^\infty(X) \mid f(x) = 0\}$, which is maximal and satisfies $\mathcal{C}^\infty(X)/\mathfrak{M}_x \simeq \mathbf{R}$, because it is the kernel of the following morphism (evaluation at x):

$$\begin{aligned} \omega_x\colon \mathcal{C}^\infty(X) &\to \mathbf{R} \\ f &\mapsto f(x). \end{aligned}$$

Lemma 4.1. *Let $x = (x^1, \ldots, x^n)$ be a point of the euclidean space $\mathbf{R}^n$. The maximal ideal $\mathfrak{M}_x$ of $\mathcal{C}^\infty(\mathbf{R}^n)$ is generated by $(X^1 - x^1, \ldots, X^n - x^n)$, where $(X^1, \ldots, X^n)$ are the canonical coordinates in $\mathbf{R}^n$.*

Proof. Let $f \in \mathfrak{M}_x$ be a differentiable function on $\mathbf{R}^n$ vanishing at x. Given an arbitrary point $(X^1, \ldots, X^n)$, let us consider the differentiable function on the closed interval $[0,1]$ given by $\psi(t) = f(t\,X^1 + (1-t)\,x^1, \ldots, t\,X^n + (1-t)\,x^n)$. By integrating one obtains

$$
\begin{aligned}
f(X^1, \ldots, X^n) &= \psi(1) - \psi(0) = \int_0^1 \frac{d\psi}{dt}\,dt \\
&= \sum_{i=1}^n (X^i - x^i) \int_0^1 \frac{\partial f}{\partial X^i}(t\,X^1 + (1-t)\,x^1, \ldots, t\,X^n + (1-t)\,x^n)\,dt
\end{aligned}
$$

thus proving the claim, because the integrals in the last line are differentiable functions of $X^1, \ldots, X^n$. ∎

We now need to recall two results, namely, that given a closed subset $Y \subset X$ there is a differentiable function $f: X \to \mathbf{R}$ such that $Y = f^{-1}(0)$, and that for any differentiable manifold X there exists a closed differentiable immersion $X \hookrightarrow \mathbf{R}^N$ of X in some euclidean space (Whitney immersion theorem; see e.g. [DR]).

Proposition 4.1. *The map*

$$
\begin{aligned}
\beta: X &\to \operatorname{Spec}_{\mathbf{R}} \mathcal{C}^\infty(X) \\
x &\mapsto \mathfrak{M}_x
\end{aligned}
$$

is a homeomorphism. That is, X is the real spectrum of its ring $\mathcal{C}^\infty(X)$ of differentiable functions.

Proof. As differentiable functions separate points, β is injective. Moreover, β is a homeomorphism of X onto $\beta(X)$ because, given a differentiable function $f: X \to \mathbf{R}$, one has $\beta^{-1}(V(f)) = f^{-1}(0)$; conversely, every closed subset $Y \subset X$ is the vanishing locus of some differentiable function, as we have pointed out.

It only remains to prove that β is surjective. Let $\mathfrak{M} \in \operatorname{Spec}_{\mathbf{R}} \mathcal{C}^\infty(X)$ be a point of the real spectrum. We consider two cases:

1) $X = \mathbf{R}^n$. Then, if $\omega: \mathcal{C}^\infty(X) \to \mathcal{C}^\infty(X)/\mathfrak{M} = \mathbf{R}$ is the quotient morphism, and $x^i = \omega(X^i)$ are the images of the global coordinates, one has

$(X^i - x^i) \in \mathfrak{M}$ for every i, and hence $\mathfrak{M}_x \subset \mathfrak{M}$ by Lemma 4.1, which means that $\mathfrak{M}_x \equiv \mathfrak{M}$ because of the maximality of the first ideal.

2) General case. By the Whitney theorem, there is a closed immersion $X \hookrightarrow \mathbf{R}^N$ in some euclidean space. Thus, X is a closed subset of $\mathbf{R}^N \simeq \operatorname{Spec}_{\mathbf{R}} \mathcal{C}^\infty(\mathbf{R}^N)$, which can be identified with $V(\mathfrak{J})$, where $\mathfrak{J}$ is the ideal of differentiable functions on $\mathbf{R}^N$ vanishing at X. As $\mathcal{C}^\infty(\mathbf{R}^N)/\mathfrak{J} \xrightarrow{\sim} \mathcal{C}^\infty(X)$, Eq. (4.1) gives a homeomorphism $\operatorname{Spec}_{\mathbf{R}} \mathcal{C}^\infty(X) \xrightarrow{\sim} V(\mathfrak{J}) \simeq X$. ∎

We shall take advantage of this theorem in the Section devoted to graded manifolds, for which a similar result still holds true.

In the sequel, a differentiable manifold X and the space $\operatorname{Spec}_{\mathbf{R}} \mathcal{C}^\infty(X) \simeq \operatorname{Hom}_{\mathbf{R}\text{-alg}}(\mathcal{C}^\infty(X), \mathbf{R})$ will be identified via the homeomorphism β, so that we shall sometimes confuse a point $x \in X$ with an ideal $\mathfrak{M}_x \in \operatorname{Spec}_{\mathbf{R}} \mathcal{C}^\infty(X)$ or with a morphism $\omega_x \in \operatorname{Hom}_{\mathbf{R}\text{-alg}}(\mathcal{C}^\infty(X), \mathbf{R})$, as best suits us.

Let X, Y be differentiable manifolds. To every differentiable map $f: X \to Y$ there corresponds a ring morphism $f^*: \mathcal{C}^\infty(Y) \to \mathcal{C}^\infty(X)$ defined by composition, $f^*(g) = g \circ f$. Moreover, for every $\mathbf{R}$-algebra morphism $\psi: \mathcal{C}^\infty(Y) \to \mathcal{C}^\infty(X)$, there is a continuous map $\psi^*: X \simeq \operatorname{Spec}_{\mathbf{R}} \mathcal{C}^\infty(X) \to \operatorname{Spec}_{\mathbf{R}} \mathcal{C}^\infty(Y) \simeq Y$ (Eq. (4.3)), which is in fact differentiable, since its composition with any differentiable function $g: Y \to \mathbf{R}$ is a differentiable function on X because $g \circ \psi^* = \psi(g) \in \mathcal{C}^\infty(X)$. This follows from the fact that $\psi(g)(x) = \omega_x(\psi(g)) = \omega_{\psi^*(x)}(g) = g \circ \psi^*(x)$ for every point $x \in X$.

Corollary 4.1. *Let* $\operatorname{Hom}(X, Y)$ *be the set of differentiable maps from* X *to* Y. *The maps:*

$$\operatorname{Hom}(X, Y) \to \operatorname{Hom}_{\mathbf{R}\text{-alg}}(\mathcal{C}^\infty(Y), \mathcal{C}^\infty(X)), \quad f \mapsto f^*,$$
$$\operatorname{Hom}_{\mathbf{R}\text{-alg}}(\mathcal{C}^\infty(Y), \mathcal{C}^\infty(X)) \to \operatorname{Hom}(X, Y), \quad \psi \mapsto \psi^*$$

are the inverse of each other.

Proof. Let $f: X \to Y$ be a differentiable map, and let $\psi = f^*: \mathcal{C}^\infty(Y) \to \mathcal{C}^\infty(X)$. The induced continuous map $\psi^*: X \simeq \operatorname{Spec}_{\mathbf{R}} \mathcal{C}^\infty(X) \to \operatorname{Spec}_{\mathbf{R}} \mathcal{C}^\infty(Y) \simeq Y$ is in fact equal to f, because $\psi^*(x) = \psi^*(\mathfrak{M}_x) = \psi^{-1}(\mathfrak{M}_x) = (f^*)^{-1}(\mathfrak{M}_x) = \mathfrak{M}_{f(x)} = f(x)$. On the other hand, if $\psi: \mathcal{C}^\infty(Y) \to \mathcal{C}^\infty(X)$ is an $\mathbf{R}$-algebra morphism, by letting $f = \psi^*: X \simeq \operatorname{Spec}_{\mathbf{R}} \mathcal{C}^\infty(X) \to \operatorname{Spec}_{\mathbf{R}} \mathcal{C}^\infty(Y) \simeq Y$, one has $f^* = \psi$; this is because for any differentiable function $g \in \mathcal{C}^\infty(Y)$ and every point $x \in X$, the equalities $(f^*g)(x) = g(f(x)) = \omega_{f(x)}(g) = \omega_x(\psi(g)) = \psi(g)(x)$ hold. ∎

Thus, there are as many differentiable maps $X \to Y$ as there are $\mathbf{R}$-algebra morphisms $\mathcal{C}^\infty(Y) \to \mathcal{C}^\infty(X)$.

The previous results allow us to develop an approach to the theory of differentiable manifolds as locally ringed spaces, as we have already hinted.

Let us start by considering the ringed spaces $(U, \mathcal{C}^\infty_U)$, where $U \subset \mathbf{R}^n$ is an open subset of euclidean space, and $\mathcal{C}^\infty_U$ denotes the sheaf of germs of differentiable functions on U. If V is another open subset of $\mathbf{R}^n$, every differentiable map $f: U \to V$ induces a morphism of ringed spaces $(f, f^*): (V, \mathcal{C}^\infty_V) \to (U, \mathcal{C}^\infty_U)$, where for any open subet $W \subset V$, the ring morphism $f^*: \mathcal{C}^\infty(W) \to \mathcal{C}^\infty(f^{-1}(W))$ is defined as above.

We now prove that there is a one-to-one correspondence between morphisms of locally ringed spaces from $(U, \mathcal{C}^\infty_U)$ to $(V, \mathcal{C}^\infty_V)$ and differentiable functions $f: U \to V$.

Proposition 4.2. *If $(f, \phi): (U, \mathcal{C}^\infty_U) \to (V, \mathcal{C}^\infty_V)$ is a morphism of locally ringed spaces, then $\phi = f^*$.*

Proof. By virtue of Corollary 4.1, it suffices to prove that for any open subset $W \subset V$ one has $f = \phi^*_W$, where $\phi_W: \mathcal{C}^\infty(W) \to \mathcal{C}^\infty(f^{-1}(W))$ is the morphism induced by ϕ. But if $x \in f^{-1}(W)$ and $y = f(x)$, then $\phi_W(\mathfrak{M}_y) \subset \mathfrak{M}_x$, since $\phi: \mathcal{C}^\infty_V \to f_* \mathcal{C}^\infty_U$ is a local morphism. This implies that $\mathfrak{M}_y = \phi^{-1}(\mathfrak{M}_x) = \phi^*_W(x)$, namely, $\phi^*_W = f$. ∎

Corollary 4.2. *Let X be a Hausdorff paracompact topological space and let $(X, \mathcal{A})$ be a locally ringed space, locally isomorphic with $(\mathbf{R}^n, \mathcal{C}^\infty_{\mathbf{R}^n})$. Then X is an n-dimensional differentiable manifold and there is a natural isomorphism of locally ringed spaces $(X, \mathcal{A}) \simeq (X, \mathcal{C}^\infty_X)$.*

Proof. By definition, there exist open covers $\{U_i\}_{i \in I}$ of X and $\{V_i\}_{i \in \mathbf{R}^n}$ of $\mathbf{R}^n$ and a family of isomorphisms $(f_i, \psi_i): (U_i, \mathcal{A}_{|U_i}) \simeq (V_i, \mathcal{C}^\infty_{V_i})$ of locally ringed spaces. Then, $(f_i, \psi_i) \circ (f_j^{-1}, \psi_j^{-1}): (V_i \cap V_j, \mathcal{C}^\infty_{V_i \cap V_j}) \to (V_i \cap V_j, \mathcal{C}^\infty_{V_i \cap V_j})$ are isomorphisms of locally ringed spaces, hence induced by the diffeomorphisms $f_i \circ f_j^{-1}: V_i \cap V_j \simeq V_i \cap V_j$ (Lemma 4.1). The claim is now easily proved. ∎

This result provides an alternative definition of differentiable manifolds in terms of locally ringed spaces.

Corollary 4.3. *Let X and Y be differentiable manifolds. There is a one-to-one* *correspondence*

$$\begin{aligned} \mathrm{Hom}(X,Y) &\simeq \mathrm{Hom}((X,\mathcal{C}^\infty_X),(Y,\mathcal{C}^\infty_Y)) \\ f &\mapsto (f,f^*) \end{aligned}$$

between the set of differentiable maps $X \to Y$ and the set of morphisms of locally ringed spaces $(X,\mathcal{C}^\infty_X) \to (Y,\mathcal{C}^\infty_Y)$.

Proof. Straightforward. ∎

Glueing of graded locally ringed spaces. Let $\{(X_i,\mathcal{A}_i)\}$ be a family of graded locally ringed spaces. Let us assume that for every pair (i,j) of indices there are an open subset $X_{ij} \subset X_i$ and an isomorphism of graded locally ringed spaces

$$(f_{ij},\phi_{ij})\colon (X_{ji},\mathcal{A}_{j|X_{ji}}) \simeq (X_{ij},\mathcal{A}_{i|X_{ij}})$$

such that $X_{ii} = X_i$ and (f_{ii},ϕ_{ii}) is the identity for every i.

Let us suppose, furthermore, that the restriction (f'_{ij},ϕ'_{ij}) of (f_{ij},ϕ_{ij}) to $X_{ij} \cap X_{ik}$ is an isomorphism of graded locally ringed spaces

$$(f'_{ij},\phi'_{ij})\colon (X_{ij}\cap X_{ik},\mathcal{A}_{i|X_{ij}\cap X_{ik}}) \simeq (X_{ji}\cap X_{jk},\mathcal{A}_{j|X_{ji}\cap X_{jk}}),$$

and that these isomorphisms fulfill the *glueing condition* ([GroD], Ch.0, 4.1.7):

$$(f'_{ik},\phi'_{ik}) = (f'_{ij},\phi'_{ij}) \circ (f'_{jk},\phi'_{jk}). \tag{4.5}$$

We can define an equivalence relation on the disjoint sum $\tilde{X} = \coprod_i X_i$ by identifying points by means of the f_{ij}'s. If we denote by X the quotient topological space, the projection map $f_i\colon \tilde{X} \to X$ induces homeomorphisms f_i of X_i with open subsets U_i of X such that $\{U_i\}$ is a cover of X. Moreover, the glueing condition (4.5) implies that the sheaves $(f_i)_*(\mathcal{A}_i)$ on the open subsets $\{U_i\}$ fulfill the sheaf glueing condition (1.2). Thus, there is a sheaf $\mathcal{A}$ on X, and sheaf isomorphisms $\theta_i\colon \mathcal{A}_{|U_i} \simeq (f_i)_*(\mathcal{A}_i)$, as in Proposition 1.2. Then, $(X,\mathcal{A})$ is a graded locally ringed space, and there are isomorphisms of graded locally ringed spaces

$$(f_i,\phi_i)\colon (X_i,\mathcal{A}_i) \simeq (U_i,\mathcal{A}_{|U_i}),$$

for every index i.

Definition 4.7. *The graded locally ringed space $(X, \mathcal{A})$ is called the graded locally ringed space obtained by glueing the $(X_i, \mathcal{A}_i)$ by means of the isomorphisms (f_{ij}, ϕ_{ij}).*

One can easily see that $(X, \mathcal{A})$ and the isomorphisms (f_i, ϕ_i) are determined up to an isomorphism. $(X, \mathcal{A})$ inherits all the local properties of the graded locally ringed spaces $(X_i, \mathcal{A}_i)$. In particular, $(X, \mathcal{A})$ is, respectively, a differentiable manifold, an analytic space, etc., if the $(X_i, \mathcal{A}_i)$'s also are.

Let us consider another family $\{(X_i, \mathcal{B}_i)\}$ of graded locally ringed spaces, endowed with isomorphisms

$$(f_{ij}, \psi_{ij}): (X_{ji}, \mathcal{A}_{j|X_{ji}}) \simeq (X_{ij}, \mathcal{A}_{i|X_{ij}}),$$

fulfilling all the above conditions, so that there exists a graded locally ringed space $(X, \mathcal{B})$ and isomorphisms

$$(f_i, \psi_i): (X_i, \mathcal{B}_i) \simeq (U_i, \mathcal{B}_{|U_i}),$$

obtained by glueing. Then, as in Lemma 1.1, one has:

Lemma 4.2. *Given sheaf morphisms $\delta_i: \mathcal{A}_i \to \mathcal{B}_i$ such that the diagram*

$$\begin{array}{ccc} \mathcal{A}_{i|X_{ij}} & \xrightarrow{\delta_i} & \mathcal{B}_{i|X_{ij}} \\ {\scriptstyle \phi_{ij}}\downarrow & & \downarrow{\scriptstyle \psi_{ij}} \\ (f_{ij})_*(\mathcal{A}_{j|X_{ji}}) & \xrightarrow{\psi_{ij}} & (f_{ij})_*(\mathcal{B}_{j|X_{ji}}) \end{array}$$

commutes, there exists a sheaf morphism $\delta: \mathcal{A} \to \mathcal{B}$ such that $\delta_i \circ \phi_i = \psi_i \circ \delta_{|U_i}$ for every i.

Sheaves of derivations. If $(X, \mathcal{A})$ is a graded ringed space and $\mathcal{M}, \mathcal{N}$ are sheaves of graded $\mathcal{A}$-modules, the homomorphism sheaf $\mathcal{H}om_{\mathcal{A}}(\mathcal{M}, \mathcal{N})$ is introduced as in Definition 1.7. If $\mathcal{B}$ is a subsheaf of graded algebras of $\mathcal{A}$, one can define the *sheaf of derivations* $\mathcal{D}er_{\mathcal{B}}(\mathcal{A}, \mathcal{M})$ as the subsheaf of $\mathcal{H}om_{\mathcal{B}}(\mathcal{A}, \mathcal{M})$ whose sections on an open subset $U \subset X$ are $\mathcal{B}$-linear graded derivations $D: \mathcal{A}_{|U} \to \mathcal{M}_{|U}$, that is, morphisms of sheaves of $\mathcal{B}_{|U}$-modules which for every open subset $V \subset U$ are graded derivations of $\mathcal{A}(V)$ over $\mathcal{B}(V)$ with values in $\mathcal{M}(V)$. It should be noticed that in general one cannot define the sheaf of derivations by letting $U \to \mathrm{Der}_{\mathcal{B}(U)}(\mathcal{A}(U), \mathcal{M}(U))$ since, given an open subset

$V \subset U$, a restriction map $\mathrm{Der}_{\mathcal{B}(U)}(\mathcal{A}(U), \mathcal{M}(U)) \to \mathrm{Der}_{\mathcal{B}(V)}(\mathcal{A}(V), \mathcal{M}(V))$ may fail to exist; complex manifolds are an example of this situation.

It is customary to denote the sheaf $\mathcal{D}er_{\mathcal{B}}(\mathcal{A}, \mathcal{A})$ simply by $\mathcal{D}er_{\mathcal{B}}\mathcal{A}$.

Chapter III
Categories of supermanifolds

Nous avons vu tant de monstres de cette espèce
que nous sommes un peu blasés, et qu'il faut accumuler
les caractères tératologiques les plus biscornus
pour arriver encore à nous étonner.

N. Bourbaki

The category of G-supermanifolds [**BB1,BBH**] provides a consistent and concrete model for the development of supergeometry. In order to supply proper motivations for the introduction of these objects, and also for historical reasons, we shall start with a brief description of graded manifolds; these were originally introduced by Berezin and Leĭtes [**BL,Leĭ**], although the most extensive treatment can be found in Kostant [**Kos**] and Manin [**Ma2**]. Graded manifolds also play a direct role in the theory developed in this book, in that some results holding in that category can be either reformulated or applied as they are in the context of G-supermanifolds.

On the other hand, the 'geometric' approach to supermanifolds due to DeWitt and Rogers [**Bch1,Bch2,DW,Rs1,Rs2**], which is our starting point to define G-supermanifolds, will be reviewed and discussed in Sections 2 and 3.

It should be pointed out that this survey of supermanifolds is by no means exhaustive; for instance, we do not dwell upon the work by Vladimirov and Volovich [**VlV**]. Besides, throughout this book we shall limit ourselves to the case where the ground graded algebra, and the geometric spaces involved, are finite-dimensional over the real (or complex) field, thus leaving aside the interesting contributions by Jadczyk and Pilch [**JP**], Matsumoto and Kakazu [**MK**], Molotkov [**Mol**], and Schmitt [**Scm**]. More specific bibliography will be cited where appropriate.

The discussion of the relationship between G-supermanifolds and the axiomatics for supermanifolds proposed by Rothstein [**Rt2**] will be postponed to the next Chapter, since it involves some constructions which will be developed there.

1. Graded manifolds

It is convenient to introduce graded manifolds as a particular case of a more general category, namely, that of graded spaces (cf. the treatment given in [**Ma2**]).

Graded spaces. Let k be a commutative field and $(X, \mathcal{S})$ a locally ringed space in commutative k-algebras; thus, X is a topological space, and $\mathcal{S}$ a sheaf of commutative k-algebras on X.

Definition 1.1. *A graded space of odd dimension n with underlying space $(X, \mathcal{S})$ is a pair $(X, \mathcal{A})$, where $\mathcal{A}$ is a sheaf of graded-commutative k-algebras, such that:*

(1) *there is an exact sheaf sequence*

$$0 \to \mathfrak{J} \to \mathcal{A} \xrightarrow{\pi} \mathcal{S} \to 0, \tag{1.1}$$

where π is a surjective morphism of graded k-algebras, and $\mathfrak{J} = \mathcal{A}_1 + (\mathcal{A}_1)^2$.

(2) *$\mathfrak{J}/\mathfrak{J}^2$ is a locally free module of rank n over $\mathcal{S} = \mathcal{A}/\mathfrak{J}$, and $\mathcal{A}$ is locally isomorphic, as a sheaf of graded-commutative algebras, to the exterior bundle $\bigwedge_{\mathcal{S}}(\mathfrak{J}/\mathfrak{J}^2)$.*

The second condition in the above definition implies that $\mathfrak{J}^{n+1} = 0$. In the case where $\mathcal{S}$ has no nilpotents, which will be relevant in what follows, $\mathfrak{J}$ coincides with the sheaf $\mathfrak{N}$ of nilpotents of $\mathcal{A}$.

This definition also implies that a graded space $(X, \mathcal{A})$ is a graded locally ringed space (in the sense of Definition II.4.1), for the unique maximal ideal of a stalk $\mathcal{A}_x$ is π^{-1} of the maximal ideal of $\mathcal{S}_x$. Therefore, one can define *morphisms of graded spaces* merely as morphisms of graded locally ringed spaces (Definition II.4.2).

Let $(X, \mathcal{A})$ and $(Y, \mathcal{B})$ be graded spaces with underlying spaces $(X, \mathcal{S})$ and $(Y, \mathcal{T})$ respectively. Given a morphism $(f, \phi): (X, \mathcal{A}) \to (Y, \mathcal{B})$, the morphism

$\phi: \mathcal{B} \to f_*\mathcal{A}$ maps $\mathcal{B}_1$ into $f_*(\mathcal{A}_1)$, and then $\mathcal{B}_1 + (\mathcal{B}_1)^2$ into $f_*(\mathcal{A}_1 + (\mathcal{A}_1)^2)$), so that it induces a sheaf morphism $\bar{\phi}: \mathcal{T} \to f_*\mathcal{S}$, such that the diagram

$$\begin{array}{ccc} \mathcal{B} & \xrightarrow{\phi} & f_*\mathcal{A} \\ {\scriptstyle\pi}\downarrow & & \downarrow{\scriptstyle f_*(\pi)} \\ \mathcal{T} & \xrightarrow{\bar{\phi}} & f_*\mathcal{S} \end{array}$$

commutes. Namely, any graded space morphism induces a morphism $(f, \bar{\phi})$: $(X, \mathcal{S}) \to (Y, \mathcal{T})$ between the underlying spaces.

Graded manifolds. A graded manifold is simply a graded space over $\mathbb{R}$ whose underlying space is a smooth manifold.

Definition 1.2. *A graded manifold of dimension (m, n) is a graded space in $\mathbb{R}$-algebras of odd dimension n whose underlying space is an m-dimensional differentiable manifold $(X, \mathcal{C}_X^\infty)$.*

Analogously, one can define *complex analytic graded manifolds* by taking $k = \mathbb{C}$ and $(X, \mathcal{S})$ as a complex manifold, or *graded analytic spaces*, or *graded schemes*, and so on.

From the exact sheaf sequence (1.1), that now reads

$$0 \to \mathfrak{N} \to \mathcal{A} \to \mathcal{C}_X^\infty \to 0,$$

one obtains, for any open subset $U \subset X$, an exact sequence of graded algebras

$$0 \to \mathfrak{N}(U) \to \mathcal{A}(U) \xrightarrow{\pi} \mathcal{C}^\infty(U).$$

A section f of $\mathcal{A}$ will be called a *graded function*. The image of a graded function $f \in \mathcal{A}(U)$ by the structural morphism $\pi: \mathcal{A}(U) \to \mathcal{C}^\infty(U)$ will be denoted by $\tilde{f}$.

Definition 1.3. *A morphism of graded manifolds is a morphism of graded spaces $(f, \psi): (X, \mathcal{A}) \to (Y, \mathcal{B})$.*

Like all graded space morphisms, a graded manifold morphism (f, ψ): $(X, \mathcal{A}) \to (Y, \mathcal{B})$ induces a morphism of locally ringed spaces $(f, \bar{\psi}): (X, \mathcal{C}_X^\infty) \to (Y, \mathcal{C}_Y^\infty)$. Now, Corollary II.4.3 entails that $f: X \to Y$ should be a differentiable map, and that $\bar{\psi}$ equals the pullback morphism $f^*: \mathcal{C}_Y^\infty \to f_*\mathcal{C}_X^\infty$. Therefore, graded manifold morphisms can be alternatively described as morphisms of graded locally ringed spaces $(f,\psi)\colon(X,\mathcal{A})\to(Y,\mathcal{B})$ such that $f\colon X\to Y$ is a differentiable map, and there is a commutative diagram

$$\begin{array}{ccc} \mathcal{B} & \xrightarrow{\psi} & f_*\mathcal{A} \\ \downarrow & & \downarrow \\ \mathcal{C}^\infty_Y & \xrightarrow[f^*]{} & f_*\mathcal{C}^\infty_X \end{array} .$$

A morphism $(f,\psi)\colon(X,\mathcal{A})\to(Y,\mathcal{B})$ of graded manifolds is therefore a differentiable map $f\colon X\to Y$ and, for every open subset $V\subset X$, an even morphism of graded algebras $\psi\colon\mathcal{B}(V)\to\mathcal{A}(f^{-1}(V))$ compatible with the restriction maps, such that the diagram

$$\begin{array}{ccc} \mathcal{B}(V) & \xrightarrow{\psi} & \mathcal{A}(f^{-1}(V)) \\ \downarrow & & \downarrow \\ \mathcal{C}^\infty(V) & \xrightarrow[f^*]{} & f_*\mathcal{C}^\infty(f^{-1}(V)) \end{array}$$

is commutative.

Isomorphisms of graded manifolds can be now defined in the obvious way. It is clear that an isomorphism of graded manifolds induces a diffeomorphism between the underlying differentiable manifolds, but the converse is not true.

It is known (cf. e.g. [Ser] and [Wel]) that the category of rank r locally free $\mathcal{C}^\infty_X$-modules and the category of rank r smooth vector bundles on X are equivalent. In particular, any locally free $\mathcal{C}^\infty_X$-module determines uniquely a smooth vector bundle, and, *vice versa*, any smooth vector bundle yields a locally free $\mathcal{C}^\infty_X$-module; namely, the sheaf of its sections. Thus, the locally free $\mathcal{C}^\infty_X$-module $\mathfrak{N}/\mathfrak{N}^2$ defines a rank n-vector bundle $E\to X$, and every point of X has an open neighbourhood $U\subset X$ such that

$$\mathcal{A}(U)\simeq\Gamma(U,\textstyle\bigwedge E)$$

as graded-commutative $\mathbf{R}$-algebras.

Definition 1.4. *A splitting neighbourhood for a graded manifold is an open subset $U\subset X$ such that $E_{|U}$ is a trivial bundle and $\mathcal{A}_{|U}\xrightarrow{\sim}\bigwedge_{\mathcal{C}^\infty_{|U}}(\mathfrak{N}/\mathfrak{N}^2)$.*

If U is a splitting neighbourhood for $(X, \mathcal{A})$, there is a basis $\{y^1, \ldots, y^n\}$ of sections of $E_{|U}$, and an isomorphism

$$\mathcal{A}(U) \simeq \mathcal{C}^\infty(U) \otimes_{\mathbf{R}} \bigwedge(E_n)$$

where $E_n = \langle y^1, \ldots, y^n \rangle$ denotes the $\mathbf{R}$-vector space generated by $\{y^1, \ldots, y^n\}$. The existence of a section $\epsilon: \mathcal{C}^\infty(U) \hookrightarrow \mathcal{A}(U)$ of the projection π follows. A graded function $f \in \mathcal{A}(U)$ can be now expressed as

$$f = \sum_{\mu \in \Xi_n} f_\mu y^\mu , \tag{1.2}$$

where the coefficients f_μ are elements of $\epsilon(\mathcal{C}^\infty(U))$, and Ξ_n is as in Example I.1.3.

Definition 1.5. *If U is a splitting neighbourhood, a family $(x, y) \equiv (x^1, \ldots, x^m, y^1, \ldots, y^n)$ of graded functions ($|x^i| = 0$, $|y^\alpha| = 1$) is called a graded coordinate system if*

(1) *$(\tilde{x}^1, \ldots, \tilde{x}^m)$ is an ordinary coordinate system in U and $x^i = \epsilon(\tilde{x}^i)$ for every i.*

(2) *$(y^1, \ldots, y^n)$ is a basis of sections of $E_{|U}$, that is, $y^1, \ldots, y^n$ are elements of $\bigwedge E$ and $\prod_{\alpha=1}^n y^\alpha \neq 0$.*

The elements $f_\mu \in \epsilon(\mathcal{C}^\infty(U))$ in the local expression (1.2) may be considered as differentiable functions of $(x^1, \ldots, x^m)$ and will be written as $f_\mu(x^1, \ldots, x^m)$.

Lemma 1.1. (Graded partitions of unity) *Let $(X, \mathcal{A})$ be a graded manifold and $W \subset X$ an open set. One has:*

(1) *if $\tilde{f} = 1$ for some $f \in \mathcal{A}(W)$, then f is invertible in $\mathcal{A}(W)$.*

(2) *if $\mathcal{V} = \{V_j\}_{j \in J}$ is an open cover of W, there exists a locally finite refinement $\{U_i\}_{i \in I}$ of $\mathcal{V}$, and even elements $f_i \in \mathcal{A}(W)$, such that* $\operatorname{Supp} f_i \subset U_i$ *and $1 = \sum_{i \in I} f_i$ in $\mathcal{A}(W)$.*

Proof. (1) If $\tilde{f} = 1$, $h = f - 1$ is nilpotent, and $f = 1 + h$ is invertible.

(2) (See [Kos], Lemma 2.4) By paracompactness, there exists a locally finite cover of W by splitting neighbourhoods U_i such that $\mathcal{A}(U_i) \simeq \epsilon_i(\mathcal{C}^\infty(U_i)) \otimes_{\mathbf{R}} \bigwedge(E_n^i)$. Let $1 = \sum_{i \in I} \tilde{\tau}_i$ be a differentiable partition of unity on W such that $\operatorname{Supp} \tilde{\tau}_i \subset U_i$. If $\tau_i = \epsilon_i(\tilde{\tau}_i) \in \epsilon_i(\mathcal{C}^\infty(U_i)$, one has that $\operatorname{Supp} \tau_i \subset U_i$, and τ_i can be extended to an even function $\tau_i \in \mathcal{A}(W)$ with the same support. Now, the

sum $h = \sum_{i \in I} \tau_i$ exists, because it is locally finite, and fulfills $\tilde{h} = 1$. By (1), h is invertible, and one concludes by taking $f_i = h^{-1}\tau_i$. ∎

In accordance with the definitions given in Section II.3, the second part of Lemma 1.1 means that $\mathcal{A}$ is a fine sheaf, so that any $\mathcal{A}$-module is soft, and therefore acyclic (Corollary II.3.4).

Corollary 1.1. *Let $(X, \mathcal{A})$ be a graded manifold and $W \subset X$ an open subset.*

(1) *The sequence*

$$0 \to \mathfrak{N}(W) \to \mathcal{A}(W) \to \mathcal{C}^\infty(W) \to 0$$

is exact.

(2) *If $\tilde{f}$ is invertible in $\mathcal{C}^\infty(W)$, then f is invertible in $\mathcal{A}(W)$.*

(3) *There is a natural homeomorphism*

$$\begin{aligned} W &\xrightarrow{\sim} \operatorname{Spec}_{\mathbf{R}} \mathcal{A}(W) \\ x &\mapsto \mathfrak{m}_x \equiv \{f \in \mathcal{A}(W) \mid \tilde{f}(x) = 0\} \end{aligned}$$

where $\operatorname{Spec}_{\mathbf{R}} \mathcal{A}(W)$ is endowed with the Zariski topology.

Proof. (1) It suffices to show that the last arrow is surjective; this follows from the exact cohomology sequence associated with (1.1), since $\mathfrak{N}$ is an $\mathcal{A}$-module and hence is acyclic. (2) is trivial. To prove (3), let us notice that, $\mathfrak{N}(W)$ being the ideal of nilpotents of $\mathcal{A}(W)$, the surjective morphism $\mathcal{A}(W) \to \mathfrak{N}(W)$ induces a homeomorphism $\operatorname{Spec}_{\mathbf{R}} \mathcal{C}^\infty(W) \xrightarrow{\sim} \operatorname{Spec}_{\mathbf{R}} \mathcal{A}(W)$. The thesis then follows from Proposition II.4.1. ∎

Topologies of the structure rings of a graded manifold and localization. In order to develop the differential geometry of graded manifolds (e.g. the definition of products) Kostant exploited the coalgebra of finitely-supported distributions over the sheaf $\mathcal{A}$ [**Kos**]. A more direct approach, that we shall adopt here, can be pursued provided the rings $\mathcal{A}(U)$ are suitably topologized [**HeM1**].

Lemma 1.2. *Let $(X, \mathcal{A})$ be a graded manifold. The derivations of $\mathcal{A}(X)$ are local operators, that is, if $U \subset X$ is open and $f_{|U} = 0$ for some $f \in \mathcal{A}(X)$, then $D(f)_{|U} = 0$ for every derivation $D \in \operatorname{Der}_{\mathbf{R}} \mathcal{A}(X)$.*

Proof. It is sufficient to prove that for every point x there is an open neighbourhood $V \subset U$ such that $D(f)_{|V} = 0$. To do that, let us take V such that

$\overline{V} \subset U$. By the existence of partitions of unity, one can write $1 = \phi + \psi$, with $\phi, \psi \in \mathcal{A}(X)$, $\operatorname{Supp}\phi \subset U$, and $\operatorname{Supp}\psi \subset X - \overline{V}$. Then, $f\phi = 0$, and so $0 = D(\phi) f + \phi D(f)$ and $0 = f_{|V} D(f)_{|V}$. Since $\phi_{|V} = 0$, one concludes. ∎

The locality property of the derivations of $\mathcal{A}$ implies that, if $V \subset U$ are open sets, there is a restriction morphism $\operatorname{Der}_{\mathbf{R}} \mathcal{A}(U) \to \operatorname{Der}_{\mathbf{R}} \mathcal{A}(V)$, which for an arbitrary ringed space may fail to exist, as pointed out in Section II.4. Thus in our case one has:

Corollary 1.2. *Let $(X, \mathcal{A})$ be a graded manifold; $U \to \operatorname{Der}_{\mathbf{R}} \mathcal{A}(U)$ is a sheaf of graded $\mathcal{A}$-modules, which coincides with the sheaf $\mathcal{D}er_{\mathbf{R}}\mathcal{A} \equiv \mathcal{D}er_{\mathbf{R}}(\mathcal{A}, \mathcal{A})$ defined as in Section II.4.*

Proposition 1.1. *Let U be a coordinate neighbourhood for a graded manifold $(X, \mathcal{A})$ with graded coordinates $(x^1, \ldots, x^m, y^1, \ldots, y^n)$. There exist even derivations $\frac{\partial}{\partial x^1}, \ldots, \frac{\partial}{\partial x^m}$ and odd derivations $\frac{\partial}{\partial y^1}, \ldots, \frac{\partial}{\partial y^n}$ of $\mathcal{A}(U)$ uniquely characterized by the conditions*

$$\frac{\partial x^h}{\partial x^i} = \delta_i^h; \quad \frac{\partial y^\alpha}{\partial x^i} = 0; \qquad \frac{\partial x^h}{\partial y^\beta} = 0; \quad \frac{\partial y^\alpha}{\partial y^\beta} = \delta_\beta^\alpha$$

$(i, h = 1, \ldots, m;\ \alpha, \beta = 1, \ldots n)$ and such that every derivation $D \in \operatorname{Der}_{\mathbf{R}} \mathcal{A}(U)$ can be written as

$$D = \sum_{i=1}^{m} D(x^i) \frac{\partial}{\partial x^i} + \sum_{\alpha=1}^{n} D(y^\alpha) \frac{\partial}{\partial y^\alpha}.$$

In particular, $\operatorname{Der}_{\mathbf{R}} \mathcal{A}(U)$ is a free $\mathcal{A}(U)$-module with basis $\frac{\partial}{\partial x^1}, \ldots, \frac{\partial}{\partial x^m}, \frac{\partial}{\partial y^1}, \ldots, \frac{\partial}{\partial y^n}$ (cf. [Kos] *Theorem 2.8).*

Proof. It is enough to prove that the conditions $D(x^i) = D(y^\alpha) = 0$ for $i = 1, \ldots, m$, $\alpha = 1, \ldots, n$ imply $D = 0$. But $\mathcal{A}(U) \simeq \mathcal{C}^\infty(U) \otimes_{\mathbf{R}} \bigwedge(\langle y^1, \ldots, y^n \rangle)$, and under this isomorphism one has $f = \sum_{\mu \in \Xi_n} f_\mu(x^1, \ldots, x^m) y^\mu$. Then, $D(f) = \sum_{\mu \in \Xi_n} D(f_\mu) y^\mu$ because $D(y^\alpha) = 0$ for every index α, and $D(f_\mu) = 0$ because $D_{|\mathcal{C}^\infty(U)}$ is an ordinary derivation from $\mathcal{C}^\infty(U)$ into $\mathcal{A}(U)$ vanishing on the coordinates $(x^1, \ldots, x^m)$. ∎

Now, let $(X, \mathcal{A})$ be a graded manifold of dimension (m, n). The next step is to endow the rings $\mathcal{A}(W)$, where $W \subset X$ is an open subset, with a structure of a graded-commutative Fréchet algebra (let us recall that a Fréchet space is a complete locally convex metrizable topological real vector space. A Fréchet algebra is an algebra over the real numbers whose underlying vector space is Fréchet and whose product is continuous [**RR**]).

REMARK 1.1. If $(x^1,\dots,x^m,y^1,\dots,y^n)$ is a graded coordinate system, for any multi-index $J=(j^1,\dots,j^m)\in\mathbf{N}^m$, whose length is $|J|=\sum_{k=1}^m j^k$, and any multi-index $\mu\in\Xi_n$, we shall write

$$\left(\frac{\partial}{\partial x}\right)^J\left(\frac{\partial}{\partial y}\right)_\mu\equiv\left(\frac{\partial}{\partial x^1}\right)^{j_1}\circ\cdots\circ\left(\frac{\partial}{\partial x^m}\right)^{j_m}\circ\frac{\partial}{\partial y^{\mu(1)}}\circ\cdots\circ\frac{\partial}{\partial y^{\mu(d(\mu))}}.$$

▲

For every compact subset K contained in a coordinate neighbourhood $U\subset W$ with graded coordinates $(x^1,\dots,x^m,y^1,\dots,y^n)$, every $f\in\mathcal{A}(W)$ and every positive integer $r\geq 0$, let us define

$$p_K^r(f)=\max_{\substack{x\in K\\ |J|\leq r,\,\mu\in\Xi_n}}\left|\left[\left(\frac{\partial}{\partial x}\right)^J\left(\frac{\partial}{\partial y}\right)_\mu f\right]^\sim(x)\right|.$$

Then one has:

Proposition 1.2.

(1) *The functions $p_K^r\colon\mathcal{A}(W)\to\mathbf{R}$ are submultiplicative seminorms, in that*

$$p_K^r(f\,g)\leq 2^{nr}\,p_K^r(f)\,p_K^r(g).$$

(2) *$\mathcal{A}(W)$, equipped with the topology induced by the seminorms $\{p_K^r\}$, where $r\geq 0$ and K is an arbitrary compact coordinate subset of W, is a Fréchet algebra.*

Proof. (1) One has:

$$\left[\left(\frac{\partial}{\partial x}\right)^J\left(\frac{\partial}{\partial y}\right)_\mu(f\,g)\right]^\sim=(-1)^{\binom{d(\mu)}{2}}\left(\frac{\partial}{\partial\tilde{x}}\right)^J\left(\sum_{\gamma\leq\mu}\varsigma(\gamma,\mu)\,\tilde{f}_\gamma\,\tilde{g}_{\mu-\gamma}\right)$$

where $\gamma\in\Xi_n$ and $\varsigma(\gamma,\mu)$ is the sign determined by $y^\mu=\varsigma(\gamma,\mu)y^\gamma\,y^{\mu-\gamma}$. A straightforward computation yields the required inequality.

(2) The topology defined in $\mathcal{A}(W)$ is locally convex (by construction) and metrizable, because X can be covered by a countable family of coordinate neighbourhoods $\{U_i\}$ and every U_i can be covered by a countable family of compacts $K_h^i \subset K_{h+1}^i$, and hence the seminorms $p^r_{K_h^i}$ define the topology. The completeness of $\mathcal{A}(W)$ is a local question, and thus one can assume that $\mathcal{A}(W) \simeq \mathcal{C}^\infty(W) \otimes \bigwedge(\langle y^1, \ldots, y^n \rangle)$. One concludes, since this is a metric isomorphism from $\mathcal{A}(W)$ onto the free $\mathcal{C}^\infty(W)$-module $\mathcal{C}^\infty(W) \otimes \bigwedge(\langle y^1, \ldots, y^n \rangle)$. ∎

This topology can be completely characterized by the requirement that for every splitting neighbourhood $U \subset W$, for which the isomorphism $\mathcal{A}(U) \simeq \mathcal{C}^\infty(U) \otimes_{\mathbb{R}} \bigwedge(\langle y^1, \ldots, y^n \rangle)$ holds, a sequence $\{f_i = \sum_{\mu \in \Xi_n} f_{\mu,i}\, y^\mu\}_{i \in \mathbb{N}}$ converges to an element $f = \sum_{\mu \in \Xi_n} f_\mu\, y^\mu$ of $\mathcal{A}(U)$ if and only if, for every $\mu \in \Xi_n$, the sequence of differentiable functions $(f_{\mu,i})_{i \in \mathbb{N}}$ converges to f_μ in the weak topology of the ring $\mathcal{C}^\infty(U)$ (that is, uniformly with all its derivatives on every compact $K \subset U$).

Corollary 1.3.

(1) *If $(f, \psi) \colon (X, \mathcal{A}) \to (Y, \mathcal{B})$ is a morphism of graded manifolds, the induced morphism $\mathcal{B}(Y) \to \mathcal{A}(X)$ is continuous.*

(2) *If $(X, \mathcal{A})$ is a graded manifold, each derivation $D \in \operatorname{Der}_{\mathbb{R}} \mathcal{A}(X)$ is continuous.*

Proof. (1) We may assume that X and Y are coordinate neighbourhoods with graded coordinates $(x^1, \ldots, x^m, y^1, \ldots, y^n)$ and $(\bar{x}^1, \ldots, \bar{x}^p, \bar{y}^1, \ldots, \bar{y}^q)$, respectively. From Proposition 1.1 one obtains:

$$\frac{\partial}{\partial x^i} \circ \psi = \sum_{j=1}^{p} \frac{\partial \psi(\bar{x}^j)}{\partial x^i} \frac{\partial}{\partial \bar{x}^j} + \sum_{\beta=1}^{q} \frac{\partial \psi(\bar{y}^\beta)}{\partial x^i} \frac{\partial}{\partial \bar{y}^\beta}$$

$$\frac{\partial}{\partial y^\alpha} \circ \psi = \sum_{j=1}^{p} \frac{\partial \psi(\bar{x}^j)}{\partial y^\alpha} \frac{\partial}{\partial \bar{x}^j} + \sum_{\beta=1}^{q} \frac{\partial \psi(\bar{y}^\beta)}{\partial y^\alpha} \frac{\partial}{\partial \bar{y}^\beta}$$

Using these formulas, the seminorms in $\mathcal{A}(X)$ are majorated in terms of those in $\mathcal{B}(Y)$ and of the maxima over compact sets of the derivatives of the various orders of the quantities $\psi(\bar{x}^j)$ and $\psi(\bar{y}^\beta)$ with respect to $(x^1, \ldots, x^m, y^1, \ldots, y^n)$.

(2) is trivial in view of the definition of the seminorms. ∎

Let $(X, \mathcal{A})$ be a graded manifold and $U \subset X$ an open subset. Let us consider the *ring of fractions* $S_U^{-1}\, \mathcal{A}(X)$ of $\mathcal{A}(X)$ with respect to the multiplicative system S_U of the elements $f \in \mathcal{A}_0(X)$ such that $\tilde{f}(x) \neq 0$ for every point

$x \in U$.[1] If an element $g \in \mathcal{A}_0(X)$ is such that $\tilde{g}$ is invertible in $\mathcal{C}^\infty(U)$, the restriction of g to U is invertible in $\mathcal{A}(U)$, so that one obtains a graded **R**-algebra morphism

$$\begin{aligned} S_U^{-1}\,\mathcal{A}(X) &\to \mathcal{A}(U) \\ f/g &\mapsto f_{|U}(g_{|U})^{-1}. \end{aligned}$$

From this one can deduce a localization property for graded manifolds.

Proposition 1.3. *The above morphism is bijective.*

Proof. 1) Injectivity. Let us take $(f/1) \in S_U^{-1}\,\mathcal{A}(X)$ such that $f_{|U} = 0$ in $\mathcal{A}(U)$. One has to find an element $g \in S_U$ such that $gf = 0$. Owing to the existence of partitions of unity, one can assume that X is a coordinate neighbourhood. Now, $f = \sum_{\mu\in\Xi_n} f_\mu y^\mu$ with $f_{\mu|U} = 0$ for every μ. Let $g \in \mathcal{C}^\infty(X)$ be a function such that $g \equiv 0$ on $X - U$, but $g > 0$ on U. Then, $gf_\mu = 0$ on U for every μ and so $gf = 0$. Furthermore, since X is assumed to be a coordinate neighbourhood, we can regard g as an element in S_U.

2) Surjectivity. Let $p_1 \leq p_2 \leq \ldots$ be an increasing countable sequence of seminorms which defines the topology of $\mathcal{A}(X)$, and let $\{U_i\}$ be a cover of U by graded coordinate neighbourhoods such that $\overline{U}_i \subset U$ and, finally, for any i let $\phi_i \in \mathcal{A}(X)$ be an even section such that $\operatorname{Supp}\phi_i \subset U_i$ and $\tilde{\phi}_i > 0$ on U_i. Given an element $f \in \mathcal{A}(U)$, let us consider the series

$$\sum_{i\geq 0} \frac{1}{2^i}\,\frac{\phi_i\, f}{1 + p_i(\phi_i\, f) + p_i(\phi_i)}, \qquad \sum_{i\geq 0} \frac{1}{2^i}\,\frac{\phi_i}{1 + p_i(\phi_i\, f) + p_i(\phi_i)};$$

if these converge in $\mathcal{A}(X)$ to sections g and h, one has $h \in S_U$ and $g = h\,f$ on $\mathcal{A}(U)$, which allows us to conclude.

Proving the convergence is similar for the two series, so that we shall only consider the first one. For every $\varepsilon > 0$ and every index j, there exists $s \geq j$

[1] Since the elements of the multiplicative system commute with any other element, the relation defined in $S_U \times \mathcal{A}(X)$ by

$$(s, f) \sim (s', f') \quad \text{if there exists an element } s'' \in S_U \text{ such that } s''\,(s\,f' - s'\,f) = 0\,,$$

is an equivalence relation. Thus, the ring of fractions is defined as $S_U^{-1}\,\mathcal{A}(X) = (S_U \times \mathcal{A}(X))/\sim$ (see [**AtM**] for the commutative case).

such that $(1/2^{s-1}) < \varepsilon$. If $k \geq s$, for any $r \geq 0$ one obtains:

$$p_j\left(\sum_{i=k}^{k+r}\frac{1}{2^i}\frac{\phi_i f}{1+p_i(\phi_i f)+p_i(\phi_i)}\right) \leq \sum_{i=k}^{k+r}\frac{1}{2^i}\frac{p_j(\phi_i f)}{1+p_i(\phi_i f)+p_i(\phi_i)}$$

$$\leq \sum_{i=k}^{k+r}\frac{1}{2^i} \leq \frac{1}{2^{k-1}} < \varepsilon$$

because $p_j \leq p_i$ for $i \geq k$ since $i \geq k \geq s \geq j$. ■

Proposition 1.3 means that any graded function $f \in \mathcal{A}(U)$ defined on an open set $U \subset X$ can be expressed as a quotient $f = g_{|U}/h_{|U}$ where $g \in \mathcal{A}(X)$ and $h \in \mathcal{A}(X)$ are *globally* defined graded functions. As a matter of fact, the structure sheaf of a graded manifold can be reconstructed from the ring of its global sections:

Corollary 1.4. *If $(X, \mathcal{A})$ is a graded manifold, the presheaf $U \to S_U^{-1}\mathcal{A}(X)$ is a sheaf of graded* **R***-algebras, canonically isomorphic with the structure sheaf $\mathcal{A}$.* ■

An important consequence of these results is that a morphism of graded manifolds $(f, \phi): (X, \mathcal{A}) \to (Y, \mathcal{B})$ is characterized only by the **R**-algebra morphism $\phi: \mathcal{B}(Y) \to \mathcal{A}(X)$, or — using the terminology of algebraic geometry — graded manifolds are 'affine' (cf. Corollary II.4.1 for the case of differentiable manifolds).

Corollary 1.5. *Let $(X, \mathcal{A})$ and $(Y, \mathcal{B})$ be graded manifolds. The natural map:*

$$\operatorname{Hom}((X, \mathcal{A}), (Y, \mathcal{B})) \to \operatorname{Hom}_{\mathbf{R}\text{-alg}}(\mathcal{B}(Y), \mathcal{A}(X))_0, \tag{1.3}$$

where the right hand side denotes the even morphisms of graded **R***-algebras, is bijective.*

Proof. The injectivity follows from the previous Corollary. As far as surjectivity is concerned, if $\phi: \mathcal{B}(Y) \to \mathcal{A}(X)$ is an even morphism of graded **R**-algebras, ϕ sends the nilpotents of $\mathcal{B}(Y)$ into those of $\mathcal{A}(X)$, so that ϕ induces a ring morphism $\tilde{\phi}: \mathcal{C}^\infty(Y) \to \mathcal{C}^\infty(X)$; passing to real spectra (see Section II.4), one obtains a differentiable map $f : X \equiv \operatorname{Spec}_{\mathbf{R}} \mathcal{C}^\infty(X) \to Y \equiv \operatorname{Spec}_{\mathbf{R}} \mathcal{C}^\infty(Y)$, such that $\tilde{\phi} = f^*$; the pair (f, f^*) provides a morphism of locally ringed spaces $(X, \mathcal{C}_X^\infty) \to (Y, \mathcal{C}_Y^\infty)$. We should observe that ϕ determines morphisms

$$S_U^{-1}\mathcal{B}(Y) \to S_{f^{-1}(U)}^{-1}\mathcal{A}(X)$$

for every open $U \subset Y$. Hence, by the previous Corollary, ϕ induces a sheaf morphism $\hat{\phi}: \mathcal{B} \to f_* \mathcal{A}$ such that the diagram

$$\begin{array}{ccc} \mathcal{B} & \longrightarrow & \mathcal{C}^\infty_Y \\ {\scriptstyle\hat{\phi}}\downarrow & & \downarrow{\scriptstyle f^*} \\ f_*\mathcal{A} & \longrightarrow & f_*\mathcal{C}^\infty_X \end{array}$$

commutes. Thus $(f, \hat{\phi})$ is a morphism of graded manifolds $(X, \mathcal{A}) \to (Y, \mathcal{B})$ which is mapped to ϕ by the morphism (1.3). ■

Products of graded manifolds. [HeM1] In the case of differentiable manifolds, the notion of the product of manifolds is usually introduced by describing the manifolds in terms of atlases and then defining a product atlas on the cartesian product of the underlying topological spaces. However, this pattern of definition is not appropriate for graded manifolds, since not all the information concerning a graded manifold is encoded in the underlying topological space. We must therefore proceed on the analogy of the definition of product in algebraic geometry [GroD], which on the other hand is still valid for smooth or complex manifolds.

If X and Y are differentiable manifolds, a classical result is that the natural morphism $\mathcal{C}^\infty(X) \otimes_{\mathbf{R}} \mathcal{C}^\infty(Y) \to \mathcal{C}^\infty(X \times Y)$, given by $f(x) \otimes g(y) \mapsto f(x)\, g(y)$, induces an isomorphism of Fréchet $\mathbf{R}$-algebras

$$\mathcal{C}^\infty(X)\hat{\otimes}_\pi \mathcal{C}^\infty(Y) \stackrel{\sim}{\to} \mathcal{C}^\infty(X \times Y), \tag{1.4}$$

where the left hand side is the completion of $\mathcal{C}^\infty(X) \otimes_{\mathbf{R}} \mathcal{C}^\infty(Y)$ with respect to Grothendieck's π topology.

For the reader's convenience, we offer some details about the isomorphism (1.4) (cf. [Gro1] Part II, p. 81). If E and F are locally convex real vector spaces, there is a unique locally convex topology on $E \otimes_{\mathbf{R}} F$ such that, for every locally convex vector space G, the continuous linear maps $E \otimes_{\mathbf{R}} F \to G$ are in a natural one-to-one correspondence with the continuous bilinear maps $E \times F \to G$ (see [Gro1],§I.1, Proposition 2). This is the so-called Grothendieck π topology, and one denotes by $E \otimes_\pi F$ the corresponding locally convex vector space. The image of the immersion $\mathcal{C}^\infty(X) \otimes_\pi \mathcal{C}^\infty(Y) \hookrightarrow \mathcal{C}^\infty(X \times Y)$ is dense, since it separates the points of $X \times Y$ and the tangent vectors at a point. As a consequence, one obtains an isomorphism $\mathcal{C}^\infty(X)\hat{\otimes}_\pi \mathcal{C}^\infty(Y) \stackrel{\sim}{\to} \mathcal{C}^\infty(X \times Y)$.

We should recall that, according to Proposition 1.2, the space of sections of the structure sheaf of a graded manifold $(X, \mathcal{A})$ over a coordinate neighbourhood $U \subset X$ is a Fréchet algebra. Now, let $(X, \mathcal{A})$ and $(Y, \mathcal{B})$ be graded manifolds, of dimension (m, n) and (p, q), respectively. Let $\mathcal{A}\hat{\otimes}_\pi\mathcal{B}$ be the sheaf associated with the presheaf characterized by $U \times V \to \mathcal{A}(U)\hat{\otimes}_\pi\mathcal{B}(V)$.

Proposition 1.4. *$(X \times Y, \mathcal{A}\hat{\otimes}_\pi\mathcal{B})$ is a graded manifold of dimension $(m+p, n+q)$, which will be called the product graded manifold of $(X, \mathcal{A})$ and $(Y, \mathcal{B})$.*

Proof. For any product of open sets $U \times V \subset X \times Y$, the surjective continuous morphisms $\mathcal{A}(U) \to \mathcal{C}^\infty(U) \to 0$ and $\mathcal{B}(V) \to \mathcal{C}^\infty(V) \to 0$ induce surjective continuous morphisms $\mathcal{A}(U) \otimes_\pi \mathcal{B}(V) \to \mathcal{C}^\infty(U) \otimes_\pi \mathcal{C}^\infty(V) \to 0$ and, by completing, $\mathcal{A}(U)\hat{\otimes}_\pi\mathcal{B}(V) \to \mathcal{C}^\infty(U)\hat{\otimes}_\pi\mathcal{C}^\infty(V) \simeq \mathcal{C}^\infty(U \times V) \to 0$. These, in turn, induce an (even) surjective continuous morphism of sheaves of graded $\mathbf{R}$-algebras

$$\mathcal{A}\hat{\otimes}_\pi\mathcal{B} \to \mathcal{C}^\infty_{X\times Y} \to 0\,. \tag{1.5}$$

One has to prove that the kernel of this morphism is the sheaf $\mathfrak{N}$ of (locally) nilpotent elements and that $\mathcal{A}\hat{\otimes}_\pi\mathcal{B}$ is locally isomorphic with $\bigwedge(\mathfrak{N}/\mathfrak{N}^2)$. These are local matters, and one can then suppose that $\mathcal{A} \simeq \mathcal{C}^\infty_X \otimes_{\mathbf{R}} \bigwedge(E)$ and $\mathcal{B} \simeq \mathcal{C}^\infty_Y \otimes_{\mathbf{R}} \bigwedge(F)$ for certain vector spaces E and F of dimensions n and q. Hence, $\mathcal{A}\hat{\otimes}_\pi\mathcal{B} \simeq (\mathcal{C}^\infty_X\hat{\otimes}_\pi\mathcal{C}^\infty_Y)\otimes_{\mathbf{R}}\bigwedge(E\oplus F)$ because $E\oplus F$ is a finite-dimensional vector space. This enables us to conclude. ■

As a consequence of the previous Proposition, one obtains an analogue of the isomorphism (1.4) for graded manifolds:

$$\mathcal{A}(X)\hat{\otimes}_\pi\mathcal{B}(Y) \simeq (\mathcal{A}\hat{\otimes}_\pi\mathcal{B})(X \times Y)\,. \tag{1.6}$$

REMARK 1.2. The product defined above is in fact *the product in the category of graded manifolds*, in the sense that one has morphisms $\mathrm{pr}_1\colon (X \times Y, \mathcal{A}\hat{\otimes}_\pi\mathcal{B}) \to (X, \mathcal{A})$ and $\mathrm{pr}_2\colon (X \times Y, \mathcal{A}\hat{\otimes}_\pi\mathcal{B}) \to (Y, \mathcal{B})$ such that for any graded manifold $(Z, \mathcal{C})$ and every pair of morphisms $\phi_1\colon (Z, \mathcal{C}) \to (X, \mathcal{A})$ and $\phi_2\colon (Z, \mathcal{C}) \to (Y, \mathcal{B})$, there is a unique morphism $\phi\colon (Z, \mathcal{C}) \to (X \times Y, \mathcal{A}\hat{\otimes}_\pi\mathcal{B})$ fulfilling $\phi_1 = \phi \circ \mathrm{pr}_1$ and $\phi_2 = \phi \circ \mathrm{pr}_2$. ▲

Global structure of graded manifolds. The structure sheaf of a graded space $(X, \mathcal{A})$ over a field k with underlying space $(X, \mathcal{S})$ is by definition locally isomorphic, as a sheaf of graded-commutative k-algebras, with the exterior algebra sheaf $\mathcal{B} = \bigwedge_{\mathcal{S}}(\mathfrak{J}/\mathfrak{J}^2)$ (cf. Definition 1.1). We restrict our attention to the case where the field k has characteristic 0 (as a matter of fact, in future applications k will be either $\mathbf{R}$ or $\mathbf{C}$), and the reduced space $(X, \mathcal{S})$ has no nilpotents, so that $\mathfrak{J} = \mathfrak{N}$.

It should be ascertained under which conditions $\mathcal{A}$ is isomorphic, as a sheaf of graded-commutative algebras, with $\mathcal{B}$. When $(X, \mathcal{A})$ is a graded manifold, we recover Batchelor's representation theorem [**Bch1**]:

the sheaf $\mathcal{A}$ can be identified with the sheaf of sections of the exterior bundle of a vector bundle over X.

Since the proof given in [**Bch1**] employs non-Abelian cohomology, we prefer to follow [**BlR**].

The sheaf $\mathcal{B} = \bigwedge_{\mathcal{S}}(\mathfrak{N}/\mathfrak{N}^2)$ has a canonical structure of graded $\mathcal{S}$-algebra, since the projection $\mathcal{B} \to \mathcal{S}$ admits a canonical section $\mathcal{S} \hookrightarrow \mathcal{B}$. Accordingly, our first aim is to determine whether the projection $\mathcal{A} \to \mathcal{S}$ admits a graded k-algebra section $\sigma: \mathcal{S} \hookrightarrow \mathcal{A}$.

We start by looking for graded algebra sections of the induced surjective morphisms $\pi_h: \mathcal{A}/\mathfrak{N}^h \to \mathcal{S}$. First, we need the algebraic results expressed by the following two Lemmas.

Lemma 1.3. *Let $(X, \mathcal{A})$ be a graded space with underlying space $(X, \mathcal{S})$ over a field k of characteristic 0, such that there is a graded k-algebra section $\sigma: \mathcal{S} \hookrightarrow \mathcal{A}$. If $\theta: \mathcal{A} \xrightarrow{\simeq} \mathcal{A}$ is a unipotent $\mathcal{S}$-algebra isomorphism (that is, $(\theta - \mathrm{Id})^p = 0$ for some integer p), the sheaf morphism defined by*

$$\log \theta = \sum_{1 \leq p} \frac{(-1)^{p-1}}{p} (\theta - \mathrm{Id})^p$$

is an even nilpotent derivation of $\mathcal{A}$ over $\mathcal{S}$. Conversely, if D is an even nilpotent derivation of $\mathcal{A}$ over $\mathcal{S}$, the sheaf morphism

$$\exp D = \mathrm{Id} + \sum_{1 \leq p} \frac{D^p}{p!}$$

is a unipotent $\mathcal{S}$-algebra automorphism. These transformations are inverse to each other.

Proof. Straightforward. ■

Lemma 1.4. *Every graded $\mathcal{S}$-algebra section $\delta_h: \mathcal{S} \hookrightarrow \mathcal{B}/\mathfrak{N}^h$ of the surjective morphism $\pi_h: \mathcal{B}/\mathfrak{N}^h \to \mathcal{S}$ has a lift to $\mathcal{B}/\mathfrak{N}^{h+1}$, that is, a graded $\mathcal{S}$-algebra* *section* $\delta_{h+1}: \mathcal{S} \hookrightarrow \mathcal{B}/\mathfrak{N}^{h+1}$ *such that* $\delta_h = p_h \circ \delta_{h+1}$, *where* $p_h: \mathcal{B}/\mathfrak{N}^{h+1} \to \mathcal{B}/\mathfrak{N}^h$ *is the natural projection.*

Proof. We introduce the exterior algebra $B = \bigwedge k^n$ and call N its nilpotent ideal, so that $\mathcal{B}/\mathfrak{N}^h \simeq B/N^h \otimes_k \mathcal{S}$; we define a graded k-algebra morphism $\theta: \mathcal{B}/\mathfrak{N}^h \to \mathcal{B}/\mathfrak{N}^h$ by letting, for every open subset $U \subset X$, $\theta(g \otimes f) = (g \otimes 1)\tilde{f}$, where $g \in \mathcal{B}/\mathfrak{N}^h$ and $f \in \mathcal{S}(U)$. As $(\theta - \mathrm{Id})(\mathfrak{N}/\mathfrak{N}^h) \subset \mathfrak{N}^2/\mathfrak{N}^h$, one has $(\theta - \mathrm{Id})^h = 0$, that is, θ is unipotent. By the above lemma, $D = \log\theta: \mathcal{B}/\mathfrak{N}^h \to \mathcal{B}/\mathfrak{N}^h$ is a nilpotent graded derivation. It follows that the morphism $\bar{D}: \mathcal{B}/\mathfrak{N}^{h+1} \to \mathcal{B}/\mathfrak{N}^{h+1}$, described on $U \subset X$ by $\bar{D}(g \otimes f) = (g \otimes 1)(\alpha_h \otimes 1)D(1 \otimes f)$, where $\alpha_h: B/N^h \simeq \bigoplus_{j<h} \bigwedge^j k^n \to B/N^{h+1} \simeq \bigoplus_{j<h+1} \bigwedge^j k^n$ is the natural immersion, is also a nilpotent derivation, thus inducing a unipotent k-algebra isomorphism $\bar{\theta}: \mathcal{B}/\mathfrak{N}^{h+1} \to \mathcal{B}/\mathfrak{N}^{h+1}$ such that $p_{h+1} \circ \bar{\theta} = \theta \circ p_h$. The algebra morphism $\delta_{h+1}: \mathcal{S} \hookrightarrow \mathcal{B}/\mathfrak{N}^{h+1}$ defined as $\delta_{h+1}(F) = \bar{\theta}(1 \otimes f)$ is the desired lift of δ_h. ∎

Corollary 1.6. *If the surjective morphism* $\pi_h: \mathcal{A}/\mathfrak{N}^h \to \mathcal{S}$ *has a section* $\sigma_h: \mathcal{S} \hookrightarrow \mathcal{A}/\mathfrak{N}^h$, *then there exist an open cover* $\{U_i\}$ *of* X *and local lifts* $\sigma_{h+1,i}: \mathcal{S}_{|U_i} \hookrightarrow (\mathcal{A}/\mathfrak{N}^{h+1})_{|U_i}$ *of* σ_h.

Proof. There is an open cover $\{U_i\}$ of X such that one has graded algebra sheaf isomorphisms $\tau_i: \mathcal{A}_{|U_i} \simeq \mathcal{B}_{|U_i} \simeq B \otimes_k \mathcal{S}_{|U_i}$ commuting with the projections onto $\mathcal{S}_{|U_i}$. One concludes by the previous Lemma. ∎

Corollary 1.7. *If the surjective morphism* $\pi_h: \mathcal{A}/\mathfrak{N}^h \to \mathcal{S}$ *has a section* $\sigma_h: \mathcal{S} \hookrightarrow \mathcal{A}/\mathfrak{N}^h$, *there is a cohomology class in* $\check{H}^1(X, \mathcal{D}er_k(\mathcal{S}, \mathfrak{N}^h/\mathfrak{N}^{h+1}))$ *which vanishes if and only if there is a global lift* $\sigma_{h+1}: \mathcal{S} \hookrightarrow \mathcal{A}/\mathfrak{N}^{h+1}$ *of* σ_h.

Proof. From the exact sequence

$$0 \to \mathfrak{N}^h/\mathfrak{N}^{h+1} \to \mathcal{A}/\mathfrak{N}^{h+1} \xrightarrow{p_h} \mathcal{A}/\mathfrak{N}^h \to 0$$

one obtains another exact sequence:

$$0 \to \mathfrak{N}^h/\mathfrak{N}^{h+1} \to \mathcal{E} \xrightarrow{p} \mathcal{S} \to 0,$$

having denoted by $\mathcal{E} = \mathcal{A}/\mathfrak{N}^{h+1} \times_{\mathcal{A}/\mathfrak{N}^h} \mathcal{S}$ the subalgebra of $\mathcal{A}/\mathfrak{N}^{h+1} \times \mathcal{S}$ whose sections on an open subset $U \subset X$ are the pairs (g, f) such that $p_h(g) = \sigma_h(f)$. It is easy to check that, on any open subset $V \subset X$, the sections of $p_{|V}: \mathcal{E}_{|V} \to \mathcal{S}_{|V}$ are in a one-to-one correspondence with the lifts of $\sigma_{h|V}$. Thus, we have to study the conditions for the existence of a global k-algebra section $\rho: \mathcal{S} \hookrightarrow \mathcal{E}$ of $p: \mathcal{E} \to \mathcal{S}$, which is routine work since $\operatorname{Ker} p = \mathfrak{N}^h/\mathfrak{N}^{h+1}$ is a square zero ideal.[2] By Corollary 1.6, there are a cover $\{U_i\}$ of X and local lifts $\sigma_{h+1,i}$ of σ_h which determine, according to the previous remark, sections $\rho_i: \mathcal{S}_{|U_i} \hookrightarrow \mathcal{E}_{|U_i}$. As $(\mathfrak{N}^h/\mathfrak{N}^{h+1})^2 = 0$ in $\mathcal{E}$, the family of maps $\rho_{ij} = \rho_{i|U_i\cap U_j} - \rho_{j|U_i\cap U_j}: \mathcal{S}_{|U_i\cap U_j} \to (\mathfrak{N}^h/\mathfrak{N}^{h+1})_{|U_i\cap U_j}$ is a Čech 1-cocycle of the sheaf $\mathcal{D}er_k(\mathcal{S}, \mathfrak{N}^h/\mathfrak{N}^{h+1})$ with respect to the cover $\{U_i\}$. The resulting cohomology class $[\rho] \in \check{H}^1(X, \mathcal{D}er_k(\mathcal{S}, \mathfrak{N}^h/\mathfrak{N}^{h+1}))$, which is independent of the choice of the local lifts $\sigma_{h+1,i}$, obviously vanishes if these are induced by a global one. Conversely if the class $[\rho]$ vanishes, then, possibly after refining the cover, there exist derivations $D_i: \mathcal{S}_{|U_i} \to (\mathfrak{N}^h/\mathfrak{N}^{h+1})_{|U_i}$, such that $D_{i|U_i\cap U_j} - D_{j|U_i\cap U_j} = \rho_{ij} = \rho_{i|U_i\cap U_j} - \rho_{j|U_i\cap U_j}$. Now, $\phi_i = \rho_i + D_i: \mathcal{S}_{|U_i} \hookrightarrow \mathcal{E}_{|U_i}$ are k-algebra sections of p which agree on $U_i \cap U_j$, thus defining a global section $\phi: \mathcal{S} \hookrightarrow \mathcal{E}$. ∎

From this result one obtains:

Proposition 1.5. *If $\check{H}^1(X, \mathcal{D}er_k(\mathcal{S}, \bigoplus_{h=0}^n \mathfrak{N}^h/\mathfrak{N}^{h+1})) = 0$, there is a global section $\sigma: \mathcal{S} \hookrightarrow \mathcal{A}$ of $\pi: \mathcal{A} \to \mathcal{S}$.* ∎

Corollary 1.8. *Let $(X, \mathcal{A})$ be a graded manifold. There is a global section $\sigma: \mathcal{C}_X^\infty \hookrightarrow \mathcal{A}$ of $\pi: \mathcal{A} \to \mathcal{C}_X^\infty$.*

Proof. In this case Čech cohomology groups coincide with sheaf cohomology groups (Proposition II.3.5). The sheaf $\mathcal{D}er_k(\mathcal{S}, \bigoplus_{h=0}^n \mathfrak{N}^h/\mathfrak{N}^{h+1})$ is soft by Corollary II.3.4, and hence acyclic. ∎

One should notice that there are graded spaces (like graded analytic spaces or schemes) which may have no global sections (cf. [**Ma2**, p. 191]).

Now we restrict our attention to those graded spaces $(X, \mathcal{A})$ with reduced space $(X, \mathcal{S})$ which admit a section $\sigma: \mathcal{S} \hookrightarrow \mathcal{A}$ of the structure morphism $\pi: \mathcal{A} \to \mathcal{S}$. In this case, by means of the morphism σ, we can give $\mathcal{A}$ an $\mathcal{S}$-algebra structure, which obviously depends on the choice of σ. By definition of graded space, there are an open cover $\{U_i\}$ of X and graded-commutative k-algebra isomorphisms $\tau_i: \mathcal{A}_{|U_i} \xrightarrow{\sim} \mathcal{B}_{|U_i} \simeq B \otimes_k \mathcal{S}_{|U_i}$ where, as above, $B = \bigwedge k^n$. In terms of these data we can construct a set of $\mathcal{S}$-algebra isomorphisms $\phi_i: \mathcal{A}_{|U_i} \xrightarrow{\sim} \mathcal{B}_{|U_i}$.

Lemma 1.5. *Let us endow $\mathcal{A}$ with the graded $\mathcal{S}$-algebra structure induced by*

[2] An analogous statement holds for extensions of Lie algebras.

the section σ. Then, $\mathcal{A}$ and $\mathcal{B}$ are locally isomorphic as sheaves of graded-commutative $\mathcal{S}$-algebras.

Proof. By means of the section σ we define maps $\theta_i: B \otimes_k \mathcal{S}_{|U_i} \to B \otimes_k \mathcal{S}_{|U_i}$ by letting $\theta_i(b \otimes f) = (b \otimes 1)\tau_i(\sigma(f))$; these are morphisms of k-algebras, and $(\theta_i - \mathrm{Id})$ maps $B \otimes_k \mathcal{S}_{|U_i}$ into $N \otimes_k \mathcal{S}_{|U_i}$; that is, $(\theta_i - \mathrm{Id})^{n+1} = 0$. It follows that θ_i is unipotent, and by Lemma 1.3 is invertible. The morphisms

$$\phi_i = \theta_i^{-1} \circ \tau_i : \mathcal{A}_{|U_i} \xrightarrow{\sim} B \otimes_k \mathcal{S}_{|U_i}$$

are easily shown to be graded-commutative $\mathcal{S}$-algebra isomorphisms. ∎

Proposition 1.6. *Under the same hypotheses of Lemma 1.5, there is a cohomology class in $\check{H}^1(X, \mathcal{D}er_{\mathcal{S}}(\mathcal{B}, \mathcal{B}))$ whose vanishing is equivalent to the existence of a graded-commutative $\mathcal{S}$-algebra isomorphism*

$$\mathcal{A} \xrightarrow{\sim} \mathcal{B} = \bigwedge_{\mathcal{S}}(\mathfrak{N}/\mathfrak{N}^2)\,.$$

Proof. By Lemma 1.5 there are an open cover $\{U_i\}$ of X and graded-commutative $\mathcal{S}_{|U_i}$-algebra isomorphisms $\phi_i: \mathcal{A}_{|U_i} \xrightarrow{\sim} \mathcal{B}_{|U_i} \simeq B \otimes_k \mathcal{S}_{|U_i}$. Now, the maps $\phi_{ij} = \phi_j \circ \phi_i^{-1}: B \otimes_k \mathcal{S}_{|U_i \cap U_j} \xrightarrow{\sim} B \otimes_k \mathcal{S}_{|U_i \cap U_j}$ are unipotent automorphisms, thus defining nilpotent derivations D_{ij} of $B \otimes_k \mathcal{S}_{|U_i \cap U_j}$ on $\mathcal{S}_{|U_i \cap U_j}$. One easily checks that $\{D_{ij}\}$ is a Čech 1-cocycle of the sheaf $\mathcal{D}er_{\mathcal{S}}(\mathcal{B}, \mathcal{B})$, whose cohomology class is independent of the choices of the $\mathcal{S}_{|U_i}$-algebra automorphisms ϕ_i. The vanishing of this class entails, possibly after refinement of the cover, the existence of nilpotent derivations D_i of $\mathcal{B}_{|U_i}$ over $\mathcal{S}_{|U_i}$, such that $D_{ij} = D_{i|U_i \cap U_j} - D_{j|U_i \cap U_j}$. After calling $\rho_i = \exp D_i$ the corresponding unipotent $\mathcal{S}_{|U_i}$-algebra automorphisms, the $\mathcal{S}_{|U_i}$-algebra isomorphisms $\rho_i^{-1} \circ \phi_i: \mathcal{A}_{|U_i} \xrightarrow{\sim} \mathcal{B}_{|U_i}$ coincide on the overlaps, thus yielding the global isomorphism we were looking for. The converse is trivial. ∎

As a direct application of this Proposition and Corollary 1.8, we find a result usually known as *Batchelor's theorem.*

Corollary 1.9. *Let $(X, \mathcal{A})$ be a graded manifold. There is a global graded-commutative algebra isomorphism*

$$\mathcal{A} \xrightarrow{\sim} \bigwedge_{\mathcal{S}}(\mathfrak{N}/\mathfrak{N}^2)\,. \tag{1.7}$$

∎

REMARK 1.3. It should be stressed that this proof of Batchelor's theorem does not apply in the category of complex analytic graded manifolds, since in that case the cohomology groups involved in Proposition 1.5 are generically not trivial. Actually, there are examples of complex analytic graded manifolds for which the isomorphism (1.7) does not hold [**Gre**]. One can prove that any complex analytic graded manifold is a deformation, in a sense analogous to that of the Kodaira-Spencer theory, of the exterior algebra of the sheaf $\mathfrak{N}/\mathfrak{N}^2$ [**Rt1**]. ▲

2. Supersmooth functions

The original idea of the 'geometric' approach to supermanifolds is to patch open sets in $B_L^{m,n}$ by means of transition functions which fulfill a suitable 'smoothness' condition. We wish now to define the various classes of functions (G^∞, GH^∞ and H^∞ functions) that have been devised to that end. We shall call them generically *supersmooth functions.* We shall introduce them in a unified manner, in terms of a morphism, called *Z-expansion,* which maps functions of real variables into functions of variables in $B_L^{m,n}$. Unless otherwise stated, whenever referring, explicitly or implicitly, to a topology on $B_L^{m,n}$, we shall mean its **R**-vector space topology.

In this Section we assume to have chosen integers L, m and n, with $L > 0$ and $m, n \geq 0$, subject to the condition $L \geq n$. For each integer L' such that $0 \leq L' \leq L$, the exterior algebra $B_{L'}$ is regarded as a subalgebra of B_L, so that B_L acquires a structure of a graded $B_{L'}$-module, which is not free, unless $L' = 0$ or $L' = L$. We recall that the graded vector space associated with $B_L^{m|n}$ according to the procedure of Section I.1 is simply $\mathbf{R}^m \oplus \mathbf{R}^n$; we denote by $\sigma^{m,n}: B_L^{m,n} \to \mathbf{R}^m$ the restriction of the augmentation map to $B_L^{m,n}$.

For any C^∞ differentiable manifold X, let us denote by $\mathcal{C}_{L'}^\infty(W)$ the graded algebra of $B_{L'}$-valued C^∞ functions on the open set $W \subset X$. For each integer $L' \leq L$ and any $U \subset \mathbf{R}^m$, the Z-expansion is the morphism of graded algebras

$$Z_{L'}: \mathcal{C}_{L'}^\infty(U) \to \mathcal{C}_L^\infty((\sigma^{m,0})^{-1}(U)),$$

defined by the formula (cf. [**Rs2**])

$$Z_{L'}(h)(x) = h(\sigma^{m,0}(x)) + \sum_{j=1}^{L} \frac{1}{j!} D^{(j)} h_{\sigma^{m,0}(x)}(s^{m,0}(x), \ldots, s^{m,0}(x)) \qquad (2.1)$$

for all $h \in C^{\infty}_{L'}(U)$ and all $x \in (\sigma^{m,0})^{-1}(U)$; here the j-th Fréchet differential $D^{(j)}h_{\sigma^{m,0}(x)}$ of h at the point $\sigma^{m,0}(x)$ acts on $B_L^{m,0} \times \cdots \times B_L^{m,0}$ (j times) simply by extending by $(B_L)_0$-linearity its action on $\mathbf{R}^m \times \cdots \times \mathbf{R}^m$. The mapping $s^{m,0}: B_L^{m,0} \to \mathfrak{N}_L^{m,0}$ is the projection onto the second component of the direct sum $B_L^{m,0} = \mathbf{R}^m \oplus \mathfrak{N}_L^{m,0}$.

Proposition 2.1. *The morphism (2.1) is injective.*

Proof. The restriction of $Z_{L'}(h)$ to real values of its arguments coincides with h. ∎

For each open $U \subset \mathbf{R}^m$, $(\sigma^{m,0})^{-1}(U) \subset B_L^{m,0}$ is a subset of $B_L^{m,n}$, so that we can define on the open set $(\sigma^{m,n})^{-1}(U) \subset B_L^{m,n}$ the graded algebra $\mathcal{S}_{L'}((\sigma^{m,n})^{-1}(U))$ formed by the functions having the following expression

$$f(x^1, \dots, x^m, y^1, \dots, y^n) = \sum_{\mu \in \Xi_n} f_\mu(x^1, \dots, x^m)\, y^\mu, \tag{2.2}$$

where $f_\mu \in Z_{L'}(C^{\infty}_{L'}(U))$, $(x^1, \dots, x^m, y^1, \dots, y^n) \in (\sigma^{m,n})^{-1}(U)$, and $y^\mu = y^{\mu(1)} \dots y^{\mu(r)}$ if $\mu = \{\mu(1), \dots, \mu(r)\}$.

We can therefore introduce a sheaf $\mathcal{S}_{L'}$ of graded-commutative $B_{L'}$-algebras over $B_L^{m,n}$ by letting, for each open $V \subset B_L^{m,n}$,

$$\mathcal{S}_{L'}(V) = \mathcal{S}_{L'}\big((\sigma^{m,n})^{-1}\sigma^{m,n}(V)\big)\,. \tag{2.3}$$

The sections of the sheaf $\mathcal{S}_{L'}$ on an open set V are C^∞ functions which show a kind of holomorphic behaviour in the nilpotent directions, in that the coefficients of the various powers of the y's in Eq. (2.2) are determined, at every point z of the fibre $(\sigma^{m,n})^{-1}(x)$ of $B_L^{m,n}$ over $x = \sigma^{m,n}(z) \in \mathbf{R}^m$, by their germs at x.

We denote by $\hat{\mathcal{S}}_{L'}$ the subsheaf of $\mathcal{S}_{L'}$ whose sections are functions not depending on the odd variables y^α, namely, they have only the first term in the sum (2.2). In other words, the sheaf $\hat{\mathcal{S}}_{L'}$ on $B_L^{m,n}$ is the inverse image under the projection $B_L^{m,n} \to B_L^{m,0}$ of the sheaf $\mathcal{S}_{L'}$ on $B_L^{m,0}$. Then Eq. (2.2) shows the existence, for any open $U \subset B_L^{m,n}$, of a surjective morphism

$$\begin{aligned} \lambda: \hat{\mathcal{S}}_{L'}(U) \otimes_{\mathbf{R}} \textstyle\bigwedge_{\mathbf{R}} \mathbf{R}^n &\to \mathcal{S}_{L'}(U) \\ \sum_{\mu \in \Xi_n} f_\mu \otimes y^\mu &\mapsto \sum_{\mu \in \Xi_n} f_\mu\, y^\mu, \end{aligned} \tag{2.4}$$

having identified $\bigwedge_{\mathbf{R}} \mathbf{R}^n$ with the exterior algebra generated by the y's.

Proposition 2.2. *The sheaf morphism (2.4) is injective (and therefore an isomorphism) if and only if $L - L' \geq n$.*

Proof. Let

$$\sum_{\mu \in \Xi_n} f_\mu \, y^\mu = \sum_{\mu \in \Xi_n} g_\mu \, y^\mu \qquad \text{for all} \quad (x,y) \in V \subset B_L^{m,n} \tag{2.5}$$

(henceforth (x,y) will denote the $(m+n)$-tuple $(x^1, \ldots, x^m, y^1, \ldots, y^n)$). We may assume without loss of generality that $V = (\sigma^{m,n})^{-1}\sigma^{m,n}(V)$. We consider in particular real x's and y's having values in the component of $(B_L)_1$ of lower degree (i.e. the y's are 1-forms in B_L), but otherwise arbitrary. Since for real x's the f_μ's are $B_{L'}$-valued, Eq. (2.5) implies $f_\mu = g_\mu$ provided that $L - L' \geq n$, thus showing the injectivity of λ.

To show the opposite implication, let as assume that $L - L' < n$. Let a be a top-degree form in $B_{L'}$; it can be regarded as a constant section of $\mathcal{S}_{L'}$. Since $ay^1 \cdots y^n = 0$ identically, λ is not injective. ∎

Derivatives of supersmooth functions. If $f \in \hat{\mathcal{S}}_{L'}(V)$, then $f = Z_{L'}(\hat{f})$ for some $\hat{f} \in \mathcal{C}_{L'}(\sigma^{m,n}(V))$. The partial derivatives of f can be defined according to the equation

$$\frac{\partial f}{\partial x^i} = Z_{L'}\left(\frac{\partial \hat{f}}{\partial \hat{x}^i}\right), \qquad i = 1 \ldots m, \tag{2.6}$$

where the $\hat{x}^i$'s are the canonical coordinates in $\mathbf{R}^m$. If $f \in \mathcal{S}_{L'}$ can be written as in (2.2), we set

$$\frac{\partial f}{\partial x^i} = \sum_{\mu \in \Xi_n} \frac{\partial f_\mu}{\partial x^i} \, y^\mu. \tag{2.7}$$

The reader may easily verify that, even when the representation (2.2) is not unique (i.e. when $L - L' < n$), the definition 2.7 is well posed. Partial derivatives of higher order with respect to even variables are obtained by successive applications of the operators $\dfrac{\partial}{\partial x^i}$.

The situation is more delicate when dealing with derivatives with respect to odd coordinates. It turns out that no consistent definition can be given, unless the condition $L - L' \geq n$ holds. This should be made clear by the following discussion. Let $f \in \mathcal{S}_{L'}(V)$, with V an open set in $B_L^{m,n}$, assume $(x,y) \in V$, and fix a $k \in B_L^{0,n}$ such that $(x, y+k)$ is still in V. The quantity $f(x, y+k)$ can be regarded as a supersmooth function of the variables x and k, and, assuming that $L - L' \geq n$, it may be given a unique representation of the form

$$f(x, y+k) = \sum_{\mu \in \Xi_n} \partial_\mu f(x, y)\, k^\mu . \tag{2.8}$$

The functions $\partial_\mu f \in \mathcal{S}_{L'}$, $\mu \neq \mu_0$, are by definition the derivatives of order 1 up to n of f with respect to the odd variables (derivatives of higher order vanish identically). We shall write

$$\frac{\partial f}{\partial y^{\mu(r)} \ldots \partial y^{\mu(1)}} \equiv \partial_\mu f \qquad \text{if} \qquad \mu = \{\mu(1), \ldots, \mu(r)\}.$$

Evidently, if $L - L' < n$, one can add terms to $\partial_\mu f$ without altering the right-hand side of (2.8), so that $\partial_\mu f$ is not well defined in that case.

The expansion (2.8), together with the Taylor formula for functions of real variables, yield a Taylor-like development for supersmooth functions:

Proposition 2.3. *Let V be an open set in $B_L^{m,n}$, such that $\sigma^{m,n}(V) \subset \mathbf{R}^m$ is convex, assume that (x, y) and $(x+h, y+k)$ are both in V, and let $f \in \mathcal{S}_{L'}(V)$. Fix a positive integer N, and let*

$$z^i = x^i, \quad t^i = h^i, \quad i = 1 \ldots m;$$

$$z^{m+\alpha} = y^\alpha, \quad t^{m+\alpha} = k^\alpha, \quad \alpha = 1 \ldots n.$$

If $L - L' \geq n$, there exist supersmooth functions $R^{(N)}_{A_N \ldots A_1}$ of z and t such that

$$\begin{aligned} f(x+h, y+k) = & f(x, y) + \sum_{j=1}^{N-1} \sum_{A_1 \ldots A_j = 1}^{m+n} \frac{\partial^j f}{\partial z^{A_j} \ldots \partial z^{A_1}}(z)\; t^{A_1} \ldots t^{A_j} \\ & + \sum_{A_1 \ldots A_N = 1}^{m+n} R^{(N)}_{A_N \ldots A_1}\, t^{A_1} \ldots t^{A_N} . \end{aligned} \tag{2.9}$$

■

H^∞ **functions.** If $L' = 0$ the sheaf $\mathcal{S}_{L'}$ coincides with the sheaf of H^∞ functions, first considered by M. Batchelor [**Bch2**] and B. DeWitt [**DW**]. They are a particular case of GH^∞ functions, and therefore the arguments of Section 3 apply to them. Proposition 2.2 also holds in this case. However, they have a distinguished feature too, in that $B_{L'}$ reduces in this case to the field $\mathbf{R}$. As far as the physical applications are concerned, it has sometimes been stated that H^∞ functions are not relevant, for the following reason. In the so-called superspace approach to supersymmetric field theory (cf. e.g. [**WsB**]), supersmooth functions are regarded as a bookkeeping device, in that the coefficient functions in the expansion (2.2) (called *superfield expansion*) are identified with the physical fields, of bosonic (resp. fermionic) type if they multiply an even (resp. odd) power of the y's. By restricting the arguments to real values (which physically means restricting to space-time), the physical fields are real-valued, so that they cannot be anticommuting, and supersymmetry cannot be implemented (cf. the discussion of the Wess-Zumino model in the Introduction). Also, graded manifolds are subject to this criticism; the reader may refer to [**DeS**] on this aspect.

G^∞ **functions.** These functions are obtained by letting $L' = L$, and were introduced by Rogers [**Rs1**]. While G^∞ functions yield physical fields of correct parity (i.e. the fermionic fields do anticommute), they are affected, however, by serious inconsistencies [**BoyG,Rt2**]. Indeed it is not possible to define for them a derivative with respect to odd variables, basically because the morphism (2.4) is not injective in this case. As a consequence, the sheaf of derivations of the sheaf of G^∞ functions is not locally free, as erroneously claimed in [**Rs1**] and [**BoyG**]. For this reason supermanifolds modeled by means of G^∞ functions are quite unmanageable, and any contact with ordinary differential geometry is lost. Nevertheless, G^∞ functions will play an important role in the development of supergeometry, since any G-supermanifold has an underlying G^∞ supermanifold.

G^∞ functions of even variables can be characterized more directly without resorting to the Z-expansion. We can indeed prove the following result [**BoyG**].

Proposition 2.4. *Let $U \subset B_L^{m,0}$ be of the form $U = (\sigma^{m,0})^{-1}(V)$ for some convex open set V in $\mathbf{R}^m$. A C^∞ function $f\colon U \to B_L$ is G^∞ if and only if its Fréchet differential is $(B_L)_0$-linear.*

Proof. If f is G^∞ then, since $n = 0$, Proposition 2.3 holds, and therefore Eq. (2.9) with $N = 2$ shows that for any $x \in U$ the Fréchet differential of f at x, say Df_x, is $(B_L)_0$-linear. To show the converse, we first notice that the $(B_L)_0$-linearity of Df_x implies the $(B_L)_0$-linearity of the j-th Fréchet differential $D^{(j)}f_x$ for all $j > 1$; then, the Taylor series for $f(x)$ around $f(\sigma^{m,0}(x))$ — which terminates at order L by nilpotency — coincides with the Z-expansion of the restriction of f to V, as given by Eq. (2.1). ∎

Proposition 2.4 is reminiscent of a similar property of holomorphic functions, i.e. a smooth function $f: U \subset \mathbf{C}^m \to \mathbf{C}$ is holomorphic if and only if its Fréchet differential is $\mathbf{C}$-linear, which fact is expressed by the Cauchy-Riemann conditions. It therefore comes as no surprise that for smooth functions $f: U \subset B_L^{m,0} \to B_L$ the fact of being G^∞ is equivalent to a set of conditions of Cauchy-Riemann type. Let $\{\beta_\mu, \mu \in \Xi_L\}$ be the canonical basis of B_L, and define real numbers $A^\rho_{\mu\nu}$ (with μ, ν, $\rho \in \Xi_L$) by letting $\beta_\mu \beta_\nu = \sum_{\rho\in\Xi_L} A^\rho_{\mu\nu}\beta_\rho$; the A's are obviously either 1, 0, or -1. For all $x = (x^1, \dots, x^m) \in U$ let

$$x^i = \sum_{\mu\in\Xi_L} x^{i\mu}\beta_\mu, \qquad f(x) = \sum_{\mu\in\Xi_L} f^\mu(x)\beta_\mu .$$

In particular we have $x^{i\mu_0} = \sigma(x^i)$.

Proposition 2.5. *The function f is G^∞ if and only if the following identities hold:*

$$\frac{\partial f^\nu}{\partial x^{i\mu}} = \sum_{\rho\in\Xi_L} \frac{\partial f^\rho}{\partial x^{i\mu_0}} A^\nu_{\mu\rho} . \tag{2.10}$$

Proof. We know that f is G^∞ if and only if its Fréchet differential Df is $(B_L)_0$-linear. If $\{e_1, \dots, e_m\}$ is the canonical basis of $B_L^{m,0}$, this condition can be written as

$$Df_x(u) = Df_x(e_i)\,u^i \quad \forall u \in B_L^{m,0} . \tag{2.11}$$

A direct computation shows that the conditions (2.10) and (2.11) are equivalent. ∎

All this discussion has been carried through setting to zero the odd dimension n. It turns out that for $n > 0$ the $(B_L)_0$-linearity of the Fréchet differential, or, equivalently, conditions (2.10), while being still necessary, are no longer sufficient to ensure that the function is G^∞. In [BoyG] it has indeed been shown that conditions (2.10) must be supplemented by suitable second order conditions.

***GH^∞* functions.** Whenever the condition

$$L - L' \geq n \tag{2.12}$$

is fulfilled we refer to supermooth functions as *GH^∞ functions*. These include H^∞ functions as a particular case. Since Proposition 2.2 holds in this case, these functions have interesting properties, which will be investigated in the next Section. For the moment let us only notice that Proposition 2.4 can also be stated in this case, in the following form: a smooth function $f: U \to B_L$ (where U is as in Proposition 2.4), which restricted to V is $B_{L'}$-valued, is GH^∞ if and only if its Fréchet differential is $(B_L)_0$-linear.

Supersmooth supermanifolds. We provide, following Rogers [**Rs1**, **Rs2**], the definition of supersmooth supermanifolds, where 'supersmooth' means either G^∞ or H^∞ or GH^∞, giving a few examples. Obviously, a supersmooth morphism $\varphi: U \to V$ between two open sets U and V in $B_L^{m,n}$ is a set of $m+n$ supersmooth functions.

Definition 2.1. *A Hausdorff, paracompact topological space is an (m,n) dimensional supersmooth supermanifold if it admits an atlas $\mathcal{A} = \{(U_\alpha, \varphi_\alpha) \mid \varphi_\alpha: U_\alpha \to B_L^{m,n}\}$ such that the transition functions $\varphi_\alpha \circ \varphi_\beta^{-1}$ are supersmooth morphisms.*

REMARK 2.1. Quite evidently, the preceding definition is equivalent to stating that an (m,n) supersmooth supermanifold is a graded locally ringed space, locally isomorphic with $(B_L^{m,n}, \mathcal{F})$, where $\mathcal{F}$ is one of the sheaves of supersmooth functions previously introduced.

Apparently, if M is a supersmooth supermanifold of dimension (m,n), it also carries a structure of C^∞ manifold of dimension $2^{L-1}(m+n)$.

EXAMPLE 2.1. The manifold $M = \mathbf{R} \times S^1$ can be endowed with a structure of (1,0) dimensional supersmooth supermanifold. We assume for L, L' the values $L = L' = 2$; to simplify the notation the canonical basis of B_2 is written as $\{1, \beta_1, \beta_2, \beta_3 = \beta_1\beta_2\}$. We choose two charts $(U_1 \times \mathbf{R}, u)$ and $(U_2 \times \mathbf{R}, w)$, where U_1 (U_2) is S^1 without the north pole (south pole). u and w are given, in terms of $z \in \mathbf{R}$ and the stereographic angles θ, ϕ , respectively from the north and south pole, as follows:

$$u = z + \theta\beta_3, \quad -\tfrac{\pi}{2} < \theta < \tfrac{\pi}{2}; \qquad w = -z + (\tfrac{\pi}{2} - \phi)\beta_3, \quad -\tfrac{\pi}{2} < \phi < \tfrac{\pi}{2}.$$

It is easily shown that u and w are C^∞ diffeomorphisms and that the transition functions $u(w)$ and $w(u)$ are supersmooth, since, e.g.,

$$w(u) = \tfrac{\pi}{2}\beta_3 - u. \tag{2.13}$$

Therefore M acquires a structure of a G^∞ supermanifold, which, having $n = 0$, is not subject to the criticism previously expressed. A direct calculation, which exploits Eq. (2.10), shows that a global supersmooth function on M can be expressed in the form

$$f = K + \sum_{i=1}^{3} (\mathrm{pr}_1^* f^i)\beta_i,$$

where the constant K and the C^∞ functions f^i on $\mathbf{R}$ are real valued, and $\mathrm{pr}_1 : M \to \mathbf{R}$ is the projection onto the first factor. Thus, we obtain

$$\mathcal{G}^\infty(M) \simeq \mathbf{R} \oplus C^\infty(\mathbf{R}) \otimes_{\mathbf{R}} \mathfrak{N}_L \tag{2.14}$$

as a direct sum of $\mathbf{R}$-vector spaces. Here $C^\infty(\mathbf{R})$ is the vector space of real valued functions on the real line. Obviously, the ring $\mathcal{G}^\infty(M)$ has a structure of graded B_L-algebra, as one can check directly. ▲

EXAMPLE 2.2. $M = T^2 \times \mathbf{R}^2$, where T^2 is the two-dimensional torus. T^2 can be covered by a smooth atlas $\{(U_j, (z_j, \xi_j)), j = 1 \ldots 4\}$ such that the transition functions are translations, $(z_j, \xi_j) \mapsto (z_j + a_j, \xi_j + b_j)$. M is endowed with a structure of (1,1) dimensional GH^∞ supermanifold, with $L = 2$, $L' = 1$, by considering an atlas $\{(U_j \times \mathbf{R}^2, (x_j, y_j)), j = 1 \ldots 4\}$, where $x_j = z_j + u\beta_3$ and $y_j = \xi_j\beta_1 + t\beta_2$; here u, t are the canonical real coordinates in $\mathbf{R}^2$. A direct computation shows that the global GH^∞ functions on M may be identified with functions of the form

$$f = \alpha + \gamma\,\beta_1 + [u\alpha' - t\mu]\,\beta_3 \tag{2.15}$$

where α, γ, and μ are periodic real valued functions of a real variable (to be identified with the coordinate z) and a prime denotes differentiation. ▲

EXAMPLE 2.3. The same underlying smooth real manifold as in Example 2.2, but with a different GH^∞ structure, obtained by letting $x_j = z_j + \xi_j\beta_3$, $y_j = u\beta_1 + t\beta_2$. Now, a global function on M can be identified with a function of the form

$$f = K_1 + (\alpha + K_2\, u)\beta_1 + K_2\, t\,\beta_2 + t\,\gamma\beta_3$$

where K_1, K_2 are real constants and α, γ are real valued periodic functions of a real variable (to be identified with z). This supermanifold structure is not equivalent to that of the previous example; indeed, in Chapter V we shall introduce a cohomology theory which discriminates between the two supermanifold structures. ▲

Other explicit examples of supermanifolds can be found in [**Rs3,HQ1,Ra, RC1,RC2**].

3. GH^∞ functions.

Henceforth while referring to 'GH^∞ functions' we shall understand that condition (2.12), i.e. the inequality $L - L' \geq n$, holds. The resulting function sheaf on $B_L^{m,n}$ will be denoted by $\mathcal{GH}_{L'}$, while the structure sheaf of a generic GH^∞ supermanifold M, defined in accordance with Definition 2.1, will be denoted by $\mathcal{GH}^M$. We wish now to show that under condition (2.12) the sheaf of derivations of $\mathcal{GH}^M$ is locally free.

Let M be a GH^∞ supermanifold, with structure sheaf $\mathcal{GH}^M$; if $(U,(x^1,\ldots,x^m,y^1,\ldots,y^n))$ is a coordinate chart, proceeding as usual one can define derivations

$$\left\{\frac{\partial}{\partial x^i}, \frac{\partial}{\partial y^\alpha} \,\middle|\, i = 1\ldots m,\ \alpha = 1\ldots n\right\}, \tag{3.1}$$

which are sections of $\mathcal{D}er\,\mathcal{GH}^M$.

Proposition 3.1. *$\mathcal{D}er\,\mathcal{GH}^M$ is a locally free graded $\mathcal{GH}^M$-module, whose rank equals the dimension of M. Given a coordinate chart $(U,(x^1,\ldots,x^m,y^1,\ldots,y^n))$ of M, $\mathcal{D}er\,\mathcal{GH}^M(U)$ is generated over $\mathcal{GH}^M(U)$ by the derivations (3.1).*

The proof is a direct consequence of the following rather technical but otherwise elementary Lemma.

Lemma 3.1. *Given an open set $V \subset B_L^{m,n}$, let us consider a function $f \in \mathcal{GH}(V)$ which depends only on the even variables, so that $f = Z_{L'}(\hat f)$, with $\hat f \in C^\infty_{L'}(\sigma^{m,n}(V))$. For all derivations $D \in \mathcal{D}er\,\mathcal{GH}(V)$ one has:*

$$D(f) = Z_{L'}(\hat D(\hat f))_{|V}, \tag{3.2}$$

where $\hat D$ is the derivation of $C^\infty_{L'}(\sigma^{m,n}(V))$ defined by

$$\hat D(\hat g) = [D(\,Z_{L'}(\hat g))]_{|\sigma^{m,n}(V)} \qquad \forall \hat g \in C^\infty_{L'}(\sigma^{m,n}(V)).$$

Proof. One has trivially $\hat D(\hat f) = [Z_{L'}(\hat D(\hat f))]_{|\sigma^{m,n}(V)}$. Since $Z_{L'}$ is injective (Proposition 2.1), one obtains Eq. (3.2), since its left- and right-hand sides coincide when restricted to $\sigma^{m,n}(V)$. ∎

Proof of Proposition 3.1. Since the result to be proved is of a local nature, we may assume $M = B_L^{m,n}$. Now, $\mathcal{D}er\,C^\infty_{L'}(\sigma^{m,n}(U))$ is a locally free graded $C^\infty_{L'}(\sigma^{m,n}(U))$-module, generated by the derivations $\{\partial/\partial\hat x^i\,,\ i = 1\ldots m\}$, where the $\hat{x}^i$'s are the canonical coordinates in $\mathbf{R}^m$. Thus, if f is a GH^∞ function of even variables, by virtue of Lemma 3.1 we obtain

$$D(f) = Z_{L'}(\hat{D}(\hat{f})) = Z_{L'}\left(\sum_{i=1}^{m} \hat{D}(\hat{x}^i)\frac{\partial \hat{f}}{\partial \hat{x}^i}\right) = \sum_{i=1}^{m} D(x^i)\frac{\partial f}{\partial x^i}.$$

In the case of functions of even variables, this proves that the derivations $\frac{\partial}{\partial x^i}$ defined in Eq. (3.1) generate $\mathcal{D}er\,\widehat{\mathcal{GH}}_{L'}(U)$. Since GH^∞ functions depend polynomially on the odd variables, all the derivations (3.1) generate $\mathcal{D}er\,\mathcal{GH}_{L'}(U)$. The linear independence of these derivations is proved by applying a vanishing linear combination with coefficients in $\widehat{\mathcal{GH}}_{L'}$

$$\sum_{i=1}^{m} f^i \frac{\partial}{\partial x^i} + \sum_{\alpha=1}^{n} g^\alpha \frac{\partial}{\partial y^\alpha} = 0$$

to the sections x^i, y^α. Thus, the thesis is proved. ∎

Even though their sheaf of derivations is locally free, the GH^∞ functions show some undesirable features, related to the quest for a reasonable definition of 'supervector bundle' within the category of GH^∞ supermanifolds. Supervector bundles will be dealt with in subsequent sections, where the discussion of GH^∞ bundles will find its natural collocation. Here we wish only to point out the origin of the bad behaviour of GH^∞ functions in this respect.

Let $\mathcal{V}_z \subset B_L$ be the space of values taken at a point $z \in B_L^{m,n}$ by the germs $f \in \mathcal{GH}_z$:[3]

$$\mathcal{V}_z = \left\{a \in B_L \mid a = \tilde{f}(z) \text{ for some } f \in \mathcal{GH}_z\right\},$$

where a tilde denotes evaluation of germs. If $\mathfrak{L}_z$ is the ideal of $\mathcal{GH}_z$ formed by the germs which vanish when evaluated in z, then there is an exact sequence of graded $B_{L'}$-modules:

$$0 \to \mathfrak{L}_z \to \mathcal{GH}_z \to \mathcal{V}_z \to 0. \tag{3.3}$$

Let us notice that, in accordance with Eqs. (2.1,2.2), constant GH^∞ functions take values only in $B_{L'}$, so that one cannot prove trivially that $B_L \hookrightarrow \mathcal{V}_z$, as happens, for instance, for C^∞ or analytic B_L-valued functions. Indeed,

[3] For notational simplicity, in the following discussion the sheaf $\mathcal{GH}_{L'}$ will be denoted by $\mathcal{GH}$.

$\mathcal{V}_z$ depends essentially on the point z, and in general it is not free as a $B_{L'}$-module. For instance, in the case of $B_L^{m,0}$ one has $\mathcal{V}_z = B_{L'}$ if $z \in \mathbf{R}^m$, while $B_{L'} \subset \mathcal{V}_z \subset B_L$ strictly for a suitable choice of z.

We are thus facing the strange phenomenon that the space of values taken by the class of functions under consideration changes from point to point. This is going to cause problems in the definition of GH^∞ bundles; to realize the relevance of the previous discussion in this respect, the reader should recall that, according to the usual definition of vector bundle, the fibre at z of a vector bundle on an ordinary (topological, smooth, complex, or algebraic) manifold M with structure sheaf $\mathcal{F}$ is a vector space over the field $\mathcal{F}_z/\mathfrak{M}_z$, where $\mathfrak{M}_z$ is the maximal ideal of $\mathcal{F}_z$, i.e. the ideal of germs in $\mathcal{F}_z$ whose evaluation vanishes. Let us consider now a GH^∞ supermanifold M with structure sheaf $\mathcal{GH}^M$ and a locally free graded $\mathcal{GH}^M$-module, say $\mathcal{F}$. For any $z \in M$, the ring $\mathcal{GH}_z^M$ is local, with maximal ideal

$$\mathfrak{N}_z = \left\{ f \in \mathcal{GH}_z^M \mid \tilde{f}(z) \in \mathfrak{N}_L \right\};$$

here $\mathfrak{N}_L$ is the ideal of nilpotents in B_L, and one has $\mathcal{GH}_z^M/\mathfrak{N}_z \simeq \mathbf{R}$ for all $z \in M$. Thus, in order to achieve a genuine generalization of $\mathbf{R}$-vector bundles, we must deviate from the ordinary theory (cf. [Del]) and quotient $\mathcal{GH}_z^M$ not by $\mathfrak{N}_z$, but by $\mathfrak{L}_z$, expecting that the fibre F_z of the supervector bundle F associated with $\mathcal{F}$ is isomorphic to $(\mathcal{GH}_z/\mathfrak{L}_z)^{m|n}$.[4] However, $\mathcal{GH}_z^M/\mathfrak{L}_z \simeq \mathcal{V}_z$, so that, in conformity with the preceding discussion, F_z would turn out to be a $B_{L'}$-module in general not free and depending upon the choice of z.

The consequences of this state of affairs will be further discussed in Section IV.3, where supervector bundles will be introduced.

4. G-supermanifolds

The discussion of the previous Section shows that the choice of a class of supersmooth functions which is free from inconsistencies, and yields a theory applicable to supersymmetry, is not trivial. In particular it seems rather difficult to combine the following requirements:

(i) the sheaf of derivations of the function sheaf under consideration should be locally free;

[4]Notice that $\mathfrak{L}_z \subset \mathfrak{N}_z$ strictly if $L > 0$.

(ii) the coefficients of the 'superfield expansion' (2.2), when restricted to real arguments, should take values in a graded-commutative algebra B;

(iii) there should be a good theory of superbundles, and in particular there is a sensible notion of graded tangent space.

These difficulties can be overcome by introducing a new category of supermanifolds, called *G-supermanifolds*, characterized in terms of a sheaf $\mathcal{G}$ on $B_L^{m,n}$, which is in a sense a 'completion' of $\mathcal{GH}_{L'}$ (condition (2.12) is assumed to hold). More precisely, we define the sheaf of graded-commutative B_L-algebras on $B_L^{m,n}$

$$\mathcal{G}_{L'} \equiv \mathcal{GH}_{L'} \otimes_{B_{L'}} B_L \tag{4.1}$$

(cf. the definition of tensor product of two graded algebras in Section I.2). It is convenient to introduce an evaluation morphism $\delta\colon \mathcal{G}_{L'} \to \mathcal{C}_L$ (we denote by $\mathcal{C}_L$ the sheaf of B_L-valued continuous functions on $B_L^{m,n}$), by extending by additivity the mapping

$$\delta(f \otimes a) = fa\,. \tag{4.2}$$

Proposition 4.1. *The image of δ is isomorphic to the sheaf $\mathcal{G}^\infty$ of G^∞ functions on $B_L^{m,n}$. The morphism δ is injective when restricted to the subsheaf $\hat{\mathcal{G}}_{L'} = \widehat{\mathcal{GH}_{L'}} \otimes_{B_{L'}} B_L$.*

Proof. The first claim is evident in view of the definition of the sheaf of G^∞ functions (cf. Section 2). In order to prove that $\delta\colon \hat{\mathcal{G}}_{L'} \to \hat{\mathcal{G}}^\infty$ is an isomorphism, we exhibit the inverse morphism $\lambda\colon \hat{\mathcal{G}}^\infty \to \hat{\mathcal{G}}_{L'}$. Given an open set $U \subset B_L^{m,n}$, every $f \in \hat{\mathcal{G}}^\infty(U)$, can written, in accordance with Eq. (2.1), in the form

$$f = \sum_{\mu\in\Xi_n} Z_0(\hat{f}^\mu)_{|U}\,\beta_\mu\,, \tag{4.3}$$

where the $\hat{f}^\mu$'s are suitable sections of $C^\infty_{\mathbf{R}^m}(\sigma^{m,n}(U))$. After letting $\lambda(f) = Z_0(\hat{f}^\mu)_{|U} \otimes \beta_\mu$, one verifies that $\lambda \circ \delta = id = \delta \circ \lambda$. ∎

Proposition 4.1 has an important consequence.

Corollary 4.1. *Given two integers L', L'' satisfying the condition (2.12), there is a canonical isomorphism of sheaves of graded commutative B_L-algebras $\mathcal{G}_{L'} \simeq \mathcal{G}_{L''}$.*

Proof. Proposition 4.1 entails the isomorphism $\hat{\mathcal{G}}_{L'} \simeq \hat{\mathcal{G}}_{L''}$. On the other hand, the isomorphism (2.4) gives

$$\mathcal{G}_{L'} \simeq \hat{\mathcal{G}}_{L'} \otimes_{\mathbf{R}} \bigwedge_{\mathbf{R}} \mathbf{R}^n, \tag{4.4}$$

so that our claim is proved. ∎

Therefore, it is possible to introduce on $B_L^{m,n}$ a canonical sheaf of graded commutative B_L-algebras $\mathcal{G}$, formally defined as the isomorphism class of the sheaves $\mathcal{G}_{L'}$ while L' varies among the non-negative integers such that $L - L' \geq n$. Alternatively, one can assume $L \geq 2n$ and take once for all $L' = [L/2]$, the biggest integer less then $L/2$ (cf. [Rs2]). A subsheaf $\hat{\mathcal{G}}$ of germs of sections of $\mathcal{G}$ 'not depending on the odd variables' is defined in the same fashion, and one obtains the isomorphism

$$\mathcal{G} \simeq \hat{\mathcal{G}} \otimes_{\mathbf{R}} \bigwedge_{\mathbf{R}} \mathbf{R}^n . \tag{4.5}$$

Let us now investigate what is the analogue of the exact sequence (3.3) in the case of G-supermanifolds. The evaluation morphism[5]

$$\sim : \mathcal{G}_z \to B_L , \tag{4.6}$$

defined by the composition of $\delta : \mathcal{G}_z \to \mathcal{G}_z^\infty$ with the usual germ evaluation morphism, gives rise to the exact sequence of graded B_L-modules

$$0 \to \mathfrak{L}_z \to \mathcal{G}_z \to B_L \to 0 . \tag{4.7}$$

The graded B_L-module $\mathfrak{L}_z$ appearing above is evidently the ideal of the germs $f \in \mathcal{G}_z$ such that $\tilde{f} = 0$; comparing the sequence (4.7) with (3.3), we see that one of the drawbacks of the GH^∞ functions has been disposed of, in that the space of values taken by the sections of $\mathcal{G}$ at a point $z \in B_L^{m,n}$ is B_L, regardless of the choice of z.

The sheaf $\mathcal{D}er\,\mathcal{G}$ of graded derivations of $\mathcal{G}$ inherits the nice algebraic properties of $\mathcal{D}er\,\mathcal{GH}$.

Proposition 4.2. *There is an isomorphism of sheaves of graded B_L-modules $\mathcal{D}er\,\mathcal{G} \simeq \mathcal{D}er\,\mathcal{GH} \otimes_{B_{L'}} B_L$.*

Proof. By virtue of the isomorphism (2.4), it is enough to show that $\mathcal{D}er\,\hat{\mathcal{G}} \simeq \mathcal{D}er\,\widehat{\mathcal{GH}} \otimes_{B_{L'}} B_L$. By identifying $\hat{\mathcal{G}}$ with $\hat{\mathcal{G}}^\infty$, we define a morphism $\eta : \mathcal{D}er\,\hat{\mathcal{G}}^\infty \to$

[5] The reader will notice that the symbol '$\sim$' has here a different meaning than in the context of graded manifolds.

$\mathcal{D}er\,\widehat{\mathcal{GH}} \otimes_{B_{L'}} B_L$ given by

$$\eta(D)(f) = \sum_{\mu \in \Xi_n} D(Z_0(\hat{f}^\mu)) \otimes \beta_\mu ,$$

where f has been factorized according to Eq. (4.3). It easily verified that η is an isomorphism. ■

The previous Proposition, together with Proposition (3.1), proves the following claim.

Proposition 4.3. *$\mathcal{D}er\,\mathcal{G}$ is a locally free graded $\mathcal{G}$-module on $B_L^{m,n}$, of rank (m,n). On every open set $U \subset B_L^{m,n}$, $\mathcal{D}er\,\mathcal{G}(U)$ is generated over $\mathcal{G}(U)$ by the derivations*

$$\left\{\frac{\partial}{\partial x^i}, \frac{\partial}{\partial y^\alpha} \,\middle|\, i = 1 \dots m\,, \alpha = 1 \dots n\right\}$$

defined as follows:

$$\frac{\partial}{\partial x^i}(f \otimes a) = \frac{\partial f}{\partial x^i} \otimes a,\ i = 1 \dots m; \quad \frac{\partial}{\partial y^\alpha}(f \otimes a) = \frac{\partial f}{\partial y^\alpha} \otimes a,\ \alpha = 1 \dots n. \tag{4.8}$$

■

The reader can easily check that these derivations satisfy a graded version of the usual Schwarz theorem:

$$\frac{\partial}{\partial x^i}\frac{\partial}{\partial x^j} = \frac{\partial}{\partial x^j}\frac{\partial}{\partial x^i}; \qquad \frac{\partial}{\partial x^i}\frac{\partial}{\partial y^\alpha} = \frac{\partial}{\partial y^\alpha}\frac{\partial}{\partial x^i}; \qquad \frac{\partial}{\partial y^\alpha}\frac{\partial}{\partial y^\beta} = -\frac{\partial}{\partial y^\beta}\frac{\partial}{\partial y^\alpha}.$$

We give now our definition of G-supermanifold.

Definition 4.1. *An (m,n) dimensional G-supermanifold is a graded locally ringed B_L-space $(M, \mathcal{A})$ satisfying the following conditions:*

1. *M is a Hausdorff, paracompact topological space;*
2. *$(M, \mathcal{A})$ is locally isomorphic with $(B_L^{m,n}, \mathcal{G})$;*
3. *denoting by $\mathcal{C}_L^M$ the sheaf of continuous B_L-valued functions on M, there exists a morphism of sheaves of B_L-algebras $\delta^M : \mathcal{A} \to \mathcal{C}_L^M$ which is locally compatible with the evaluation morphism (4.2) and with the isomorphisms ensuing from condition (2).*

Thus, rephrasing the previous assumptions, any point $z \in M$ has a neighbourhood U such that:

(i) there is an isomorphism of graded locally ringed spaces

$$(\bar{\varphi}, \varphi) \colon (U, \mathcal{A}_{|U}) \xrightarrow{\sim} (\bar{\varphi}(U), \mathcal{G}_{|\bar{\varphi}(U)}), \tag{4.9}$$

(ii) the diagram

$$\begin{array}{ccc} \mathcal{G}_{|\bar{\varphi}(U)} & \xrightarrow{\varphi} & \mathcal{A}_{|U} \\ {\scriptstyle\delta}\downarrow & & \downarrow{\scriptstyle\delta^M} \\ \mathcal{C}_{L|\bar{\varphi}(U)} & \xrightarrow[\bar{\varphi}^*]{} & \mathcal{C}_{L|U}^M \end{array}, \tag{4.10}$$

where $\bar{\varphi}^*$ is the ordinary pull-back associated with the mapping $\bar{\varphi}$, commutes.

When no confusion can arise, the evaluation morphism δ^M will be denoted simply by δ. The image of the sheaf $\mathcal{A}$ through δ is a sheaf on M of graded-commutative B_L-algebras, denoted by $\mathcal{A}^\infty$. The next result establishes a relationship between G-supermanifolds and G^∞ supermanifolds. Let $(M, \mathcal{A})$ be a G-supermanifold, and let $\{(U_i, (\bar{\varphi}_i, \varphi_i)),\ i \in \mathbb{N}\}$ be an atlas of local isomorphisms as in condition (2) of Definition 4.1.

Proposition 4.4.

(1) *The atlas $\mathfrak{A}^\infty = \{(U_i, \bar{\varphi}_i), i \in \mathbb{N}\}$ endows M with a structure of G^∞ supermanifold, of the same dimension as $(M, \mathcal{A})$.*

(2) *The G^∞ structure sheaf of M coincides with $\mathcal{A}^\infty$.*

Proof. The only non-trivial aspect of Part (1) to be proved is that the transition functions $\bar{\varphi}_i \circ \bar{\varphi}_j^{-1}$ are G^∞. Since (taking into account the diagram (4.10))

$$\begin{aligned} \bar{\varphi}_i \circ \bar{\varphi}_j^{-1} = (\bar{\varphi}_j)^{-1*}(\bar{\varphi}_i) = (\bar{\varphi}_j)^{-1*}(\delta^M \circ \varphi_i) \\ = \delta \circ \varphi_j^{-1}(\varphi_i) \in \mathcal{C}_{L|\bar{\varphi}_i(U_i) \cap \bar{\varphi}_j(U_j)}, \end{aligned}$$

the claim is proved (in the notation $\varphi_j^{-1}(\varphi_i)$, the symbol φ_i stands for the set of local coordinates on the chart U_i regarded as sections of $\mathcal{A}$). Part (2) is a direct consequence of the commutativity of (4.10) and of Proposition 4.1. ■

It is clear that G-supermanifolds generalize the notion of GH^∞ supermanifolds; indeed, if $(M, \mathcal{GH}^M)$ is a GH^∞ supermanifold, the pair $(M, \mathcal{A})$, with $\mathcal{A} = \mathcal{GH}^M \otimes_{B_{L'}} B_L$, is a G-supermanifold (the evaluation morphism is globally defined as $\delta^M(f \otimes a) = fa$). The resulting G-supermanifold will be called the *trivial extension* of the original GH^∞ supermanifold.

Graded tangent space. As a consequence of Proposition 4.3, the sheaf $\mathcal{D}er\mathcal{A}$ of graded derivations on a G-supermanifold $(M, \mathcal{A})$ is locally free, with local bases given by the derivations

$$\{\frac{\partial}{\partial x^i}, \frac{\partial}{\partial y^\alpha} \mid i = 1 \ldots m, \alpha = 1 \ldots n\}$$

associated with a local coordinate system $(x^1, \ldots, x^m, y^1, \ldots, y^n)$.

Definition 4.2. *The graded tangent space $T_z(M, \mathcal{A})$ at a point $z \in M$ is the graded B_L-module whose elements are the graded derivations $X: \mathcal{A}_z \to B_L$.*

The graded tangent space $T_z(M, \mathcal{A})$ is quite evidently free of rank (m, n), and the elements $(\frac{\partial}{\partial x^i})_z$, $(\frac{\partial}{\partial y^\alpha})_z$ defined by

$$\left(\frac{\partial}{\partial x^i}\right)_z (f) = \widetilde{\frac{\partial f}{\partial x^i}}(z), \qquad \left(\frac{\partial}{\partial y^\alpha}\right)_z (f) = \widetilde{\frac{\partial f}{\partial y^\alpha}}(z) \qquad \text{for all} \quad f \in \mathcal{A}_z,$$

yield a graded basis for it. Furthermore, there is a canonical isomorphism of graded B_L-modules

$$T_z(M, \mathcal{A}) \simeq (\mathcal{D}er\mathcal{A})_z / (\mathfrak{L}_z \cdot (\mathcal{D}er\mathcal{A})_z)\,,$$

where $\mathfrak{L}_z$ is the ideal of germs in $\mathcal{A}_z$ which vanish when evaluated, i.e.

$$\mathfrak{L}_z = \{f \in \mathcal{A}_z \mid \tilde{f}(z) = 0\}.$$

Topologies of rings of G-functions. In order to introduce the notions of morphisms and products of G-supermanifolds, and to discuss Rothstein's axiomatics, we need to topologize in a suitable way the rings of sections of the structure sheaves of G-supermanifolds. This will parallel the analogous study performed in the case of graded manifolds in Section III.1.

Let $(M, \mathcal{A})$ be a G-supermanifold, and let $\| \;\|$ denote the l^1 norm in B_L; for every open subset $U \subset M$ the rings $\mathcal{A}(U)$ of $\mathcal{A}$ can be topologized by means of the seminorms $p_{L,K}: \mathcal{A}(U) \to \mathbb{R}$ defined by

$$p_{L,K}(f) = \max_{z \in K} \| \delta(L(f))(z) \|$$

where L runs over the differential operators of $\mathcal{A}$ on U, and $K \subset U$ is compact. The above topology is also given by the family of seminorms

$$p_K^I(f) = \max_{\substack{z \in K \\ |J| \leq I,\, \mu \in \Xi_n}} \left\| \delta\left(\left(\frac{\partial}{\partial x}\right)^J \left(\frac{\partial}{\partial y}\right)_\mu f\right)(z) \right\| \quad , \tag{4.11}$$

where K runs over the compact subsets of a coordinate neighbourhood W with coordinates $(x^1, \ldots, x^m, y^1, \ldots, y^n)$ (see Remark III.1.1 for notation). Under this form it is clear that this topology makes $\mathcal{A}(U)$ into a locally convex metrizable graded algebra.

The next results will allow to prove that $\mathcal{A}(U)$ is *complete*, so that it is in fact a graded Fréchet algebra. Without loss of generality, we may assume that $(M, \mathcal{A}) = (B_L^{m,n}, \mathcal{G})$. With reference to the isomorphism (4.5), we topologize the rings $\hat{\mathcal{G}}(U)$ by means of the seminorms

$$\hat{p}_K^I(f) = \max_{\substack{z \in K \\ |J| \leq I}} \left\| \delta\left(\left(\frac{\partial}{\partial x}\right)^J f\right)(z) \right\| \quad . \tag{4.12}$$

The tensor product $\hat{\mathcal{G}}(U) \otimes_{\mathbf{R}} \bigwedge_{\mathbf{R}} \mathbf{R}^n$ is in turn given its natural topology, which is induced by the seminorms

$$p_K^{I,\mu}(f) = \hat{p}_K^I(f^\mu)$$

having set $f = \sum_{\mu \in \Xi_n} f_\mu \otimes y^\mu$.

Lemma 4.1. *The isomorphism (4.5), $\mathcal{G}(U) \simeq \hat{\mathcal{G}}(U) \otimes_{\mathbf{R}} \bigwedge_{\mathbf{R}} \mathbf{R}^n$, is a metric isomorphism.*

Proof. A direct majoration argument shows that

$$p_K^I \leq \sum_{\mu \in \Xi_n} c_\mu \hat{p}_K^{I,\mu} \qquad \text{where} \qquad c_\mu = \max_{\substack{z \in K \\ \nu \in \Xi_n}} \left\| \delta\left(\left(\frac{\partial}{\partial y}\right)_\nu y^\mu\right)(z) \right\| \quad .$$

This shows the continuity of the inverse morphism. We now display the opposite majoration. The seminorm p_K^I is explicitly written as

$$p_K^I(f) = \max_{\substack{z \in K \\ |J| \leq I,\, \nu \in \Xi_n}} \left\| \sum_{\mu \in \Xi_n} \varepsilon_{\mu\nu} \frac{\partial f^\mu}{\partial x^J}(z) \delta\left(\left(\frac{\partial}{\partial y}\right)_\nu y^\mu\right)(z) \right\| \quad , \tag{4.13}$$

with $\varepsilon_{\mu\nu}$ a suitable sign. The seminorms $p_K^{I,\mu}$ are majorated by descending recurrence, starting from the last one, i.e. from $p_K^{I,\omega}$, where ω is the sequence $\{1,2,\ldots,n\}$. Indeed, from (4.13) we obtain $p_K^{I,\omega} \leq p_K^I$, since $p_K^{I,\omega}$ is one of the terms over which the maximum (4.13) is taken. For the same reason, if we consider the seminorms p_K^{I,ω_i}, $i = 1,\ldots,n$, with $\omega_i = \{1,2,\ldots,\hat{i},\ldots,n\}$, we obtain

$$\begin{aligned} p_K^{I,\omega_i}(f) &= \max_{\substack{z\in K \\ |J|\leq I}} \left\| \frac{\partial f^{\omega_i}}{\partial x^J}(z) + \frac{\partial f^{\omega}}{\partial x^J}(z)\delta(y^i)(z) - \frac{\partial f^{\omega}}{\partial x^J}(z)\delta(y^i)(z) \right\| \\ &\leq p_K^I(f) + \max_{\substack{z\in K \\ |J|\leq I}} \left\| \frac{\partial f^{\omega}}{\partial x^J}(z)\delta(y^i)(z) \right\| \\ &\leq (1 + c_{iK})\, p_K^I(f), \end{aligned}$$

where $c_{iK} = \max_{z\in K} \| \delta(y^i)(z) \|$. The remaining majorations are performed in the same way. ■

For any open $W \subset \mathbf{R}^m$, the space $\mathcal{C}^\infty(W)\otimes_{\mathbf{R}} B_{L'}$ is equipped with the usual topology of uniform convergence of derivatives of any order, which is induced by the family of seminorms

$$q_K^I(h) = \max_{\substack{z\in K \\ |J|\leq I}} \left\| \left(\frac{\partial}{\partial x}\right)^J h(z) \right\|$$

where K is a compact in W, and the norm is taken in $B_{L'}$. Moreover, since δ is injective when restricted to $\hat{\mathcal{G}}$, we may identify the sheaves $\hat{\mathcal{G}}$ and $\hat{\mathcal{G}}^\infty$.

Lemma 4.2. *For any open $U \subset B_L^{m,n}$, and all L' such that $0 \leq L' \leq L$, the Z-expansion*

$$Z_{L'}\colon \mathcal{C}^\infty(\sigma^{m,n}(U)) \otimes B_{L'} \to \hat{\mathcal{G}}(U) \tag{4.14}$$

is an isometry onto its image. In particular, when $L' = L$, we obtain a metric isomorphism $\mathcal{C}^\infty(\sigma^{m,n}(U)) \otimes B_L \xrightarrow{\sim} \hat{\mathcal{G}}(U)$, while, for $L' = 0$, we obtain a metric isomorphism $\mathcal{C}^\infty(\sigma^{m,n}(U)) \xrightarrow{\sim} \hat{\mathcal{H}}^\infty(U)$.

Proof. One easily shows that the seminorms which defines the topology in the right-hand side are majorated in terms of the relevant seminorms on the left-hand side. To show the converse, let K be a compact subset of an open W in

$\mathbf{R}^m$, and I a nonnegative integer; for any $h \in \mathcal{C}^\infty_{\mathbf{R}^m}(W)$, we have

$$q^I_K(h) \leq \max_{\substack{z\in\tilde{K}\\ |J|\leq I}} \left\| \left(\frac{\partial}{\partial x}\right)^J Z_{L'}(h)(z) \right\| = \hat{p}^I_{\tilde{K}}(Z_{L'}(h)) \,,$$

where $\tilde{K}$ is a compact in $(\sigma^{m,n})^{-1}(W)$ containing K. It is clear that the previous minoration implies the thesis. ∎

Reasoning as in Proposition 1.2, one proves that the topological algebra $\hat{\mathcal{G}}(U)$ is complete, whence, using Lemma 4.1 and reasoning as in Proposition 1.2 again, the algebra $\mathcal{G}(U)$ is complete as well.

We eventually obtain the result we were looking for.

Proposition 4.5. *Let $(M, \mathcal{A})$ be a G-supermanifold. For every open $U \subset M$, the space $\mathcal{A}(U)$, endowed with the topology induced by the seminorms (4.11), is a graded Fréchet algebra.* ∎

The previous Lemmas also imply a further result, which will be useful when dealing with morphisms of G-supermanifolds. For any open $W \subset \mathbf{R}^m$, we topologize the space

$$\mathcal{C}^\infty(W) \otimes B_L \otimes \bigwedge \mathbf{R}^n \simeq \mathcal{C}^\infty(W) \otimes \bigwedge \mathbf{R}^{L+n}$$

as in Proposition 1.2. Then,

Corollary 4.2. *The spaces $\mathcal{G}(U)$, $\mathcal{H}^\infty(U)\otimes_{\mathbf{R}} B_L$ and $\mathcal{C}^\infty(\sigma^{m,n}(U))\otimes B_L \otimes \bigwedge \mathbf{R}^n$ are isometrically isomorphic for any open $U \subset B^{m,n}_L$.* ∎

Complex G-supermanifolds. Holomorphic G-supermanifolds are defined in exactly the same way as the G-supermanifolds based on B_L. One considers a Z-expansion mapping $C_{L'}$-valued holomorphic functions defined on open sets in $\mathbf{C}^m$ into C_L-valued functions on $C^{m,0}_L$; the functions in the image of this map are called OH functions of even variables, and the corresponding sheaf is denoted by $\widehat{\mathcal{OH}}_{L'}$. The OH functions on $C^{m,n}_L$ are functions which can be written as

$$f(z^1,\ldots,z^m,\zeta^1,\ldots,\zeta^n) = \sum_{\mu\in\Xi_n} f_\mu(z^1,\ldots,z^m)\zeta^\mu \,,$$

where $\{z^1,\ldots,z^m,\zeta^1,\ldots,\zeta^n\}$ are the canonical coordinates on $C^{m,n}_L$, and the f_μ's are OH. The relevant sheaf is denoted by $\mathcal{OH}_{L'}$.

The holomorphic counterpart of the sheaf $\mathcal{G}$ is obviously the sheaf

$$\mathcal{O}\mathcal{G} = \mathcal{O}\mathcal{H}_{L'} \otimes_{C_{L'}} C_L .$$

Complex G-supermanifolds $(M, \mathcal{B})$ are defined in the obvious way. Quite evidently, any (m, n) dimensional complex G-supermanifold has an underlying complex manifold of dimension $2^{L-1}(m+n)$. The relevant evaluation morphism δ now takes values in the sheaf of C_L-valued holomorphic functions on M.

An (m, n) dimensional complex G-supermanifold $(M, \mathcal{B})$ can be also regarded as a $(2m, 2n)$ dimensional real G-supermanifold, which we denote by $(M, \mathcal{A})$. The complexified sheaf $\mathcal{I} = \mathcal{A} \otimes_{\mathbf{R}} \mathrm{C}$ is apparently both a $\mathcal{B}$-module and a $\bar{\mathcal{B}}$-module (where a bar denotes complex conjugation), and we can define the sheaf of graded differential form of type (p, q) as

$$\Omega_{\mathcal{I}}^{p,q} = \mathcal{A} \otimes_{\mathcal{B}} \otimes \Omega_{\mathcal{B}}^{p} \otimes_{\bar{\mathcal{B}}} \overline{\Omega_{\mathcal{B}}^{q}} .$$

Here $\Omega_{\mathcal{B}}^{p}$ is the sheaf of graded holomorphic forms on $(M, \mathcal{B})$, i.e.

$$\Omega_{\mathcal{B}}^{p} = \textstyle\bigwedge_{\mathcal{B}}^{p} \mathcal{D}er_{C_L}^{*} \mathcal{B} .$$

We have

$$\Omega_{\mathcal{I}}^{r} = \bigoplus_{p+q=r} \Omega_{\mathcal{I}}^{p,q} ,$$

with projections $\pi^{p,q} \colon \Omega_{\mathcal{I}}^{r} \to \Omega_{\mathcal{I}}^{p,q}$, and from the exterior differential $d \colon \Omega_{\mathcal{I}}^{r} \to \Omega_{\mathcal{I}}^{r+1}$ we may define morphism of graded C_L-modules

$$\partial \colon \Omega_{\mathcal{I}}^{p,q} \to \Omega_{\mathcal{I}}^{p+1,q}, \qquad \partial = \pi^{p+1,q} \circ d ;$$

$$\bar{\partial} \colon \Omega_{\mathcal{I}}^{p,q} \to \Omega_{\mathcal{I}}^{p,q+1}, \qquad \bar{\partial} = \pi^{p,q+1} \circ d .$$

Since the complex structure of $(M, \mathcal{B})$ is integrable, we have

$$d = \partial + \bar{\partial}$$

(cf. [**Wel**] on this topic in the ordinary case), and the usual identities

$$\partial^2 = \bar{\partial}^2 = \partial \circ \bar{\partial} + \bar{\partial} \circ \partial = 0$$

are easily recovered.

Chapter IV
Basic geometry of G-supermanifolds

They explore the new field and bring back their spoils — a few simple generalizations — to apply them to the practical world of three dimensions. Some guiding light will be given to the attempts to build a scheme of things entire.

A.S. EDDINGTON

The first five Sections of this Chapter, the technical core of this book, are dedicated to set down the basic differential geometry of G-supermanifolds, by introducing the fundamental objects one needs: morphisms, products, super-vector bundles, and differential forms. It should be pointed out that the relevant definitions are quite different from the usual ones, and rather in the spirit of the algebraic geometry. This is a consequence of the fact that part of the information conveyed by the structure sheaf of a G-supermanifold is not otherwise embodied in the associated topological space.

In Section 6, an important class of supermanifolds — the so-called DeWitt supermanifolds — is studied, and the investigation of their global geometric structure is initiated. The completion of this analysis requires the use of some cohomological properties of supermanifolds and is therefore relegated to Chapter V.

In the last Section we discuss Rothstein's axiomatics for supermanifolds and show that, in order to provide a generalization of graded manifolds and an extension of G^{∞} supermanifolds (in a sense that will be specified later), it must be supplemented by a further axiom which assumes the topological completeness of the rings of sections of the structural sheaves of the supermanifolds. We also show that, when this additional axioms is imposed, and a finite-dimensional exterior algebra is chosen as ground algebra, the resulting category coincides with that of G-supermanifolds.

1. Morphisms

In order to devise a proper definition of a morphism, we consider as a guiding principle the requirement that, given a G-supermanifold $(M, \mathcal{A})$, the sheaf of germs of morphisms $(M, \mathcal{A}) \to (B_L, \mathcal{G})$ (where B_L is regarded as $B_L^{1,1}$) is canonically isomorphic with the structure sheaf $\mathcal{A}$.

Definition 1.1. *Given two G-supermanifolds $(M, \mathcal{A})$ and $(N, \mathcal{B})$, a G-morphism $(f, \phi): (M, \mathcal{A}) \to (N, \mathcal{B})$ is a morphism of graded locally ringed B_L-spaces, where $f: M \to N$ is a G^∞ morphism such that the diagram*

$$\begin{array}{ccc} \mathcal{B} & \xrightarrow{\phi} & f_*\mathcal{A} \\ {\scriptstyle \delta^N}\downarrow & & \downarrow{\scriptstyle \delta^M} \\ \mathcal{B}^\infty & \xrightarrow[f^*]{} & f_*\mathcal{A}^\infty \end{array} \tag{1.1}$$

commutes.

Thus, a G-morphism $(f, \phi): (M, \mathcal{A}) \to (N, \mathcal{B})$ preserves the underlying G^∞ structure; on the other hand, the morphism $f: M \to N$ is not sufficient on its own to determine the G-morphism (f, ϕ) (cf. Example 1.1 below), so that a separate specification of ϕ is needed. We recall from Definition II.4.2 that the morphism ϕ is even by definition.

As we remarked in the previous Chapter, by tensoring the structure sheaf of a GH^∞ supermanifold by B_L, we obtain a G-supermanifold. Thus, if $(M, \mathcal{GH}^M)$ and $(N, \mathcal{GH}^N)$ are GH^∞ supermanifolds, and we let $\mathcal{A} = \mathcal{GH}^M \otimes_{B_{L'}} B_L$ and $\mathcal{B} = \mathcal{GH}^N \otimes_{B_{L'}} B_L$, any GH^∞ map $f: M \to N$ defines a G-morphism $(f, \phi): (M, \mathcal{A}) \to (N, \mathcal{B})$ by letting $\phi(g \otimes \lambda) = f^* g \otimes \lambda$. However, not all G-morphisms between 'trivially extended' G-supermanifolds are of this kind, as the following examples show.

EXAMPLE 1.1. Considering the case $M = N = B_L \equiv B_L^{1,1}$, both with the structure sheaf $\mathcal{G}$, we can define two different G-morphisms (f, ϕ) and (f, ψ), both of which have the same 'topological' part. Let $f: B_L \to B_L$ be the GH^∞ map $f(x, y) = (x, 0)$, and let a be an even top-degree element in B_L (obviously, we assume that L is even). We have just noticed that the condition $\phi(g \otimes \lambda) = f^*g \otimes \lambda$ defines a G-morphism $(f, \phi): (B_L, \mathcal{G}) \to (B_L, \mathcal{G})$. A second morphism $\psi: \mathcal{G} \to f_*\mathcal{G}$ can be defined by

$$\psi(g \otimes \lambda) = \alpha \otimes \lambda + \hat{\beta} \otimes a\lambda,$$

having set $g(x, y) = \alpha(x) + y\beta(x)$ and $\hat{\beta}(x, y) = y\beta(x)$. A simple direct calculation shows that $\delta \circ \psi = f^* \circ \delta$, and that ψ is a morphism of graded B_L-algebras. Thus, (f, ψ) is another G-morphism, with the same underlying G^∞ (actually, GH^∞) map as (f, ϕ); however, (f, ψ) is not a 'trivially extended' G-morphism. ▲

EXAMPLE 1.2. We now offer another example of a G-morphism which is not a trivial extension of a GH^∞ one; this time, the underlying 'topological' morphism is G^∞. Let $f: B_L^{1,n} \to B_L^{1,n}$ be the G^∞ map $f(x, y^1, \ldots, y^n) = (ax, y^1, \ldots, y^n)$ with $a \in (B_L)_0 - (B_{L'})_0$; notice that f is not GH^∞ (we have set the even dimension m to 1 for simplicity, but any value of m would do). We define a sheaf morphism $\phi: \mathcal{G} \to f_*\mathcal{G}$ by letting

$$\phi(g \otimes \lambda) = \sum_{k=0}^{L} g_k \otimes a^k \lambda \qquad \text{with} \qquad g_k(x, y^1, \ldots y^n) = \frac{1}{k!} \left(\frac{\partial^k g}{\partial x^k} \right)_{(0, y^1, \ldots, y^n)} x^k .$$

The commutativity of the diagram (1.1) corresponds to the equality

$$\delta(\phi(g \otimes \lambda)) = f^*(\delta(g \otimes \lambda)), \qquad \text{i.e.} \qquad (f^*g)\lambda = f^*(g\lambda),$$

which is trivially verified since f is a G^∞ map. We therefore deduce that (f, ϕ): $(B_L^{1,n}, \mathcal{G}) \to (B_L^{1,n}, \mathcal{G})$ — which certainly is not GH^∞ — is a G-morphism. ▲

As in the case of smooth, complex, and graded manifolds, one can prove that a G-morphism induces a continuous morphism between the rings of sections of the structure sheaves.

Proposition 1.1. *Let $(f, \phi): (M, \mathcal{A}) \to (N, \mathcal{B})$ be a G-morphism; for any $U \subset N$ the graded B_L-algebra morphism $\phi: \mathcal{B}(U) \to \mathcal{A}(f^{-1}(U))$ is continuous.*

Proof. We may assume that $M = B_L^{m,n}$ and $N = B_L^{p,q}$ with their standard G-supermanifold structures, denoting by $\mathcal{G}_{m,n}$ and $\mathcal{G}_{p,q}$ the corresponding structure sheaves. By Corollary III.4.2, there are metric isomorphisms

$$\mathcal{G}_{m,n}(V) \simeq C^\infty(\sigma^{m,n}(V)) \otimes B_L \otimes \bigwedge \mathbb{R}^n \simeq C^\infty(\sigma^{m,n}(V)) \otimes \bigwedge \mathbb{R}^{L+n}$$
$$\mathcal{G}_{p,q}(U) \simeq C^\infty(\sigma^{p,q}(U)) \otimes B_L \otimes \bigwedge \mathbb{R}^q \simeq C^\infty(\sigma^{p,q}(U)) \otimes \bigwedge \mathbb{R}^{L+q}$$

where $V = f^{-1}(U)$. The G^∞ morphism $f: V \to U$ induces a (not uniquely determined) smooth map $\hat{f}: \sigma^{m,n}(V) \to \sigma^{p,q}(U)$ such that $\sigma^{p,q} \circ f = \hat{f} \circ \sigma^{m,n}$ as follows: f admits a (in general not unique) representation $f(x,y) = \sum_{\mu \in \Xi_L} Z_L(f_\mu)(x)\, y^\mu$. One then lets $\hat{f} = \sigma^{p,q} \circ f_0$. Now,

$$(\hat{f}, \phi): (\sigma^{m,n}(V), \mathcal{C}^\infty{}_{|\sigma^{m,n}(V)} \otimes \bigwedge \mathbb{R}^{L+n}) \to (\sigma^{p,q}(U), \mathcal{C}^\infty{}_{|\sigma^{p,q}(U)} \otimes \bigwedge \mathbb{R}^{L+q})$$

is a morphism of graded manifolds; then Corollary III.1.3 allows us to conclude. ■

This implies that, if $U \subset B_L^{p,q}$ is an open subset with coordinates $(x^1, \ldots, x^p, y^1, \ldots, y^q)$, and $(f, \phi): (M, \mathcal{A}) \to (U, \mathcal{G}_{p,q})$ is a G-morphism, ϕ is characterized by the values $\phi(x^i)$, $\phi(y^\alpha)$, that is:

Lemma 1.1. *If $(f, \phi): (M, \mathcal{A}) \to (U, \mathcal{G}_{p,q})$ and $(f, \phi'): (M, \mathcal{A}) \to (U, \mathcal{G}_{p,q})$ are G-morphisms, and $\phi(x^i) = \phi'(x^i)$ for $i = 1, \ldots, p$, $\phi(y^\alpha) = \phi'(y^\alpha)$ for $\alpha = 1, \ldots, q$, then $\phi = \phi'$.*

Proof. ϕ and ϕ' coincide over $B_L[x^1, \ldots, x^p] \otimes \bigwedge\langle y^1, \ldots, y^q\rangle$ and by continuity, they also coincide over its completion $\mathcal{G}_{p,q}(U) \simeq \mathcal{C}^\infty(\sigma^{p,q}(U)) \otimes B_L \otimes \bigwedge \mathbb{R}^q$. ■

Let us state the definitions of injective and surjective morphism in the category of G-supermanifolds.

Definition 1.2. *A G-morphism $(f, \phi): (M, \mathcal{A}) \to (N, \mathcal{B})$ is said to be*

(1) *injective (or to be a monomorphism) if f is injective, and ϕ is surjective;*
(2) *surjective (or to be an epimorphism) is f is surjective, and ϕ is injective.*

We now come to one of the main results of this Section. The counterpart of this property in the theory of differentiable manifolds is somewhat trivial and states that, given a differentiable manifold X, the sheaf of differentiable maps $X \to \mathbb{R}$, where $\mathbb{R}$ is regarded as a differentiable manifold, is isomorphic with the structure sheaf of X. With a slight abuse of language, we denote by $\mathcal{H}om\,(M, N)$ the sheaf $\mathcal{H}om\,((M, \mathcal{A}), (N, \mathcal{B}))$ of germs of G-morphisms $(M, \mathcal{A}) \to (N, \mathcal{B})$; in particular, $\mathcal{H}om\,(M, B_L)$ is the sheaf of germs of G-morphisms $(M, \mathcal{A}) \to (B_L, \mathcal{G})$.

Proposition 1.2. *The morphism $\gamma: \mathcal{H}om\,(M, B_L) \to \mathcal{A}$, defined, for any open $U \subset M$, by*

$$\begin{aligned} \gamma_U: \mathcal{H}om\,(M, B_L)(U) &\to \mathcal{A}(U) \\ (f, \phi) &\mapsto \phi(j \otimes 1) \end{aligned} \tag{1.2}$$

(where j denotes the natural inclusion $f(U) \hookrightarrow B_L$), is an isomorphism of sheaves of graded-commutative B_L-algebras.

Proof. We can limit ourselves to the case where $M = B_L^{m,n}$ and $\mathcal{A}$ is the canonical sheaf (III.4.1) over it since, if the statement is proved to be true locally, then it is also so globally. For any open $U \subset B_L^{m,n}$, an element $h \in \mathcal{A}(U)$ can be written as $h = \sum_i h_i \otimes \xi_i$, where $h_i \in \mathcal{GH}(U)$ and $\xi_i \in B_L$. By means of h we can determine a G-morphism $(\delta(h), h^\flat): (U, \mathcal{A}_{|U}) \to (B_L, \mathcal{G})$, where $\delta(h) = \sum_i h_i \xi_i: U \to B_L$ is the G^∞ morphism obtained by evaluating h, while $h^\flat: \mathcal{G} \to \mathcal{A}_{|U}$ is the morphism defined by

$$h^\flat(g \otimes \lambda) = \sum_i (-1)^{|h_i||\lambda|}(g \circ h_i) \otimes \lambda \xi_i .$$

In this way we have defined a sheaf morphism

$$\vartheta: \mathcal{A}_{|U} \to \mathcal{H}om(U, B_L)$$
$$h \mapsto (\delta(h), h^\flat),$$

which fulfills the condition $\gamma \circ \vartheta = id$. In order to prove the claim, we need only to show that $\vartheta \circ \gamma$ is the identity morphism, or, equivalently, to prove that each element $(f, \phi) \in \mathcal{H}om(U, B_L)$ is determined by the morphism $\phi(j \otimes 1)$. In fact, by Lemma 1.1, ϕ is determined by $\phi(x \otimes 1)$ and $\phi(y \otimes 1)$, where x, y are the canonical coordinates in B_L. On the other hand, the obvious identity $j \otimes 1 = x \otimes 1 + y \otimes 1$ shows that $\phi(x \otimes 1)$ and $\phi(y \otimes 1)$ are the even and odd parts of $\phi(j \otimes 1)$, respectively. ∎

Glueing of G-supermanifolds. G-supermanifolds are graded locally ringed spaces, so that we can glue G-supermanifolds together by means of a family of isomorphisms fulfilling the glueing condition (II.4.5) to obtain a new graded locally ringed space. In this section we shall see that this graded locally ringed space is actually a G-supermanifold.

Let $\{(M_i, \mathcal{A}_i)\}$ be a family of G-supermanifolds of dimension (m, n), such that for every pair (i, j) there are an open subset $M_{ij} \subset M_i$ and an isomorphism of G-supermanifolds

$$(f_{ij}, \phi_{ij}): (M_{ji}, \mathcal{A}_{j|M_{ji}}) \xrightarrow{\sim} (X_{ij}, \mathcal{A}_{i|M_{ij}}),$$

fulfilling the glueing condition of Section II.4. Then, the graded locally ringed space $(M, \mathcal{A})$ obtained from the spaces $(M_i, \mathcal{A}_i)$ by glueing is, by its very construction, locally isomorphic with $(B_L^{m,n}, \mathcal{G})$. In order to prove that it is a G-supermanifold one has only to show that there exists a sheaf morphism $\delta^M: \mathcal{A} \to \mathcal{C}_L^M$ as in (3) of Definition III.4.1.

However, the glueing condition for the spaces $(M_i, \mathcal{A}_i)$ implies the corresponding glueing condition for the locally ringed spaces $(M_i, \mathcal{C}_L^{M_i})$, and $(M, \mathcal{C}_L^M)$ is exactly the ringed space obtained by glueing. Thus, by Lemma II.4.2, the morphisms $\delta^{M_i}: \mathcal{A}_i \to \mathcal{C}_L^{M_i}$ define a sheaf morphism $\delta^M: \mathcal{A} \to \mathcal{C}_L^M$ such that $\delta^{M_i} \circ \phi_i = (f_i)^* \circ \delta^M{}_{|M_i}$ for every i, as claimed.

In conclusion, one arrives at the following result.

Lemma 1.2. *The graded locally ringed space obtained by glueing of G-supermanifolds is also a G-supermanifold.* ■

2. Products

To give a proper definition of the product of two G-supermanifolds, we have to proceed, for analogous motivations, as in the case of graded manifolds (cf. Section III.1).

For fixed values of L, m and n, the structure sheaf of the canonical G-supermanifold over $B_L^{m,n}$ is again denoted by $\mathcal{G}_{m,n}$. Given open sets $U \subset B_L^{m,n}$, $V \subset B_L^{p,q}$, we consider the presheaf defined by the correspondence

$$U \times V \to \mathcal{G}_{m,n}(U) \hat{\otimes}_{L,\pi} \mathcal{G}_{p,q}(V), \tag{2.1}$$

where $\hat{\otimes}_{L,\pi}$ denotes the tensor product over B_L completed in the Grothendieck π topology (cf. Section 1 and [**Gro1,Pie**]).

Proposition 2.1. *The structure sheaf $\mathcal{G}_{m+p,n+q}$ of the canonical G-supermanifold over $B_L^{m+p,n+q}$ is isomorphic with the sheaf associated with the presheaf defined by the assignment (2.1).*

Proof. In accordance with Corollary III.4.2, there is a metric isomorphism of graded B_L-algebras

$$\mathcal{G}_{m,n}(U) \simeq \mathcal{C}^\infty(\sigma^{m,n}(U)) \otimes B_L \otimes \wedge \mathbf{R}^n \tag{2.2}$$

for every open subset $U \subset B_L^{m,n}$. Thus, given an open $V \subset B_L^{p,q}$, we obtain a metric isomorphism

$$
\begin{aligned}
&\mathcal{G}_{m,n}(U)\hat{\otimes}_{L,\pi}\mathcal{G}_{p,q}(V) \\
&\qquad \simeq [\mathcal{C}^\infty(\sigma^{m,n}(U)) \otimes B_L \otimes \wedge \mathbb{R}^n] \,\hat{\otimes}_{L,\pi}\, [\mathcal{C}^\infty(\sigma^{p,q}(V)) \otimes B_L \otimes \wedge \mathbb{R}^q] \\
&\qquad \simeq \left[\mathcal{C}^\infty(\sigma^{m,n}(U))\hat{\otimes}_\pi \mathcal{C}^\infty(\sigma^{p,q}(V))\right] \otimes B_L \otimes \wedge \mathbb{R}^{n+q} \\
&\qquad \simeq \mathcal{C}^\infty(\sigma^{m+p,n+q}(U \times V)) \otimes B_L \otimes \wedge \mathbb{R}^{n+q} \\
&\qquad \simeq \mathcal{G}_{m+p,n+q}(U \times V) \quad . \qquad\qquad (2.3)
\end{aligned}
$$

■

Let us observe that the evaluation morphism $\delta\colon \mathcal{G}_{m,n} \to \mathcal{C}_L^\infty$ yields, for any open $U \subset B_L^{m,n}$, continuous morphisms between the spaces of sections, so that one obtains the following commutative diagram, whose arrows are morphisms of Fréchet algebras:

$$
\begin{array}{ccc}
\mathcal{G}_{m+p,n+q}(U \times V) & \xrightarrow{\sim} & \mathcal{G}_{m,n}(U)\hat{\otimes}_\pi \mathcal{G}_{p,q}(V) \\
\delta \downarrow & & \downarrow \delta \\
\mathcal{C}_L^\infty(U \times V) & \xrightarrow{\sim} & \mathcal{C}_L^\infty(U)\hat{\otimes}_\pi \mathcal{C}_L^\infty(V)
\end{array} \qquad (2.4)
$$

We now generalize this construction to the case of two generic G-supermanifolds $(M, \mathcal{A})$ and $(N, \mathcal{B})$, of dimension (m, n) and (p, q) respectively.

Definition 2.1. *The product $(M, \mathcal{A}) \times (N, \mathcal{B})$ is the graded locally ringed B_L-space $(M \times N, \mathcal{A}\hat{\otimes}_{L,\pi}\mathcal{B})$, where $\mathcal{A}\hat{\otimes}_{L,\pi}\mathcal{B}$ is the sheaf associated with the assignment*

$$U \times V \to \mathcal{A}(U)\hat{\otimes}_{L,\pi}\mathcal{B}(V)$$

for any pair of open subsets $U \subset M$, $V \subset N$.

Proposition 2.2. *The graded locally ringed B_L-space thus defined is a G-supermanifold of dimension $(m+p, n+q)$; moreover, there is a pair of canonical G-epimorphisms $\pi_1\colon (M, \mathcal{A}) \times (N, \mathcal{B}) \to (M, \mathcal{A})$ and $\pi_2\colon (M, \mathcal{A}) \times (N, \mathcal{B}) \to (N, \mathcal{B})$, such that, for any G-supermanifold $(Q, \mathcal{D})$, a G-morphism $\Phi\colon (Q, \mathcal{D}) \to (M, \mathcal{A}) \times (N, \mathcal{B})$ is uniquely characterized by the compositions $\pi_1 \circ \Phi$ and $\pi_2 \circ \Phi$.*

Proof. The space $(M \times N, \mathcal{A}\hat{\otimes}_{L,\pi}\mathcal{B})$ is locally isomorphic with the G-supermanifold $(B_L^{m+p,n+q}, \mathcal{G}_{B_L^{m+p,n+q}})$, as a consequence of Proposition 2.1. The morphism

$$\delta^M \otimes \delta^N : \mathcal{A} \otimes \mathcal{B} \to \mathcal{C}_L^{\infty M} \otimes \mathcal{C}_L^{\infty N},$$

defined in the natural way, is continuous, and induces, as a consequence of the commutativity of (2.4), an evaluation morphism on the completion of the tensor products in the π topology:[1]

$$\delta^{M\times N} \equiv \delta^M \hat{\otimes}_\pi \delta^N : \mathcal{A} \hat{\otimes}_\pi \mathcal{B} \to \mathcal{C}_L^{\infty M\times N}.$$

This demonstrates the first part of the claim. Concerning the second part, the morphism $\pi_1 \equiv (p_1, \underline{\pi}_1)$ (the case of π_2 is obviously identical) is defined by the canonical topological projection $p_1 : M \times N \to M$ and by the morphism of graded locally ringed B_L-spaces $\underline{\pi}_1 : \mathcal{A} \to (p_1)_*(\mathcal{A}\hat{\otimes}_\pi \mathcal{B})$ determined by the natural monomorphisms $\mathcal{A}(U) \hookrightarrow (\mathcal{A} \otimes \mathcal{B})((p_1)^{-1}(U))$, where U is an open set in M. In this way one obtains a commutative diagram like (2.4) and π_1 is obviously an epimorphism. The thesis follows, like the corresponding result for graded manifolds, by the universal property of the topological tensor product. ■

REMARK 2.1. The universality property stated in Proposition 2.2 entails that the product introduced in Definition 2.1 should actually be *the* product in the category of G-supermanifolds (cf. Remark III.1.2 and [**GroD**]). ▲

Product supermanifolds as free modules. Given two G-supermanifolds $(M, \mathcal{A})$ and $(N, \mathcal{B})$, we may consider — loosely speaking — the product $(M, \mathcal{A}) \times (N, \mathcal{B})$ as a fibration over $(M, \mathcal{A})$, and can define the sections of this fibration as the G-morphisms $s : (U, \mathcal{A}_{|U}) \to (M, \mathcal{A}) \times (N, \mathcal{B})$ such that $\pi_1 \circ s = \mathrm{id}$ (here U is any open subset of M). These sections define a sheaf of sets on M. We consider in particular the case where N is the free graded B_L-module $B_L^{p|q}$, equipped with its standard G-supermanifold structure (see below). In this case, the sheaf of sections previously introduced is a free $\mathcal{A}$-module; it is interesting to establish the relationship between this sheaf and the structure sheaf of the product supermanifold. This will be important in next Section in order to provide a proper definition of vector bundle within the category of G-supermanifolds. An analogous result holds in the smooth ordinary case, as well as in the category of graded manifolds [**HeM1**]; in the case of smooth manifolds, it can be briefly described as follows. The smooth functions on a vector bundle can be regarded as smooth functions of the fibre coordinates with coefficients in the

[1] The fact that this morphism exists and is uniquely defined, albeit seemingly, is not entirely trivial; for a proof, see [**Gro1**].

ring of smooth functions on the base manifold. In this way, the ring of smooth functions of the total space is no more than the completion of the polynomial ring of the fibre coordinates with coefficients in the smooth functions on the base manifold.

Firstly, let us recall from Section II.2 that the *graded symmetric algebra* of a rank (p,q) free graded R-module F, denoted by $S(F)$, is the quotient of the graded tensor algebra $\bigoplus_{h\geq 0}\bigotimes^h F$ by the ideal generated by the elements of the form $a\otimes b-(-1)^{|a||b|}\,b\otimes a$. We also define the *total graded symmetric algebra* of F:

$$ST(F)=S(F\oplus\Pi F)\,.$$

Here Π denotes the *parity change functor* (cf. [**Ma2**]), which is defined by stating that $\Pi(F)$ is the abelian group $F_1\oplus F_0$ endowed with the R-module structure given by $a(\varpi(f))=(-1)^{|a|}\varpi(af)$ for any $a\in R$ and $f\in F$, where $\varpi\colon F_0\oplus F_1\to F_1\oplus F_0$ is the map $a_0\oplus a_1\to a_1\oplus a_0$.

Once a homogeneous basis $\{e_1,\dots,e_p,f_1,\dots,f_q\}$ for F has been fixed, $S(F)$ can be identified with the algebra $R[e_1,\dots,e_p]\otimes_R\bigwedge_R\langle f_1,\dots,f_q\rangle$, while $ST(F)$ is identified with the algebra

$$R[e_1,\dots,e_p,\varpi(f_1),\dots,\varpi(f_q)]\otimes_R\bigwedge\nolimits_R\langle f_1,\dots,f_q,\varpi(e_1),\dots,\varpi(e_p)\rangle\,;$$

here $R[\dots]$ denotes the graded-commutative R-algebra generated by the elements within the bracket, while $\langle\dots\rangle$ is the graded R-module generated by the elements within the triangular brackets.

In particular we are interested in the case $F\equiv B_L^{p|q}$; since the $(B_L)_0$-modules $B_L^{p|q}$ and $(B_L)^{p+q,p+q}$ are isomorphic, $B_L^{p|q}$ has a natural structure of a G-supermanifold of dimension $(p+q,p+q)$. In order to have a coherent notation, we denote its structure sheaf by $\mathcal{G}_{p|q}$. The sheaf $\mathcal{F}$ of sections of the product G-supermanifold $(M,\mathcal{A})\times(B_L^{p|q},\mathcal{G}_{p|q})$ is obviously a rank (p,q) graded $\mathcal{A}$-module; we denote by

$$\{(\omega_i,\eta_\alpha)\;i=1,\dots,m,\;\alpha=1,\dots,n\}$$

a local basis of the dual $\mathcal{A}$-module $\mathcal{F}^*$, say on an open $U\subset M$. The total graded symmetric algebra of $\mathcal{F}^*(U)$ admits the following characterization:

$$ST(\mathcal{F}^*(U)) \simeq$$
$$\mathcal{A}(U)[\omega_1,\dots,\omega_p,\varpi(\eta_1),\dots,\varpi(\eta_q)] \otimes_{\mathbf{R}} \bigwedge\nolimits_{\mathbf{R}}\langle\varpi(\omega_1),\dots,\varpi(\omega_p),\eta_1,\dots,\eta_q\rangle \simeq$$
$$\mathcal{A}(U) \otimes_{B_L} (B_L[\omega_1,\dots,\omega_p,\varpi(\eta_1),\dots,\varpi(\eta_q)] \otimes_{\mathbf{R}}$$
$$\bigwedge\nolimits_{\mathbf{R}}\langle\varpi(\omega_1),\dots,\varpi(\omega_p),\eta_1,\dots,\eta_q\rangle). \quad (2.5)$$

We equip the space

$$B_L[\omega_1,\dots,\omega_p,\varpi(\eta_1),\dots,\varpi(\eta_q)] \otimes_{\mathbf{R}} \bigwedge\nolimits_{\mathbf{R}}\langle\varpi(\omega_1),\dots,\varpi(\omega_p),\eta_1,\dots,\eta_q\rangle$$

with the topology that it inherits as a subring of $\mathcal{G}_{p|q}(B_L^{p|q})$. The metric structure of $ST(\mathcal{F}^*(U))$ is independent of the choice of the basis $\{(\omega_i,\eta_\alpha)\}$. Finally, we denote by $\widehat{ST}(\mathcal{F}^*)$ the sheaf of B_L-algebras on $M \times B_L^{p|q}$, whose sections on the open set $U \times B_L^{p|q}$ are the completion of (2.5) with respect to the Grothendieck topology; by reasoning as in Proposition 2.2, we can prove the following result:

Proposition 2.3. *The sheaf $\widehat{ST}(\mathcal{F}^*)$ is canonically isomorphic with the structure sheaf of the product G-supermanifold $(M,\mathcal{A}) \times (B_L^{p|q},\mathcal{G}_{p|q})$.* ∎

All this can be summarized as follows: the sheaf of sections of a product G-supermanifold $(M,\mathcal{A}) \times (B_L^{p|q},\mathcal{G}_{p|q})$ is a free graded $\mathcal{A}$-module of rank (p,q); conversely, given a G-supermanifold $(M,\mathcal{A})$, and a free graded $\mathcal{A}$-module $\mathcal{F}$ of rank (p,q), we can construct a product G-supermanifold whose sheaf of sections is isomorphic with $\mathcal{F}$.

The graded tangent space of the product. Let $(M,\mathcal{A})$ and $(N,\mathcal{B})$ be G-supermanifolds of dimension (m,n) and (p,q). Let us consider the product G-supermanifold $(M,\mathcal{A})\times(N,\mathcal{B}) = (M\times N,\mathcal{A}\hat{\otimes}_\pi\mathcal{B})$ and the natural projections

$$\pi_1 = (p_1,\underline{\pi}_1)\colon (M \times N,\mathcal{A}\hat{\otimes}_\pi\mathcal{B}) \to (M,\mathcal{A})$$
$$\pi_2 = (p_2,\underline{\pi}_2)\colon (M \times N,\mathcal{A}\hat{\otimes}_\pi\mathcal{B}) \to (N,\mathcal{B})\,.$$

Every graded derivation $D \in \mathcal{D}er\mathcal{A}(U)$ on an open subset $U \subset M$ induces a graded derivation $D \otimes \mathrm{Id}$ of $\mathcal{A}(U) \otimes \mathcal{B}(V)$ for every open subset $V \subset N$. Since D is linear and continuous, it induces a graded derivation of $\mathcal{A}(U)\hat{\otimes}_\pi\mathcal{B}(V)$. In this way one obtains morphisms of sheaves of $\mathcal{A}\hat{\otimes}_\pi\mathcal{B}$-modules $\pi_1^*\colon \pi_1^*(\mathcal{D}er\mathcal{A}) \to$

$\mathcal{D}er(\mathcal{A}\hat{\otimes}_\pi\mathcal{B})$ and $\pi_2^*\colon \pi_2^*(\mathcal{D}er\mathcal{B}) \to \mathcal{D}er(\mathcal{A}\hat{\otimes}_\pi\mathcal{B})$ given respectively by $D \mapsto D\otimes\mathrm{Id}$ and $D \mapsto \mathrm{Id}\otimes D$, and then, a morphism of locally free sheaves of $\mathcal{A}\hat{\otimes}_\pi\mathcal{B}$-modules

$$\begin{array}{ccc} \pi_1^*(\mathcal{D}er\mathcal{A}) \oplus \pi_2^*(\mathcal{D}er\mathcal{B}) & \xrightarrow{\pi_1^*+\pi_2^*} & \mathcal{D}er(\mathcal{A}\hat{\otimes}_\pi\mathcal{B}) \\ D + D' & \mapsto & D\otimes\mathrm{Id} + \mathrm{Id}\otimes D' \end{array} \quad . \tag{2.6}$$

Proposition 2.4. *The previous morphism is an isomorphism.*

Proof. The question being local one can assume that $(M,\mathcal{A}) = (B_L^{m,n}, \mathcal{G}_{m,n})$ and $(N,\mathcal{B}) = (B_L^{p,q}, \mathcal{G}_{p,g})$, so that $(M,\mathcal{A})\times(N,\mathcal{B}) = (B_L^{m+p,n+q}, \mathcal{G}_{m+p,n+q})$. In this case, if $(x^1,\dots,x^m,y^1,\dots,y^n)$ are graded coordinates in $B_L^{m,n}$, then $\mathcal{D}er\,\mathcal{G}_{m,n}$ is a free $\mathcal{G}_{m,n}$-module with basis $\left(\frac{\partial}{\partial x^i},\frac{\partial}{\partial y^\alpha}\right)$ $(i=1,\dots,m,\ \alpha=1,\dots,n)$. Then, $\pi_1^*(\mathcal{D}er\,\mathcal{G}_{m,n})$ is a free $\mathcal{G}_{m+p,n+q}$-module with basis $\left(\frac{\partial}{\partial x^i}\otimes\mathrm{Id}, \frac{\partial}{\partial y^\alpha}\otimes\mathrm{Id}\right)$ $(i=1,\dots,m,\ \alpha=1,\dots,n)$. Similarly, if $(z^1,\dots,z^p,t^1,\dots,t^q)$ are graded coordinates in $B_L^{p,q}$, then $\pi_2^*(\mathcal{D}er\,\mathcal{G}_{p,q})$ is a free $\mathcal{G}_{m+p,n+q}$-module with basis $\left(\frac{\partial}{\partial z^j}\otimes\mathrm{Id}, \frac{\partial}{\partial t^\beta}\otimes\mathrm{Id}\right)$ $(j=1,\dots,p,\ \beta=1,\dots,q)$.

Now, if one writes, as customary, $x^i = \underline{\pi}_1(x^i)$, $y^\alpha = \underline{\pi}_1(y^\alpha)$, $z^j = \underline{\pi}_2(z^j)$, $t^\beta = \underline{\pi}_2(t^\beta)$, then $(x^1,\dots,x^m,z^1,\dots,z^p,y^1,\dots,y^n,t^1,\dots,t^q)$ are graded coordinates in $B_L^{m,n}\times B_L^{p,q} \simeq B_L^{m+p,n+q}$ and $\mathcal{D}er\,\mathcal{G}_{m+p,n+q}$ is the free $\mathcal{G}_{m+p,n+q}$-module with basis

$$\left(\frac{\partial}{\partial x^i} = \frac{\partial}{\partial x^i}\otimes\mathrm{Id}, \frac{\partial}{\partial y^\alpha} = \frac{\partial}{\partial y^\alpha}\otimes\mathrm{Id}, \frac{\partial}{\partial z^j} = \mathrm{Id}\otimes\frac{\partial}{\partial z^j}, \frac{\partial}{\partial t^\beta} = \mathrm{Id}\otimes\frac{\partial}{\partial t^\beta}\right)$$

thus finishing the proof. ■

Then, one has the following characterization of the graded tangent space to a product G-supermanifold.

Corollary 2.1. *For every pair of points $z \in M$, $\bar{z} \in N$, there is a natural isomorphism of free B_L-modules*

$$T_z(M,\mathcal{A}) \oplus T_{\bar{z}}(N,\mathcal{B}) \simeq T_{(z,\bar{z})}(M\times N, \mathcal{A}\hat{\otimes}_\pi\mathcal{B})\,.$$

■

3. Supervector bundles

Quite naturally, the notion of product provides the local model for the construction of superbundles. In particular, we are interested in a theory of vector bundles in the category of G-supermanifolds, that we shall call *supervector bundles.* In ordinary differential geometry, it is well known (cf. for instance [**Wel,Del**]) that, given a smooth manifold X, the category of rank r (say, smooth real) vector bundles over X is equivalent to the category of rank r locally free modules over the structure sheaf of X. This equivalence also applies to the topological, holomorphic, and algebraic cases while, on the other hand, in algebraic geometry vector bundles are *defined* as locally free modules.

Before entering the realm of supervector bundles, we should like to state explicitly the relationship existing between the fibre over $z \in X$ of a rank r vector bundle ξ over X and the structure sheaf $\mathcal{C}_X^\infty$ of X. Since vector bundles are locally trivial, and we are interested in a local matter, we may assume ξ to be trivial. After fixing a specific trivialization, the sheaf $\mathcal{F}$ of sections of ξ can be identified with $(\mathcal{C}_X^\infty)^r$, i.e. with the sheaf of smooth maps $X \to \mathbb{R}^r$. Now, it is evident that the space $\mathcal{F}_z/(\mathfrak{M}_z \cdot \mathcal{F}_z) \simeq \mathbb{R}^r$ — where $\mathfrak{M}_z$ is the maximal ideal of $(\mathcal{C}_X^\infty)_z$; i.e., the set of germs of functions which vanish at z — may be identified with the fibre of ξ over z. It is also evident that this identification is independent of the trivialization chosen.

This discussion suggests that one should tackle the construction of a theory of supervector bundles in the following way. Let $(M, \mathcal{A}, \delta^M)$ be a G-supermanifold (we recall from Section III.4 that δ^M is a B_L-algebra morphism $\mathcal{A} \to \mathcal{C}_L^M$, the latter being the sheaf of smooth B_L-valued functions on M). We require that:

(i) the category of supervector bundles over $(M, \mathcal{A}, \delta^M)$ be equivalent to the category of locally free graded $\mathcal{A}$-modules;

(ii) the fibre over $z \in M$ of a rank (r, s) supervector bundle over $(M, \mathcal{A}, \delta^M)$, whose sheaf of sections is $\mathcal{F}$, be canonically isomorphic with the graded B_L-module $\mathcal{F}_z/(\mathfrak{L}_z \cdot \mathcal{F}_z)$, where

$$\mathfrak{L}_z = \{f \in \mathcal{A}_z \mid \delta(f)(z) = 0\}. \tag{3.1}$$

It should be noticed that the ideal involved in this quotient is not the maximal ideal of $\mathcal{A}_z$, which is

$$\mathfrak{N}_z = \{f \in \mathcal{A}_z \mid \sigma(\delta^M(f)(z)) = 0\},$$

where σ is, as usual, the body map. Indeed, the quotient $\mathcal{F}_z/(\mathfrak{N}_z \cdot \mathcal{F}_z)$ is isomorphic with $\mathbf{R}^r$; therefore, by sticking with the maximal ideal of $\mathcal{A}_z$ we would obtain an inconsistency with requirement (i), in that the objects resulting from our construction would be basically ordinary vector bundles with standard fibre $\mathbf{R}^r$ (cf. discussion at the end of Section III.3).

Let $(M, \mathcal{A})$ and $(F, \mathcal{A}_F)$ be two G-supermanifolds.

Definition 3.1. *A locally trivial superbundle over $(M, \mathcal{A})$ with standard fibre $(F, \mathcal{A}_F)$ is a pair $((\xi, \mathcal{A}_\xi), \pi)$, consisting of a G-supermanifold $(\xi, \mathcal{A}_\xi)$ and a G-epimorphism $(\xi, \mathcal{A}_\xi) \xrightarrow{\pi} (M, \mathcal{A})$, such that M admits an open cover $\{U_j\}$ together with a family of local G-isomorphisms*

$$\psi_j: (\pi^{-1}(U_j), \mathcal{A}_{\xi|\pi^{-1}(U_j)}) \to (U_j, \mathcal{A}_{|U_j}) \times (F, \mathcal{A}_F) \tag{3.2}$$

fulfilling the condition $\pi_1 \circ \psi_j = id$.

If $\pi = (p, \underline{\pi})$, and $z \in M$, we denote by $\pi^{-1}(z)$ (the *fibre* over z) the G-supermanifold whose underlying topological space is $p^{-1}(z)$, and whose structure sheaf is

$$\mathcal{A}_{(z)} = \left(\mathcal{A}_\xi/\mathcal{K}_{(z)}\right)_{|p^{-1}(z)},$$

where $\mathcal{K}_{(z)}$ is the subsheaf of $\mathcal{A}_\xi$ whose sections vanish when restricted to $p^{-1}(z)$.

For any $z \in M$, $\pi^{-1}(z)$ is G-isomorphic with the standard fibre $(F, \mathcal{A}_F)$. A pair (U_j, ψ_j) is said to be a *local trivialization*; a *G-section* of the superbundle ξ on an open set $U \subset M$ is a G-morphism $\mu: (U, \mathcal{A}_{|U}) \to (\xi, \mathcal{A}_\xi)$, such that $\pi \circ \mu = \text{id}$. Given two locally trivial superbundles $((\xi, \mathcal{A}_\xi), \pi)$ and $((\xi', \mathcal{A}_{\xi'}), \pi')$ over a G-supermanifold $(M, \mathcal{A})$, a *superbundle morphism* $\phi: (\xi, \mathcal{A}_\xi) \to (\xi', \mathcal{A}_{\xi'})$ is, by definition, a G-morphism, making the following diagram commutative:

$$\begin{array}{ccc} (\xi, \mathcal{A}_\xi) & \xrightarrow{\phi} & (\xi', \mathcal{A}_{\xi'}) \\ {\scriptstyle\pi}\downarrow & & \downarrow{\scriptstyle\pi'} \\ (M, \mathcal{A}) & = & (M, \mathcal{A}) \end{array}$$

G^∞ **vector bundles.** When defining *supervector bundles*, we can restrict ourselves, with no loss of generality, to the case where the standard fibre is $B_L^{r|s}$, which can be endowed with a G-supermanifold structure as described in the previous Section. Since any G-supermanifold has an underlying G^∞

supermanifold, we can expect that any supervector bundle has an underlying 'G^∞ vector bundle,' and hence that we need to define this concept. Since the structure sheaf of a G^∞ supermanifold is a sheaf of functions, the notion of G^∞ vector bundle is a verbatim translation of the definition of ordinary vector bundles.

Definition 3.2. *A triple (M, E, p) is said to be a G^∞ vector bundle if M and E are G^∞ supermanifolds, $p: E \to M$ is a G^∞ mapping, and the following conditions are fulfilled:*

(1) *there exists a cover $\{U_j\}$ of M and G^∞ isomorphisms*

$$\psi_j: p^{-1}(U_j) \to U_j \times B_L^{r|s}$$

such that $\mathrm{pr}_1 \circ \psi_j = p$;

(2) *the morphisms $\psi_j \circ \psi_k^{-1}$, when restricted to the fibres of the space $(U_j \cap U_k) \times B_L^{r|s}$, are morphisms of graded B_L-modules.*

Supervector bundles. In the previous Section we learnt how to associate a particular kind of superbundle — i.e. a product bundle — with any free graded $\mathcal{A}$-module. This procedure can be extended to the case of locally free graded $\mathcal{A}$-modules on the G-supermanifold $(M, \mathcal{A})$.

Proposition 3.1. *With any locally free graded $\mathcal{A}$-module $\mathcal{F}$ on $(M, \mathcal{A})$ one can associate a locally trivial superbundle, whose sheaf of G-sections is isomorphic with $\mathcal{F}$.*

Proof. Let $\mathcal{F}$ be a rank (r, s) locally free graded $\mathcal{A}$-module on $(M, \mathcal{A})$; there exists a cover $\mathfrak{U} = \{U_j\}$ of M, together with a family of isomorphisms of graded $\mathcal{A}$-modules

$$\varrho_j: \mathcal{F}_{|U_j} \to (\mathcal{A}_{|U_j})^{p|q}\,. \tag{3.3}$$

The composition $h_{jk} = \varrho_j \circ \varrho_k^{-1}$ yields an isomorphism

$$h_{jk}: (\mathcal{A}_{|U_j \cap U_k})^{p|q} \to (\mathcal{A}_{|U_j \cap U_k})^{p|q}\,, \tag{3.4}$$

which is described by a matrix whose entries are sections of $\mathcal{A}_{|U_j \cap U_k}$; this matrix will be denoted by the same symbol h_{jk}. By letting

$$g_{jk} = \delta(h_{jk})\,, \tag{3.5}$$

one obtains a G^∞ morphism $g_{jk}: U_j \cap U_k \to GL_L[p|q]$, where $GL_L[p|q]$ is the general linear supergroup of rank (r,s) (cf. Section I.3). Quite obviously, the morphisms g_{jk} fulfill the *cocycle property*

$$g_{jk}(z) \cdot g_{kh}(z) \cdot g_{hj}(z) = 1 \qquad \forall z \in U_j \cap U_k \cap U_h . \tag{3.6}$$

Proceeding by analogy with the ordinary theory of fibre bundles (cf. e.g. [**KN**]), it is therefore possible to construct a G^∞ vector bundle $p: \xi \to M$, with the standard fibre $B_L^{r|s}$ and transition functions g_{jk}. As a matter of fact, for every point $z \in M$ the following isomorphism of graded B_L-modules holds:

$$p^{-1}(z) \simeq \mathcal{A}_z^{p|q}/(\mathfrak{L}_z \cdot \mathcal{A}_z^{p|q}) \simeq B_L^{p|q} ; \tag{3.7}$$

this is a direct consequence of the commutativity of the following diagram:

$$\begin{array}{ccccccccc} 0 & \longrightarrow & \mathfrak{L}_z \cdot \mathcal{A}_z^{p|q} & \longrightarrow & \mathcal{A}_z^{p|q} & \overset{\sim}{\longrightarrow} & \mathcal{A}_z^{p|q}/(\mathfrak{L}_z \cdot \mathcal{A}_z^{p|q}) & \longrightarrow & 0 \\ & & {\scriptstyle h_{jk}}\downarrow & & {\scriptstyle h_{jk}}\downarrow & & \downarrow{\scriptstyle g_{jk}(z)} & & \\ 0 & \longrightarrow & \mathfrak{L}_z \cdot \mathcal{A}_z^{p|q} & \longrightarrow & \mathcal{A}_z^{p|q} & \overset{\sim}{\longrightarrow} & \mathcal{A}_z^{p|q}/(\mathfrak{L}_z \cdot \mathcal{A}_z^{p|q}) & \longrightarrow & 0 \end{array} \tag{3.8}$$

In order to build a superbundle we simply have to introduce a sheaf $\mathcal{A}_\xi$ making ξ into a G-supermanifold, compatible with the underlying G^∞ structure. Since $\mathcal{F}$ is locally free, we may define a sheaf $\mathcal{A}_\xi = \widehat{ST}(\mathcal{F}^*)$, repeating at a local level — by means of the isomorphisms (3.3) — the same procedure followed in the case of a free $\mathcal{A}$-module. This can be done because the results obtained on the overlaps of different U_j's coincide, since the metric structure of $ST(\mathcal{F}^*)$ is independent of the choice of the isomorphisms (3.3). The pair $(\xi, \mathcal{A}_\xi)$ is a G-supermanifold, as can be deduced from Proposition 3.1 and from the fact that the isomorphisms (3.3) induce local trivializations

$$\hat{\rho}_j: \widehat{ST}(\mathcal{F}^*)_{|U_j \times B_L^{r|s}} \to \mathcal{A}_{|U_j} \hat{\otimes}_\pi \mathcal{G}_{p|q} . \tag{3.9}$$

Finally, the natural immersion $\pi: \mathcal{A} \hookrightarrow p_*(\widehat{ST}(\mathcal{F}^*))$ determines a G-epimorphism $\underline{\pi} = (p, \pi)$, and one can easily observe that the pair $((\xi, \mathcal{A}_\xi), \pi)$ is a locally trivial superbundle — with the G-supermanifold $(B_L^{r|s}, \mathcal{G}_{p|q})$ as the standard fibre – whose sheaf of G-sections coincides with $\mathcal{F}$. ■

Given two locally free graded $\mathcal{A}$-modules on $(M, \mathcal{A})$, say $\mathcal{F}$ and $\mathcal{F}'$, any morphism $\Psi: \mathcal{F} \to \mathcal{F}'$ singles out a morphism (f, ψ) between the corresponding superbundles $(\xi, \mathcal{A}_\xi)$ e $(\xi', \mathcal{A}_{\xi'})$. Indeed — referring Eqs. (3.3–7) to a fixed cover $\mathfrak{U}$ for both $\mathcal{A}$-modules — the morphism Ψ gives rise, through diagram (3.8), to the following commutative diagram

$$\begin{array}{ccccccccc} 0 & \longrightarrow & \mathfrak{L}_z\mathcal{F}_z & \longrightarrow & \mathcal{F}_z & \longrightarrow & \mathcal{F}_z/(\mathfrak{L}_z \cdot \mathcal{F}_z) & \longrightarrow & 0 \\ & & {\scriptstyle \Psi_z}\downarrow & & {\scriptstyle \Psi_z}\downarrow & & \downarrow{\scriptstyle \tilde{\Psi}_z} & & \\ 0 & \longrightarrow & \mathfrak{L}_z\mathcal{F}'_z & \longrightarrow & \mathcal{F}'_z & \longrightarrow & \mathcal{F}'_z/(\mathfrak{L}_z \cdot \mathcal{F}'_z) & \longrightarrow & 0 \end{array} \quad . \tag{3.10}$$

The map f is defined by imposing that, for any $z \in M$, its restriction to the fibre $\pi^{-1}(z)$ is merely the graded B_L-module morphism $\tilde{\Psi}_z$. On the other hand, the morphism of sheaves of graded B_L-algebras $\psi\colon \mathcal{A}_{\xi'} \to f_*(\mathcal{A}_\xi)$ is determined by evident algebraic constructions.

The analysis developed so far leads to the following definition:

Definition 3.3. *A rank (r,s) supervector bundle on a G-supermanifold $(M,\mathcal{A})$ is a locally trivial superbundle $((\xi,\mathcal{A}_\xi),\pi)$ associated, according to Proposition 3.1, with a rank (r,s) locally free graded $\mathcal{A}$-module.*

Quite naturally, we designate by the term *supervector bundle morphism* a superbundle morphism which is induced by a morphism between the corresponding $\mathcal{A}$-modules. A sequence of supervector bundle morphisms is said to be exact if the corresponding sequence of $\mathcal{A}$-modules is exact.

One can verify directly that the correspondence between supervector bundles and locally free $\mathcal{A}$-modules established by Proposition 3.1 determines a one-to-one correspondence between the respective isomorphism classes, thus yielding an equivalence of categories.

REMARK 3.1. We should like to stress that the G^∞ supermanifold ξ underlying a supervector bundle over a G-supermanifold $(M,\mathcal{A},\delta^M)$ is a G^∞ vector bundle over M, whose transition functions g_{jk} are related to the transition morphisms of the supervector bundle by Eq. (3.5). ▲

Graded tangent bundle. A very important example of supervector bundle is provided by the *graded tangent bundle* $T(M,\mathcal{A})$ to a G-supermanifold $(M,\mathcal{A})$, which is simply defined as the supervector bundle associated with the locally free $\mathcal{A}$-module $\mathcal{D}er\mathcal{A}$. If $(z_j^1,\ldots,z_j^{m+n})$ and $(z_k^1,\ldots,z_k^{m+n})$ are coordinate systems for $(M,\mathcal{A})$ on the overlapping sets U_j and U_k, then the jacobian matrix

$$(h_{jk})^A{}_B = \frac{\partial z_j^A}{\partial z_k^B}, \qquad A, B = 1, \ldots, m+n \tag{3.11}$$

provides the relevant transition morphisms for $T(M, \mathcal{A})$ (the jacobian matrix is evaluated according to prescription (III.4.8)). The discussion which has led to the definition of supervector bundle and Definition III.4.2 show that the fibre of the graded tangent bundle of $(M, \mathcal{A})$ at a point $z \in M$ is no more than the graded B_L-module $T_z(M, \mathcal{A})$ (the graded tangent space at z) with its canonical structure of a G-supermanifold. The sections of the graded tangent bundle, i.e. the graded derivations of $\mathcal{A}$, will also be called *graded vector fields*, in the sense that at any point $z \in M$ they single out an element (a vector) in $T_z(M, \mathcal{A})$.

The graded tangent bundle $T(M, \mathcal{A})$ has an underlying G^∞ vector bundle, whose transition functions are the mappings $g_{jk} = \delta^M(h_{jk})$; these functions cannot be written as jacobian matrices, since derivatives of G^∞ functions with respect to odd variables are not defined. This is consistent with the fact that the sheaf of sections of the underlying G^∞ vector bundle (which is no more than $\mathcal{D}er\mathcal{A}^\infty$) is not locally free.

Superline bundles. A particular, but important, case is that of *superline bundles,* — i.e. supervector bundles over a G-supermanifold $(M, \mathcal{A})$, having either rank (1,0) or (0,1). Since in both cases the transition morphisms of the bundle are local sections of the sheaf $\mathcal{A}_0^*$ (invertible even sections of $\mathcal{A}$), the categories of the two kinds of superline bundles are in fact equivalent. Superline bundles will be studied in some detail in Chapter VI.

Categorial operations with supervector bundles. The category of SVB's over a G-supermanifold $(M, \mathcal{A})$ is equivalent to that of locally free $\mathcal{A}$-modules, so that one can define the usual operations of direct sum, tensor product, etc. in terms of the corresponding operations for $\mathcal{A}$-modules. Henceforth, the terminology 'supervector bundle' will often be shortened into 'SVB.'

Let $(M, \mathcal{A})$ be a G-supermanifold, $((\xi, \mathcal{A}_\xi), \pi)$ and $((\xi', \mathcal{A}_{\xi'}), \pi')$ be SVB's over $(M, \mathcal{A})$ of respective rank (p, q) and (r, s), and let $\mathcal{F}$ and $\mathcal{F}'$ be the corresponding locally free $\mathcal{A}$-modules of G-sections.

Definition 3.4. *The superbundle of homomorphisms from $(\xi, \mathcal{A}_\xi)$ to $(\xi', \mathcal{A}_{\xi'})$ is the SVB $(\mathrm{Hom}(\xi, \xi'), \mathcal{A}_{\mathrm{Hom}(\xi,\xi')})$ associated with the rank $(pr + qs, ps + qr)$ locally free $\mathcal{A}$-module of homomorphisms $\mathcal{H}om_{\mathcal{A}}(\mathcal{F}, \mathcal{F}')$ according to Definition 3.3. In the same way, the direct sum and the tensor product of $(\xi, \mathcal{A}_\xi)$ and*

$(\xi', \mathcal{A}_{\xi'})$ are the SVB's $(\xi \oplus \xi', \mathcal{A}_{\xi\oplus\xi'})$ and $(\xi \otimes \xi', \mathcal{A}_{\xi\otimes\xi'})$ associated with the locally free $\mathcal{A}$-modules $\mathcal{F} \oplus \mathcal{F}'$ and $\mathcal{F} \otimes_{\mathcal{A}} \mathcal{F}'$, respectively.

All these bundles are trivial when $((\xi, \mathcal{A}_\xi), \pi)$ and $((\xi', \mathcal{A}_{\xi'}), \pi')$ are also trivial. In particular, if we denote by $\xi_z = \pi^{-1}(z)$ the fibre of $\pi: \xi \to M$ over a point $z \in M$, and similarly ξ'_z etc., one has

$$\mathrm{Hom}(\xi, \xi')_z \simeq \mathrm{Hom}_{B_L}(\xi_z, \xi'_z),$$

$$(\xi \oplus \xi')_z \simeq \xi_z \oplus \xi'_z,$$

$$(\xi \otimes \xi')_z \simeq \xi_z \otimes_{B_L} \xi'_z,$$

when we consider all fibres with their natural structures of graded B_L-modules.

Let us briefly comment upon the structure of the direct sum SVB $(p, \Phi): (\xi \oplus \xi', \mathcal{A}_{\xi\oplus\xi'}) \to (M, \mathcal{A})$. The underlying manifold $\xi \oplus \xi'$ is the fibre product $\xi \times_M \xi'$ over M of the underlying manifolds, taken with respect to the maps $\pi: \xi \to M$ and $\pi': \xi' \to M$. Actually, $(\xi \oplus \xi', \mathcal{A}_{\xi\oplus\xi'})$ would be the fibre product G-supermanifold $(\xi, \mathcal{A}_\xi) \times_{(M,\mathcal{A})} (\xi', \mathcal{A}_{\xi'})$, if this notion had been defined; in fact, the fibre product of two G-morphisms only exists when certain 'transversality' conditions are fulfilled. Although this is certainly the case for locally free superfibre bundles, as the proof involves some not entirely trivial technicalities, we shall confine ourselves to the study of $(\xi \oplus \xi', \mathcal{A}_{\xi\oplus\xi'})$ in the particular case when $(\xi, \mathcal{A}_\xi) \simeq (M \times B_L^{p|q}, \mathcal{A}\hat{\otimes}_\pi \mathcal{G}_{p|q})$ is trivial, such that $\mathcal{F} \equiv \mathcal{A}^{p|q} \simeq B_L^{p|q} \otimes_{B_L} \mathcal{A}$. Now, one has

$$ST((\mathcal{A}^{p|q} \oplus \mathcal{F}')^*) \simeq ST(B_L^{p|q}) \otimes_{B_L} ST((\mathcal{F}')^*)$$

by (2.5), and then

$$\widehat{ST}((\mathcal{A}^{p|q} \oplus \mathcal{F}')^*) \simeq \mathcal{G}_{p|q} \hat{\otimes}_\pi \widehat{ST}((\mathcal{F}')^*) .$$

This proves that there is a G-isomorphism

$$(M \times B_L^{p|q}, \mathcal{A}\hat{\otimes}_\pi \mathcal{G}_{p|q}) \oplus (\xi', \mathcal{A}_{\xi'}) \simeq (B_L^{p|q}, \mathcal{G}_{p|q}) \times (\xi', \mathcal{A}_{\xi'}) . \tag{3.12}$$

The natural morphism of $\mathcal{A}$-modules

$$\begin{aligned} \mathcal{H}om_{\mathcal{A}}(\mathcal{F}, \mathcal{F}') \oplus \mathcal{F} &\to \mathcal{F}' \\ (\zeta\, , f) &\mapsto \zeta(f) \end{aligned}$$

induces a morphism of SVB's

$$(\mathrm{Hom}(\xi,\xi'),\mathcal{A}_{\mathrm{Hom}(\xi,\xi')})\oplus(\xi,\mathcal{A}_\xi)\to(\xi',\mathcal{A}_{\xi'})\,. \tag{3.13}$$

If $(\xi,\mathcal{A}_\xi)\simeq(M\times B_L^{p|q},\mathcal{A}\hat{\otimes}_\pi\mathcal{G}_{p|q})$ is trivial, (3.12) gives rise to a G-morphism

$$(\mathrm{Hom}(M\times B_L^{p|q},\xi'),\mathcal{A}_{\mathrm{Hom}(M\times B_L^{p|q},\xi')})\times B_L^{p|q}\to(\xi',\mathcal{A}_{\xi'})\,. \tag{3.14}$$

In particular, taking $\widehat{M}=\widehat{z}=(z,B_L)$, this proves that the natural map

$$(\mathrm{Hom}_{B_L}(B_L^{p|q},B_L^{r|s}),\mathcal{G}_{pr+qs|ps+qr})\times(B_L^{p|q},\mathcal{G}_{p|q})\to(B_L^{r|s}\mathcal{G}_{r|s})\,,$$

is a G-morphism; this also follows from the fact that this morphism is GH^∞.

Furthermore, let us consider the general linear supergroup $GL_L[p|q]$ over B_L (cf. Section I.3), endowed with its natural structure of G-supermanifold as an open submanifold of $\mathrm{Hom}_{B_L}(B_L^{p|q},B_L^{p|q})$, and let us denote by $\mathcal{B}$ the corresponding structure sheaf; the above morphism induces a G-morphism

$$(GL_L[p|q],\mathcal{B})\times(B_L^{p|q},\mathcal{G}_{p|q})\to(B_L^{p|q},\mathcal{G}_{p|q})\,. \tag{3.15}$$

Let us take the SVB's $((\xi,\mathcal{A}_\xi),\pi)$, $((\xi',\mathcal{A}_{\xi'}),\pi')$ and $((\xi'',\mathcal{A}_{\xi''}),\pi'')$ associated with locally free $\mathcal{A}$-modules $\mathcal{F}$, $\mathcal{F}'$ and $\mathcal{F}''$. The composition of morphisms defines a morphism of $\mathcal{A}$-modules

$$\begin{aligned}\mathcal{H}om_{\mathcal{A}}(\mathcal{F}',\mathcal{F}'')\oplus\mathcal{H}om_{\mathcal{A}}(\mathcal{F},\mathcal{F}')&\to\mathcal{H}om_{\mathcal{A}}(\mathcal{F},\mathcal{F}'')\\ (\zeta\,,\theta)&\mapsto\zeta\circ\theta\,,\end{aligned}$$

and, thus, a morphism of SVB's

$$\begin{aligned}(\mathrm{Hom}(\xi',\xi''),\mathcal{A}_{\mathrm{Hom}(\xi',\xi'')})\oplus(\mathrm{Hom}(\xi,\xi'),\mathcal{A}_{\mathrm{Hom}(\xi,\xi')})&\\ \to(\mathrm{Hom}(\xi,\xi''),\mathcal{A}_{\mathrm{Hom}(\xi,\xi'')})\,,&\end{aligned} \tag{3.16}$$

whose effect on fibres is, of course, the composition of morphisms

$$\begin{aligned}\mathrm{Hom}(\xi',\xi'')_z\oplus\mathrm{Hom}(\xi,\xi')_z&\to\mathrm{Hom}(\xi,\xi'')\\ (\zeta_z\,,\theta_z)&\mapsto\zeta_z\circ\theta_z\,.\end{aligned}$$

Equations (3.12) and (3.16) provide G-morphisms

$$(\mathrm{Hom}_{B_L}(B_L^{r|s}, B_L^{r|s}), \mathcal{G}_{r^2+s^2|2rs}) \times (\mathrm{Hom}(\xi, M \times B_L^{r|s}), \mathcal{A}_{\mathrm{Hom}(\xi, M\times B_L^{r|s})}) \\ \to (\mathrm{Hom}(\xi, M \times B_L^{r|s}), \mathcal{A}_{\mathrm{Hom}(\xi, M\times B_L^{r|s})}) \quad (3.17)$$

and

$$(\mathrm{Hom}(M \times B_L^{r|s}, \xi''), \mathcal{A}_{\mathrm{Hom}(M\times B_L^{r|s}, \xi'')}) \times (\mathrm{Hom}_{B_L}(B_L^{r|s}, B_L^{r|s}), \mathcal{G}_{r^2+s^2|2rs}) \\ \to (\mathrm{Hom}(M \times B_L^{r|s}, \xi''), \mathcal{A}_{\mathrm{Hom}(M\times B_L^{r|s}, \xi'')}). \quad (3.18)$$

If we again take the general linear supergroup $GL_L[r|s]$ over B_L endowed, as before, with its natural structure of a G-supermanifold, the above morphisms induce G-morphisms

$$(GL_L[r|s], \mathcal{B}) \times (\mathrm{Hom}(\xi, M \times B_L^{r|s}), \mathcal{A}_{\mathrm{Hom}(\xi, M\times B_L^{r|s})}) \\ \to (\mathrm{Hom}(\xi, M \times B_L^{r|s}), \mathcal{A}_{\mathrm{Hom}(\xi, M\times B_L^{r|s})}) \quad (3.19)$$

and

$$(\mathrm{Hom}(M \times B_L^{r|s}, \xi''), \mathcal{A}_{\mathrm{Hom}(M\times B_L^{r|s}, \xi'')}) \times (GL_L[r|s], \mathcal{B}) \\ \to (\mathrm{Hom}(M \times B_L^{r|s}, \xi''), \mathcal{A}_{\mathrm{Hom}(M\times B_L^{r|s}, \xi'')}). \quad (3.20)$$

Let us now consider SVB's $((\xi, \mathcal{A}_\xi), \pi)$ and $((\xi', \mathcal{A}_{\xi'}), \pi')$ with the same rank, $(p,q) = (r,s)$. We can then talk of isomorphisms between them. The subset $\mathrm{Iso}(\xi, \xi')$ of those points in $\mathrm{Hom}(\xi, \xi')$ that are isomorphisms of ξ_z with ξ'_z is open, and we have an open sub-G-supermanifold

$$(\mathrm{Iso}(\xi, \xi'), \mathcal{A}_{\mathrm{Iso}(\xi,\xi')}),$$

where $\mathcal{A}_{\mathrm{Iso}(\xi,\xi')} = \mathcal{A}_{\mathrm{Hom}(\xi,\xi')|\,\mathrm{Iso}(\xi,\xi')}$. The restriction of the natural projection $(p, \phi)\colon (\mathrm{Hom}(\xi, \xi'), \mathcal{A}_{\mathrm{Hom}(\xi,\xi')}) \to (M, \mathcal{A})$ is a G-morphism,

$$(p, \phi)\colon (\mathrm{Iso}(\xi, \xi'), \mathcal{A}_{\mathrm{Iso}(\xi,\xi')}) \to (M, \mathcal{A}),$$

namely, it is a locally trivial G-superbundle (cf. Definition 3.1) whose standard fibre is $GL_L[p|q]$.

We can thus give the following definition:

Definition 3.5. *The superbundle of isomorphisms from* $(\xi, \mathcal{A}_\xi)$ *to* $(\xi', \mathcal{A}_{\xi'})$ *is the locally trivial superbundle with standard fibre* $GL_L[p|q]$ *described by*

$$(p, \phi): (\mathrm{Iso}(\xi, \xi'), \mathcal{A}_{\mathrm{Iso}(\xi,\xi')}) \to (M, \mathcal{A}) .$$

Now, given a SVB $(\xi, \mathcal{A}_\xi)$ of rank (r, s), Eq. (3.14) defines a G-morphism

$$(\mathrm{Iso}(M \times B_L^{p|q}, \xi), \mathcal{A}_{\mathrm{Iso}(M\times B_L^{p|q},\xi)}) \times B_L^{p|q} \to (\xi, \mathcal{A}_\xi) , \tag{3.21}$$

while (3.19) and (3.20) provide G-morphisms

$$\begin{aligned}(GL_L[r|s], \mathcal{G}'_{r^2+s^2|2rs}) \times (\mathrm{Iso}(\xi, M \times B_L^{r|s}), \mathcal{A}_{\mathrm{Iso}(\xi,M\times B_L^{r|s})})\\ \to (\mathrm{Iso}(\xi, M \times B_L^{r|s}), \mathcal{A}_{\mathrm{Iso}(\xi,M\times B_L^{r|s})})\end{aligned} \tag{3.22}$$

and

$$\begin{aligned}(\mathrm{Iso}(M \times B_L^{r|s}, \xi), \mathcal{A}_{\mathrm{Iso}(M\times B_L^{r|s},\xi)}) \times (GL_L[r|s], \mathcal{B})\\ \to (\mathrm{Iso}(M \times B_L^{r|s}, \xi), \mathcal{A}_{\mathrm{Iso}(M\times B_L^{r|s},\xi)}) .\end{aligned} \tag{3.23}$$

EXAMPLE 3.1. There is an important superbundle of isomorphisms canonically associated with a G-supermanifold $(M, \mathcal{A})$ of dimension (m, n). Taking $(\xi, \mathcal{A}_\xi) = (M \times B_L^{r|s}, \mathcal{A}\hat{\otimes}_\pi \mathcal{G}_{m|n})$ as the trivial SVB of rank (m, n), and $(\xi', \mathcal{A}_{\xi'}) = T(M, \mathcal{A})$ as the graded tangent bundle, the locally trivial G-superbundle of isomorphisms of the trivial SVB of rank (m, n) with the graded tangent bundle is called the *superbundle of graded frames* of $(M, \mathcal{A})$, and is denoted by $Fr(M, \mathcal{A})$. ▲

4. Graded exterior differential calculus

The graded tensor calculus developed in Section I.2 can be applied to the case of the spaces of sections of a locally free sheaf $\mathcal{M}$ on a graded ringed space. We are interested in the case of the sheaf of graded differential forms on a G-supermanifold; the case of graded manifolds, which is in fact very similar, is treated in detail in [**Kos**] and [**HeM1**].

Let $(M, \mathcal{A})$ be a (m, n) dimensional G-supermanifold; the sheaf $\mathcal{D}er\mathcal{A}$ of graded derivations of the sheaf of graded B_L-algebras $\mathcal{A}$ is locally free as a consequence of Proposition III.4.3, since $(M, \mathcal{A})$ and $(B_L^{m,n}, \mathcal{G})$ are locally isomorphic. In accordance with Proposition I.2.1, $\mathcal{D}er\mathcal{A}$ is a sheaf of graded B_L-algebras, with the graded Lie bracket between local sections D_1, D_2 given by Eq. (I.2.4):

$$[D_1, D_2] = D_1 \circ D_2 - (-1)^{|D_1||D_2|} D_2 \circ D_1 .$$

If $(U, (x^1, \ldots, x^m, y^1, \ldots, y^n))$ is a coordinate chart for $(M, \mathcal{A})$, the graded derivations $\left\{ \frac{\partial}{\partial x^i}, \frac{\partial}{\partial y^\alpha} \right\}$, $i = 1, \ldots, m$, $\alpha = 1, \ldots, n$, are defined as in Eq. (III.4.8) by enforcing the local identification of $(M, \mathcal{A})$ with $(B_L^{m,n}, \mathcal{G})$, and form a basis of $\mathcal{D}er\mathcal{A}(U)$:

$$D = \sum_{i=1}^{m} D(x^i) \frac{\partial}{\partial x^i} + \sum_{\alpha=1}^{n} D(y^\alpha) \frac{\partial}{\partial y^\alpha}.$$

for any $D \in \mathcal{D}er\mathcal{A}(U)$.

Definition 4.1. *The sheaves of graded differential forms on $(M, \mathcal{A})$ are the sheaves*

$$\Omega_{\mathcal{A}}^k = \bigwedge^k \mathcal{D}er^* \mathcal{A}.$$

The graded differential forms on $(M, \mathcal{A})$ will also be called simply graded forms.

Section I.2 provides the algorithm for computing the wedge product of two graded forms and the inner product between a graded vector field and a graded form: for $\omega^p \in \Omega_{\mathcal{A}}^p(U)$ and $\omega^q \in \Omega_{\mathcal{A}}^q(U)$ homogeneous graded forms, and homogeneous graded vector fields $D_1, \ldots, D_{p+q} \in \mathcal{D}er\mathcal{A}(U)$, we have

$$\begin{aligned} &(\omega^p \wedge \omega^q)(D_1, \ldots, D_{p+q}) \\ &= \frac{1}{(p+q)!} \sum_{\sigma \in \mathfrak{S}_{p+q}} (-1)^{|\sigma| + \Delta(\sigma, D, \omega^q)} \omega^p(D_1, \ldots, D_{\sigma(p)}) \omega^q(D_{\sigma(p+1)}, \ldots, D_{\sigma(p+q)}) \end{aligned} \tag{4.1}$$

where — as in Section I.2 — we have denoted by $|\sigma|$ the parity of the permutation σ, and have set

$$\Delta(\sigma, D, \omega^q) = \sum_{1 \le i < j \le p} \sum_{\sigma(i) > \sigma(j)} |D_{\sigma(i)}| |D_{\sigma(j)}| + |\omega^q| \sum_{i=1}^{p} |D_{\sigma(i)}| ;$$

$$(D_1 \rfloor \omega^p)(D_2, \ldots, D_p) = p(-1)^{|D|_1 |\omega^p|} \omega^p(D_1, \ldots, D_p). \tag{4.2}$$

We now wish to generalize the notion of Cartan exterior differential to the setting of graded forms.

Definition 4.2. *The exterior differential is the morphism of graded B_L-modules*

$$d: \Omega^p_A(U) \to \Omega^{p+1}_A(U)$$

described on a homogeneous graded form by

$$\begin{aligned} d\omega^p(D_1, \ldots, D_{p+1}) = & \\ & \tfrac{1}{p+1} \sum_{i=1}^{p+1} (-1)^{i-1+\alpha_i+|\omega^p||D_i|} D_i(\omega^p(D_1, \ldots, \widehat{D_i}, \ldots, D_p)) \\ + & \tfrac{1}{p+1} \sum_{i<j} (-1)^{\alpha_i+\alpha_j+i+j+|D_i||D_j|} \omega^p([D_i, D_j], \ldots, \widehat{D_i}, \ldots, \widehat{D_j}, \ldots, D_{p+1}) \end{aligned} \tag{4.3}$$

for homogeneous $D_1, \ldots, D_p \in \mathcal{D}er(\mathcal{A}(U))$, where $\alpha_i = |D_i| \sum_{h<i} |D_h|$ (a hat denotes omission).

In particular, one has:

$$df(D) = (-1)^{|f||D|} D(f) \tag{4.4}$$

for homogeneous $f \in \mathcal{A}(U)$, $D \in \mathcal{D}er\,\mathcal{A}(U)$.

Proposition 4.1. *The exterior differential d verifies the condition $d^2 = 0$, i.e., $(\mathcal{A}^\bullet, d)$ is a complex of sheaves of graded B_L-modules.*[2] *Moreover, d is a differential operator of bidegree $(1,0)$, that is:*

$$d(\omega^p \wedge \omega^q) = d\omega^p \wedge \omega^q + (-1)^p \omega^q \wedge d\omega^q$$

for $\omega^p \in \Omega^p_A(U)$, $\omega^q \in \Omega^q_A(U)$. ■

Obviously, the sheaf $\mathcal{D}er^*\mathcal{A} \equiv \Omega^1_A$ is locally free, and one can characterize the bases of the modules of its sections over a coordinate chart.

[2] Cf. Definition II.2.3.

Proposition 4.2. *Let $(U,(x^1,\ldots,x^m,y^1,\ldots,y^n))$ be a coordinate chart on a G-supermanifold $(M,\mathcal{A})$. The graded forms $(dx^1,\ldots,dx^m,dy^1,\ldots,dy^n)$ provide a basis for the free $\mathcal{A}(U)$-module $\Omega^1_{\mathcal{A}}(U)$, and for any homogeneous function $f\in\mathcal{A}(U)$ one has*

$$\begin{aligned} df &= \sum_{i=1}^m \frac{\partial f}{\partial x^i}\,dx^i - \sum_{\alpha=1}^n (-1)^{|f|}\frac{\partial f}{\partial y^\alpha}\,dy^\alpha \\ &= \sum_{i=1}^m dx^i\,\frac{\partial f}{\partial x^i} + \sum_{\alpha=1}^n dy^\alpha\,\frac{\partial f}{\partial y^\alpha}\,. \end{aligned}$$

Proof. Since

$$dx^i(\frac{\partial}{\partial x^j}) = \delta_{ij},\quad dx^i(\frac{\partial}{\partial y^\alpha}) = 0,\quad dy^\alpha(\frac{\partial}{\partial x^i}) = 0,\quad dy^\alpha(\frac{\partial}{\partial y^\beta}) = -\delta_{\alpha\beta}\,,$$

the collection $\{dx^1,\ldots,dx^m,-dy^1,\ldots,-dy^n\}$ is the basis of the module $\Omega^1_{\mathcal{A}}(U)$ $=$ $(\mathcal{D}er\,\mathcal{A}(U))^*$ dual to the basis $\left(\frac{\partial}{\partial x^i},\frac{\partial}{\partial y^\alpha}\right)$ of $\mathcal{D}er\,\mathcal{A}(U)$. This demonstrates the first claim. The second one is proved by letting $df = \sum_{i=1}^m f_i\,dx^i + \sum_{\alpha=1}^n f_\alpha\,dy^\alpha$ and applying both sides of this equation to the graded vector fields $\frac{\partial}{\partial x^i}$ and $\frac{\partial}{\partial y^\alpha}$, thus obtaining the first expression for df. The other is obtained from Eq.(I.1.2), which formally means that we are regarding $\Omega^1_{\mathcal{A}}(U)$ as a right module rather than a left one. ■

The second expression for df holds even if f is not homogeneous.

As a corollary, one obtains the basis of the module of graded r-forms $\Omega^r_{\mathcal{A}}(U)$. Let us recall that Ξ_r is the set of all strictly increasing sequences $\{\mu\colon\{1\ldots p\}\to\{1\ldots r\}\mid 1\le p\le r\}\cup\{\mu_0\}$, where μ_0 is the empty sequence. The number p is also denoted by $d(\mu)$. Moreover, $J=(J^1,\ldots,J^n)\in\mathbb{N}^n$ will denote a multi-index, whose length is defined as $|J|=\sum_{i=1}^n J^i$.

Corollary 4.1. *Let $(U,(x^1,\ldots,x^m,y^1,\ldots,y^n))$ be a coordinate chart on a G-supermanifold $(M,\mathcal{A})$. The module $\Omega^r_{\mathcal{A}}(U)$ is free over $\mathcal{A}(U)$ with a basis given by the graded r-forms*

$$dx^\mu\wedge dy^J = dx^{\mu(1)}\wedge\cdots\wedge dx^{\mu(p)}\wedge\underbrace{dy^1\wedge\cdots\wedge dy^1}_{J^1}\wedge\cdots\wedge\underbrace{dy^n\wedge\cdots\wedge dy^n}_{J^n}$$

where $\mu\in\Xi_p$, and p and $|J|$ are such that $p+|J|=r$. ■

All this enables us to compute the exterior differential of any graded form $\omega \in \Omega^r_A(U)$. By letting

$$\omega = \sum_{d(\mu)+|J|=r} dx^\mu \wedge dy^J \, f_{\mu J}, \qquad \text{with} \quad f_{\mu J} \in \mathcal{A}(U),$$

we obtain

$$\begin{aligned} d\omega &= \sum_{d(\mu)+|J|=r} d(dx^\mu \wedge dy^J \, f_{\mu J}) = (-1)^r \sum_{d(\mu)+|J|=r} dx^\mu \wedge dy^J \wedge df_{\mu J} \\ &= (-1)^r \sum_{d(\mu)+|J|=r} \left[\sum_{i=1}^m \varepsilon_{\mu i} dx^{\mu+i} \wedge dy^J \frac{\partial f_{\mu J}}{\partial x^i} + \sum_{\alpha=1}^n dx^\mu \wedge dy^{J+(\alpha)} \frac{\partial f_{\mu J}}{\partial y^\alpha} \right]; \end{aligned}$$

here $\mu+i$ is the strictly increasing sequence obtained by reordering the sequence $(\mu(1), \ldots, \mu(p), i)$, and $\varepsilon_{\mu i}$ is a ± 1 which takes account of the changes of sign involved in the reordering, while, if i is already contained in μ, the summation on that term is skipped; (α) is the multi-index given by $(\alpha) = (0, \ldots, 1, \ldots, 0) \in \mathbb{N}^n$ with 1 in the αth place.

It is also possible to introduce the Lie derivative of a graded differential form with respect to a graded vector field, by using as a definition a well-known property of the Lie derivative in the ordinary setting.

Definition 4.3. *The Lie derivative with respect to a graded vector field $D \in \mathcal{D}er\,\mathcal{A}(U)$ is the morphism of graded B_L-modules*

$$\mathrm{Lie}_D \colon \Omega^p_A(U) \to \Omega^p_A(U)$$

given by

$$\mathrm{Lie}_D\, \omega = D \rfloor d\omega + d(D \rfloor \omega)$$

for any $\omega \in \Omega^p_A(U)$.

It follows that if $D \in \mathcal{D}er\,\mathcal{A}(U)$ is a homogeneous graded vector field, the Lie derivative Lie_D is a graded derivation of bidegree $(0, |D|)$, that is:

$$\mathrm{Lie}_D(\omega^p \wedge \omega^q) = \mathrm{Lie}_D\, \omega^p \wedge \omega^q + (-1)^{|D||\omega^p|} \omega^q \wedge \mathrm{Lie}_D\, \omega^q \tag{4.5}$$

where $\omega^p \in \Omega^p_A(U)$ and $\omega^q \in \Omega^q_A(U)$.

The explicit expression of the Lie derivative is easily achieved from its very definition:

$$\begin{aligned}\mathrm{Lie}_D\,\omega(D_1,\dots,D_p) &= (-1)^{|D|\sum_{i=1}^{p}|D_i|}D(\omega(D_1,\dots,D_p))\\ &-(-1)^{|D||\omega|}\sum_{i=1}^{p}(-1)^{|D|\sum_{j<i}|D_j|}\omega(D_1,\dots,[D,D_i],\dots,D_p)\end{aligned} \tag{4.6}$$

for homogeneous $D \in \mathcal{D}er\mathcal{A}(U)$, $\omega \in \Omega^p_{\mathcal{A}}(U)$ and $D_1,\dots,D_p \in \mathcal{D}er\mathcal{A}(U)$.

5. Projectable graded vector fields

In this section we introduce the notion of a projectable graded derivation (graded vector field) on the total space of a superbundle. This will be particularly useful when dealing with principal superfibre bundles. Let $(p,\psi)\colon(P,\mathcal{B})\to(M,\mathcal{A})$ be a G-morphism, and let D be a graded vector field on $(P,\mathcal{B})$.

Definition 5.1. *D is (p,ψ)-projectable to $(M,\mathcal{A})$ if for every open subset $V\subset M$ it preserves the subring $\psi(\mathcal{A}(V))$ of $\mathcal{B}(p^{-1}(V))$; that is, if for every $f\in\mathcal{A}(V)$, there exists $p(D)(f)\in\mathcal{A}(V)$ such that $D(\psi(f))=\psi(p(D)(f))$. Moreover, D is said to be vertical if it is (p,ψ)-projectable and $p(D)=0$.*

The graded vector field $p(D)$ on $(V,\mathcal{A}_V)$ is called the *projection* of D. In sheaf language, if $D\colon\mathcal{B}\to\mathcal{B}$ is projectable, the diagram

$$\begin{array}{ccc}\mathcal{A} & \xrightarrow{p^*} & p_*\mathcal{B}\\ {\scriptstyle p(D)}\downarrow & & \downarrow{\scriptstyle D}\\ \mathcal{A} & \xrightarrow[p^*]{} & p_*\mathcal{B}\end{array}$$

is commutative.

Let $V\subset M$ be an open subset and $U=p^{-1}(V)$. It is clear that if D is (p,ψ)-projectable on $(U,\mathcal{A}_U)$ with projection $p(D)$, and $f\in\mathcal{A}(V)$, then $f\cdot D$, defined as $f\cdot D=\psi(f)\,D$ is also (p,ψ)-projectable with projection $f\,p(D)$. It follows that the set $\mathcal{P}ro(p_*\mathcal{B})(V)$ of (p,ψ)-projectable graded vector fields on $p^{-1}(V)$ is a graded $\mathcal{A}(V)$ module, so that we can define a sheaf of graded $\mathcal{A}$-modules $\mathcal{P}ro(p_*\mathcal{B})$, called the *sheaf of (p,ψ)-projectable graded vector fields.*

Furthermore, if D_1, D_2 are (p, ψ)-projectable on $(U, \mathcal{A}_U)$, the Lie bracket $[D_1, D_2]$ is also (p, ψ)-projectable, with projection $p([D_1, D_2]) = [p(D_1), p(D_2)]$.

If D is a vertical graded vector field on an open subset $U \subset P$, for every $g \in \mathcal{B}(U)$ the graded vector field gD is also vertical. One can thus define a sheaf $\mathcal{V}er(\mathcal{B})$ of $\mathcal{B}$-modules, called the sheaf of *vertical graded vector fields* on $(P, \mathcal{B})$. By definition, there is a sequence of sheaves of $\mathcal{A}$ modules

$$0 \longrightarrow p_*(\mathcal{V}er(\mathcal{B})) \longrightarrow \mathcal{P}ro(p_*\mathcal{B}) \xrightarrow{p} \mathcal{D}er\mathcal{A} \longrightarrow 0 \tag{5.1}$$

which in general is exact only on the left, i.e. the morphism p may fail to be surjective.

Proposition 5.1. *Let $(p, \psi)\colon (P, \mathcal{B}) \to (M, \mathcal{A})$ be a locally trivial superbundle. The sequence of sheaves of $\mathcal{A}$-modules (5.1) is exact.*

Proof. One has only to prove that there is a cover of M by open subsets V such that every graded vector field on $(V, \mathcal{A}_V)$ (where $\mathcal{A}_V = \mathcal{A}_{|V}$) is the projection of a graded vector field on $(U, \mathcal{A}_U)$, $U = p^{-1}(V)$. Let us then cover M by open subsets V such that $(U, \mathcal{A}_U) = (F, \mathcal{A}_F) \times (V, \mathcal{A}_V) = (F \times V, \mathcal{A}_F \hat{\otimes}_\pi \mathcal{A}_V)$ and p is the second projection. Thus, if D' is a graded vector field on $(V, \mathcal{A}_V)$, $D = \mathrm{Id} \otimes D'$ defines a graded vector field on $(U, \mathcal{A}_U) = (F \times V, \mathcal{A}_F \hat{\otimes}_\pi \mathcal{A}_V)$ that is (p, ψ)-projectable and projects onto D', i.e. $p(D) = D'$. ∎

Let us look at the local structure of this sequence, taking V as above, so that $(U, \mathcal{A}_U) = (F, \mathcal{A}_F) \times (V, \mathcal{A}_V) = (F \times V, \mathcal{A}_F \hat{\otimes}_\pi \mathcal{A}_V)$, and (p, ψ) is the second projection ($U = p^{-1}(V)$). We assume that there is a system $(x^1, \dots, x^m, y^1, \dots, y^n)$ of graded coordinates in $(V, \mathcal{A}_V)$, and we take graded coordinates $(z^1, \dots, z^p, t^1, \dots, t^q)$ in an open subset $W \subset F$ of the fibre. Then, $(x^1, \dots, x^m, z^1, \dots, z^p, y^1, \dots, y^n, t^1, \dots, t^q)$ are graded coordinates in $W \times V$, and the general expression of a graded vector field in $W \times V$ is

$$D = \sum_{\mu \in \Xi_L} \left(\sum_{i=1}^m f_i^\mu \frac{\partial}{\partial x^i} + \sum_{\alpha=1}^n F_\alpha^\mu \frac{\partial}{\partial y^\alpha} + \sum_{j=1}^p g_j^\mu \frac{\partial}{\partial z^j} + \sum_{\gamma=1}^q G_\gamma^\mu \frac{\partial}{\partial t^\gamma} \right) \otimes \beta_\mu ,$$

where the f's, F's, g's and G's are GH^∞ functions of the coordinates $(x^1, \dots, x^m, z^1, \dots, z^p, y^1, \dots, y^n, t^1, \dots, t^q)$, and $\{\beta_\mu,\ \mu \in \Xi_L\}$ is the canonical basis of B_L.

Now, D is *(p, ψ)-projectable* if and only if the coefficients f and F's depend only on the graded coordinates $(x^1, \dots, x^m, y^1, \dots, y^n)$ of the base supermanifold; in this case, the projection of D is

$$p(D) = \sum_{\mu \in \Xi_L} \left(f_i^\mu \frac{\partial}{\partial x^i} + \sum_{\alpha=1}^{n} F_\alpha^\mu \frac{\partial}{\partial y^\alpha} \right) \otimes \beta_\mu \quad .$$

Furthermore, D is *vertical* if it is given by:

$$D = \sum_{\mu \in \Xi_L} \left(\sum_{j=1}^{p} g_j^\mu \frac{\partial}{\partial z^j} + \sum_{\gamma=1}^{q} G_\gamma^\mu \frac{\partial}{\partial t^\gamma} \right) \otimes \beta_\mu$$

where the g's and the G's are GH^∞ functions of all the graded coordinates.

Finally, one easily obtains the following result.

Proposition 5.2. *Let $(p, \psi): (P, \mathcal{B}) \to (M, \mathcal{A})$ be a locally trivial superbundle. There is a canonical isomorphism of sheaves of $\mathcal{B}$-modules:*

$$\psi^* \mathcal{P}ro(p_* \mathcal{B}) = \mathcal{B} \otimes_{\mathcal{A}} \mathcal{P}ro(p_* \mathcal{B}) \simeq \mathcal{D}er\, \mathcal{B} \,,$$

so that the sequence 5.1 induces an exact sequence of sheaves of $\mathcal{B}$-modules

$$0 \longrightarrow \mathcal{V}er(\mathcal{B}) \longrightarrow \mathcal{D}er\, \mathcal{B} \xrightarrow{p} \psi^* \mathcal{D}er\, \mathcal{A} \longrightarrow 0 \,. \tag{5.2}$$

■

Moreover, Proposition 2.4 implies:

Proposition 5.3. *Let $(p, \psi): (F, \mathcal{A}_F) \times (M, \mathcal{A}) \to (M, \mathcal{A})$ be a trivial superbundle. If $\pi_1 = (p_1, \psi_1): (F, \mathcal{A}_F) \times (M, \mathcal{A}) \to (F, \mathcal{A}_F)$ denotes the first projection, there is a canonical isomorphism*

$$\pi_1^* (\mathcal{D}er \mathcal{A}_F) \simeq \mathcal{V}er(\mathcal{B}) \,.$$

and the exact sequence (5.2) is induced by (5.1). ■

6. DeWitt supermanifolds

There is a class of supermanifolds, customarily called *DeWitt supermanifolds* [**DW**], which has found important applications in theoretical physics. This has happened both in a proper sense — just to cite a few examples, let us mention the formulation of field theories with BRST symmetry [**BoPT1,HQ2**], the anomaly problem in supersymmetric quantum field theory [**BoPT2,BBL,BruL**], and the introduction of super Riemann surfaces in superstring theory [**Fri,RSV1, RSV2**] — and implicitly: by this we mean that most of the work in theoretical physics involving 'superspaces' with non-trivial topology deals in fact with DeWitt supermanifolds.

These supermanifolds have a much simpler geometric structure than generic supermanifolds in that they are fibrations over smooth manifolds with contractible fibres. DeWitt supermanifolds are in many respects similar to graded manifolds, and it is indeed possible to establish a precise relationship between the two categories.

DeWitt supermanifolds are most conveniently defined by introducing in $B_L^{m,n}$ a topology τ_{DW} (called *DeWitt topology*), which is coarser than the usual Euclidean topology of $B_L^{m,n}$, and is indeed the coarsest topology such that the projection $\sigma^{m,n}: B_L^{m,n} \to \mathbb{R}^m$ is continuous. Therefore, the open sets in τ_{DW} have the form $V \times \mathfrak{N}_L^{m,n}$, where V is an open set in $\mathbb{R}^m$. The topological space $(B_L^{m,n}, \tau_{DW})$ is evidently not T_1, and therefore is neither Hausdorff nor paracompact. In a sense, the topology τ_{DW} is the most natural one for considering supersmooth functions, which always admit extensions to open sets of the type $V \times \mathfrak{N}_L^{m,n}$ (cf. Section III.2).

Definition 6.1. *A G-supermanifold $(M, \mathcal{A})$ is said to be DeWitt if it admits an atlas $\mathfrak{A} = \{(U_j, (\bar{\varphi}_j, \varphi_j))\}$, where the pair $(\bar{\varphi}_j, \varphi_j)$ is an isomorphism of graded locally ringed spaces*

$$(\bar{\varphi}_j, \varphi_j): (U_j, \mathcal{A}_{|U_j}) \to (\bar{\varphi}_j(U_j), \mathcal{G}_{|\bar{\varphi}_j(U_j)}), \tag{6.1}$$

such that the sets $\bar{\varphi}_j(U_j) \subset B_L^{m,n}$ are open in the DeWitt topology.

In the same way, we may define a supersmooth (i.e., G^∞ or GH^∞ or H^∞) De Witt supermanifold by repeating Definition III.2.1, but requiring that the images of the coordinate maps be open in the DeWitt topology. Quite trivially, the G^∞ supermanifold underlying a DeWitt G-supermanifold is DeWitt itself, and, conversely, the trivial extension (in the sense of Section III.4) of a GH^∞ or H^∞ DeWitt supermanifold is a DeWitt G-supermanifold.

We wish to show that any (m, n) dimensional DeWitt supermanifold $(M, \mathcal{A})$ intrinsically defines an m-dimensional differentiable manifold M_B, usually called the *body* of M, and that M is a locally trivial fibre bundle over M_B, with typical fibre $\mathfrak{N}_L^{m,n}$ [**Rs1,DW**].[3] We regard M as a G^∞ supermanifold, and consider on it a coarse G^∞ atlas, $\mathfrak{A} = \{(U_j, \varphi_j)\}$. We define in M the following relation:

$$p_1 \approx p_2 \quad \text{if} \quad p_1, p_2 \in U_j \quad \text{for some } j \text{ and} \quad \sigma^{m,n} \circ \varphi_j(p_1) = \sigma^{m,n} \circ \varphi_j(p_2).$$

It is not hard to see that this relation is independent of the choice of the index j, and is an equivalence relation. We can therefore take the quotient $M/\approx$; we denote by M_B the quotient topological space and by $\Phi: M \to M_B$ the (continuous) projection. Moreover, we set $W_j = \Phi(U_j)$ and define mappings $\psi_j: W_j \to \mathbb{R}^m$ by letting $\psi_j(\Phi(p)) = \sigma^{m,n} \circ \varphi_j(p)$. The atlas $\mathfrak{A}_B = \{(W_j, \psi_j)\}$ endows M_B with the structure of an m-dimensional smooth real manifold, and Φ is smooth. Simple routine checks show that the construction of the body manifold is independent of the coarse atlas originally chosen. In addition, since $\varphi_j(U_j) \simeq \psi_j(W_j) \times \mathfrak{N}_L^{m,n}$, it follows that M is a locally trivial fibre bundle over M_B with typical fibre $\mathfrak{N}_L^{m,n}$, as previously mentioned. It is easy to exhibit explicitly the transition functions of this bundle, that we denote by g_{jk}; for any $p \in W_j \cap W_k$, and $u \in \mathfrak{N}_L^{m,n}$, one has

$$g_{jk}(p)(u) = s \circ \varphi_j \circ \varphi_k^{-1}(\psi_k(p) + u)\,. \tag{6.2}$$

In order to check that these functions fulfill the cocycle condition, one needs to use the identity

$$\sigma^{m,n} \circ \varphi_j \circ \varphi_k^{-1}(z) = \psi_j \circ \psi_k^{-1}(\sigma^{m,n}(z))\,,$$

where $z \in \varphi_k(U_j \cap U_k)$. In general, the g_{jk}'s take values in $\mathrm{Diff}(\mathfrak{N}_L^{m,n})$ (the group of smooth diffeomorphisms of the standard fibre) and need not be linear, so that $\Phi: M \to M_B$ is not necessarily a vector bundle.

By means of the projection Φ we can introduce a coarse (DeWitt) topology in M as well: again, this is the coarsest topology such that Φ is continuous, that is, its open sets have the form $\Phi^{-1}(W)$, with $W \subset M_B$ open. Covers of M formed by sets which are open in the DeWitt topology will be called *coarse*.

[3] The notion of body of a supermanifold is more general, and applies to a wider category of supermanifolds than DeWitt ones [**BoyG,CaRT**].

Relationship between different categories of DeWitt supermanifolds. We have so far introduced a certain number of different kinds of DeWitt supermanifolds, i.e. we have defined objects of the DeWitt type within the category of G-supermanifolds and the various categories of supersmooth supermanifolds. Actually, it can be shown that these various kinds of DeWitt supermanifolds can be identified and, moreover, that DeWitt supermanifolds having a certain manifold as body are in a one-to-one correspondence with graded manifolds based on that manifold.

Using the tools that we have so far in our hands, we can only shed light on the relationship between H^∞ DeWitt supermanifolds and graded manifolds; a complete analysis of this issue requires some knowledge of the cohomology of DeWitt supermanifolds, and will therefore be postponed to Chapter V. The ideas of the following discussion are taken from [**Bch1,Bch2**].

We start by making an analogy with vector bundles. If X is a smooth manifold, and ξ a rank r vector bundle on it, the sheaf of sections of ξ locally has the form $\mathcal{C}_X^\infty \otimes \mathbf{R}^r$; in order to glue these sections to yield a globally defined sheaf, we need a Čech cocycle of the sheaf of smooth mappings from X into $\operatorname{Aut}\mathbf{R}^r \simeq Gl(r)$, i.e., a set of transition functions. Thus, the isomorphism classes of rank r smooth vector bundles over X are the elements of the first cohomology set $H^1(X, Gl(r))$ (cf. [**Hirz**]).[4] On the other hand, if we have a graded manifold $(X, \mathcal{F})$ of odd dimension r, there are local isomorphisms

$$\mathcal{F}_{|U} \simeq \Phi^*(\mathcal{C}_X^\infty{}_{|U} \otimes \textstyle\bigwedge \mathbf{R}^r),$$

and therefore (the equivalence classes of) graded manifolds of odd dimension r over X are classified by the cohomology set $H^1(X, \operatorname{Aut} \bigwedge \mathbf{R}^r)$.[5]

Let us now consider an H^∞ DeWitt supermanifold $(M, \mathcal{H}_M)$ of odd dimension r, with body M_B. Since there are local isomorphisms

$$\mathcal{H}_{M|U} \simeq \mathcal{C}^\infty_{M_B|\Phi(U)} \otimes \textstyle\bigwedge \mathbf{R}^r,$$

the isomorphism classes of H^∞ DeWitt supermanifolds of odd dimension r and body M_B are again in correspondence with the elements of $H^1(M_B, \operatorname{Aut} \bigwedge \mathbf{R}^r)$.

---

[4] Since $Gl(r)$ is not abelian, $H^1(X, Gl(r))$ is not a group, but only a pointed set; see [**Hirz**].

[5] Even though we shall not need this fact in the sequel, let us notice that Batchelor's theorem (Corollary III.1.9) implies an isomorphism $H^1(X, \operatorname{Aut} \bigwedge \mathbf{R}^r) \simeq H^1(X, Gl(r))$; a direct proof of this fact was given in [**Bch1**].

Thus, H^∞ DeWitt supermanifolds and graded manifolds with the same body and odd dimension are in a one-to-one correspondence.

We wish to make this correspondence more transparent, by contructing explicitly a graded manifold from an H^∞ DeWitt supermanifold, and *vice versa*.

Lemma 6.1. *If $(M, \mathcal{H}_M)$ is an H^∞ DeWitt supermanifold, with body projection $\Phi: M \to M_B$, the graded locally ringed space $(M_B, \Phi_* \mathcal{H}_M)$ is a graded manifold. Moreover, the spaces $(M, \mathcal{H}_M)$ and $(M_B, \Phi_* \mathcal{H}_M)$ determine the same element in $H^1(M_B, \operatorname{Aut} \bigwedge \mathbf{R}^r)$.*

Proof. Proposition III.2.2 implies that, for any suitable open set $W \subset M_B$, one has $(\Phi_* \mathcal{H}_M)_{|W} \simeq \mathcal{C}^\infty_{M_B\,|W} \otimes_{\mathbf{R}} \bigwedge_{\mathbf{R}} \mathbf{R}^n$; thus, local triviality is ensured. The augmentation map $\sim: \Phi_* \mathcal{H}_M \to \mathcal{C}^\infty_{M_B}$ is defined by letting $\tilde{f}(\Phi(p)) = \sigma \circ f(p)$, where $\sigma: B_L \to \mathbf{R}$ is the body map. The second part of the statement is apparent. ■

Now we construct a DeWitt supermanifold starting from a graded manifold. If $(X, \mathcal{F})$ is an (m, n) dimensional graded manifold, we consider on it an atlas $\widehat{\mathfrak{A}} = \{(W_j, \psi_j)\}$; if we denote $\psi_j = (x_j^1, \ldots, x_j^m, y_j^1, \ldots, y_j^n)$, the transition functions of $\widehat{\mathfrak{A}}$ have the expression

$$
\begin{aligned}
x_j^i &= \sum_{\mu \in \Xi'_n} \hat{\vartheta}^{i\mu}_{jk}(x_k^1, \ldots, x_k^m)\, y_k^\mu, \qquad & i = 1, \ldots, m, \\
y_j^\alpha &= \sum_{\mu \in \Xi''_n} \hat{\varpi}^{\alpha\mu}_{jk}(x_k^1, \ldots, x_k^m)\, y_k^\mu, \qquad & \alpha = 1, \ldots, n,
\end{aligned}
$$

where Ξ'_n (resp. Ξ''_n) is formed by the sequences in Ξ_n with an even (resp. odd) number of elements. The functions $\hat{\vartheta}^{i\mu}_{jk}$ and $\hat{\varpi}^{\alpha\mu}_{jk}$ are real-valued and are defined on the sets $\psi_k(W_j \cap W_k) \subset \mathbf{R}^m$. By Z-expanding them, we obtain H^∞ functions

$$
\vartheta^i_{jk} = Z_0(\hat{\vartheta}^i_{jk}), \qquad \varpi^{\alpha\mu}_{jk} = Z_0(\hat{\varpi}^{\alpha\mu}_{jk})
$$

defined on the sets $\psi_k(W_j \cap W_k) \times \mathfrak{N}_L^{m,0} \subset B_L^{m,0}$.

Lemma 6.2. *It is possible to associate with any graded manifold $(X, \mathcal{F})$ an H^∞ supermanifold $(M, \mathcal{H}_M)$ whose body manifold coincides with X, and is such that $\Phi_* \mathcal{H}_M \simeq \mathcal{F}$ (here $\Phi: M \to X$ is the body projection). Moreover, the manifolds $(X, \mathcal{F})$ and $(M, \mathcal{H}_M)$ determine the same element in $H^1(X, \operatorname{Aut} \bigwedge \mathbf{R}^r)$.*

Proof. With reference to the previous discussion, we define the H^∞ functions

$$
\varphi_{jk}: \psi_k(W_j \cap W_k) \times \mathfrak{N}_L^{m,n} \to \psi_j(W_j \cap W_k) \times \mathfrak{N}_L^{m,n},
$$

$$\varphi_{jk}^i(x^1,\ldots,x^m,y^1,\ldots,y^n) = \sum_{\mu\in\Xi_n'} \vartheta_{jk}^{i\mu}(x^1,\ldots,x^m)\,y^\mu, \quad i=1,\ldots,m \tag{6.3a}$$

$$\varphi_{jk}^\alpha(x^1,\ldots,x^m,y^1,\ldots,y^n) = \sum_{\mu\in\Xi_n''} \varpi_{jk}^{\alpha\mu}(x^1,\ldots,x^m)\,y^\mu, \quad \alpha=1,\ldots,n. \tag{6.3b}$$

The collection of these functions satisfies the cocycle condition $\varphi_{jk}\circ\varphi_{kh}=\varphi_{jh}$, and allows us to glue together, in the usual way, the sets $\psi_j(W_j)\times\mathfrak{N}_L^{m,n}$. In this manner, we obtain an H^∞ supermanifold, say M, with structure sheaf $\mathcal{H}_M$. The bodies of the transition functions (6.3) coincide with the transition functions of X, so that the body manifold M_B can be identified with X while, on the other hand, it is straightforward to show that $\Phi_*\mathcal{H}_M \simeq \mathcal{F}$ (again canonically). This fact entails the last statement in the thesis. ∎

Summing up, we may say that there is a one-to-one correspondence between isomorphism classes of (m,n) dimensional DeWitt supermanifolds whose body is a fixed smooth m-dimensional manifold X, and isomorphism classes of (m,n) dimensional graded manifolds over X. The explicit relationship between the two kinds of objects is established by Lemmas 6.1 and 6.2. Along the way we have also found the following result, that we would like to state explicitly.

Corollary 6.1. *There exists a one-to-one correspondence between isomorphism classes of DeWitt supermanifolds of odd dimension n, whose body is a fixed smooth manifold X, and isomorphism classes of rank n smooth vector bundles over X.*

Proof. This is implied by Proposition 6.2 together with Batchelor's theorem (Corollary III.1.9). ∎

A direct consequence of the results we have expounded so far is that any H^∞ DeWitt supermanifold admits atlases of a rather special kind (cf. [Rs4]).

Definition 6.2. *A coarse atlas $\mathfrak{A}=\{(U_j,\varphi_j)\}$ on an (m,n) dimensional DeWitt H^∞ supermanifold M is said to be split if — denoting $\varphi_j(p)=(x_j^1(p),\ldots,x_j^m(p),y_j^1(p),\ldots,y_j^n(p))$ — its transition functions have the form*

$$\begin{aligned} x_j^i &= \vartheta_{jk}^i(x_k^1,\ldots,x_k^m), & i=1,\ldots,m; \\ y_j^\alpha &= \sum_{\beta=1}^n \zeta_{jk}^{\alpha\beta}(x_k^1,\ldots,x_k^m)\,y_k^\beta, & \alpha=1,\ldots,n, \end{aligned} \tag{6.4}$$

where the functions ϑ_{jk}^i and $\zeta_{jk}^{\alpha\beta}$ are H^∞. ∎

In particular, the transition functions of a split atlas are such that the 'new' odd coordinates depend linearly on the 'old' odd coordinates, contrary to the general case described in Eq. (6.2). We say that a DeWitt supermanifold is *split* if it admits a split atlas. Given any H^∞ DeWitt supermanifold $(M, \mathcal{H}_M)$, the associated graded manifold can be endowed, as a consequence of Batchelor's theorem, with a split atlas (in a sense analogous to Definition 6.2). Then, the construction which led to Lemma 6.2 shows that $(M, \mathcal{H}_M)$ admits itself a split atlas. Therefore:

Corollary 6.3. *Any DeWitt supermanifold is split.* ■

We wish to point out once more that this result does not imply that the fibration $\Phi: M \to M_B$ is a vector bundle: indeed the transition functions (6.4) are not linear in the soul part of the even coordinates.

REMARK 6.1. It should be noticed that Corollaries 6.2 and 6.3 do not hold true in the complex analytic case. However, Batchelor's theorem can be generalized to that case in terms of a deformation theory à la Kodaira-Spencer (cf. Remark III.1.3). ▲

REMARK 6.2. Let M be an H^∞ DeWitt supermanifold of dimension (m, n); it is not hard to construct an $(m, 0)$ dimensional H^∞ DeWitt supermanifold M_0, together with a projection $\tau: M \to M_0$, and a rank $(0, n)$ H^∞ supervector bundle on it, $p: E \to M_0$, such that $M \simeq E$ as H^∞ supermanifolds. Moreover, there are canonical isomorphisms $(TM)_0 \simeq \tau^{-1}TM_0$ and $(TM)_1 \simeq \tau^{-1}E$. ▲

7. Rothstein's axiomatics

In Sections 2 to 4 of Chapter III we have described an approach to supermanifolds essentially due to De Witt and Rogers. We have also discussed some inadequacies of their proposal, and, eventually, have suggested a modification of their approach, which aims at disposing of some undesirable features of their model. In an interesting paper [**Rt2**], Rothstein dealt with the same problem. In his paper, the terms of the question are turned upside-down, in the sense that the required properties are imposed as axioms; contact with the usual approaches is gained by means of a series of theorems.

Although [**Rt2**] contains some inexactnesses, as we shall comment presently, the framework presented in that paper appears to be very convenient for discussing certain general features of supermanifold theory. Also, it turns out that

G-supermanifolds are a particular case of Rothstein supermanifolds (which we shall call *R-supermanifolds* for brevity). More precisely, we can prove that G-supermanifolds are exactly those R-supermanifolds based on the graded algebra B_L whose rings of sections are topologically complete.

In order to state Rothstein's axioms, the following objects are needed:

1. a Hausdorff, paracompact space M;
2. a graded-commutative Banach algebra B;
3. a sheaf $\mathcal{A}$ on M of graded-commutative B-algebras with identity;
4. an 'evaluation' morphism $\delta\colon \mathcal{A} \to \mathcal{C}_B$, where $\mathcal{C}_B$ is the sheaf of continuous B-valued functions on M.

Furthermore, we denote by $\mathcal{D}er^*\mathcal{A}$ the dual sheaf to $\mathcal{D}er\,\mathcal{A}$, i.e. the sheaf $\mathcal{D}er^*\mathcal{A} = \mathcal{H}om_{\mathcal{A}}(\mathcal{D}er\,\mathcal{A}, \mathcal{A})$. A morphism of sheaves of graded B-modules $d\colon \mathcal{A} \to \mathcal{D}er^*\mathcal{A}$ (exterior differential) is defined as usual by letting

$$df(D) = (-1)^{|f|\,|D|}\, D(f)$$

for all homogeneous $f \in \mathcal{A}(U)$, $D \in \mathcal{D}er\,\mathcal{A}(U)$ and all open $U \subset M$ (cf. Section 4).

Let (m, n) be fixed, nonnegative integers. The triple $(M, \mathcal{A}, \delta)$ is said to be an (m, n) dimensional R-*supermanifold* if the following four axioms are satisfied.

Axiom 1. *$\mathcal{D}er^*\mathcal{A}$ is a locally free $\mathcal{A}$-module of rank (m, n). Any $z \in M$ has an open neighbourhood U with sections $x^1, \ldots, x^m \in \mathcal{A}(U)_0$, $y^1, \ldots, y^n \in \mathcal{A}(U)_1$ such that $\{dx^1, \ldots, dx^m, dy^1, \ldots, dy^n\}$ is a graded basis of $\mathcal{D}er^*\mathcal{A}(U)$.*

The collection $(U, (x^1, \ldots, x^m, y^1, \ldots, y^n))$ is called a *coordinate chart* for the supermanifold. This axiom evidently implies that $\mathcal{D}er\,\mathcal{A}$ is locally free of rank (m, n), and is locally generated by the derivations $\dfrac{\partial}{\partial x^i}, \dfrac{\partial}{\partial y^\alpha}$ defined by duality with the dx^i's and dy^α's.

Let us denote by a tilde the action of the evaluation morphism δ, i.e. $\tilde{f} = \delta(f)$.

Axiom 2. *If $(U, (x^1, \ldots, x^m, y^1, \ldots, y^n))$ is a coordinate chart, the mapping*

$$\begin{aligned} \psi\colon U &\to B^{m,n} \\ z &\mapsto (\tilde{x}^1(z), \ldots, \tilde{x}^m(z), \tilde{y}^1(z), \ldots, \tilde{y}^n(z)) \end{aligned}$$

is a homeomorphism onto an open set in $B^{m,n}$.

Axiom 3. (Existence of Taylor expansion) *Let* $(U,(x^1,\ldots,x^m,y^1,\ldots,y^n))$ *be a coordinate chart. For any* $z \in U$ *and any* $f \in \mathcal{A}_z$ *there are germs* $g_1,\ldots,g_m,h_1,\ldots,h_n \in \mathcal{A}_z$ *such that*

$$f = \tilde{f}(z) + \sum_{i=1}^{m} g_i\,(x^i - \tilde{x}^i(z)) + \sum_{\alpha=1}^{n} h_\alpha\,(y^\alpha - \tilde{y}^\alpha(z)). \tag{7.1}$$

Perhaps this axiom needs some explanation. Since any $\mathcal{A}(U)$ has the unit section, there is in an injection $B \hookrightarrow \mathcal{A}_z$ for all $z \in U$, and this permits us to regard the values $\tilde{f}(z)$, $\tilde{x}^i(z)$, $\tilde{y}^\alpha(z)$ as germs in $\mathcal{A}_z$. Moreover, in the case of smooth functions, Eq. (7.1) would be no more than a transcription in sheaf-theoretic language of the zeroth-order Taylor formula with a Lagrange remainder.

Axiom 4. *Let* $\mathcal{D}(\mathcal{A})$ *denote the graded* $\mathcal{A}$*-module generated multiplicatively by* $\mathcal{D}er\,\mathcal{A}$ *over* $\mathcal{A}$*, i.e. the sheaf of differential operators over* $\mathcal{A}$*, and let* $f \in \mathcal{A}_z$*, with* $z \in M$*. If* $L(f) = 0$ *for all* $L \in \mathcal{D}(\mathcal{A})_z$*, then* $f = 0$.

Definition 7.1. *A morphism of R-supermanifolds is a graded ringed space morphism* $(f,\psi)\colon(M,\mathcal{A},\delta) \to (N,\mathcal{B},\delta')$ *such that there is a commutative diagram*

$$\begin{array}{ccc} \mathcal{B} & \xrightarrow{\psi} & f_*\mathcal{A} \\ {\scriptstyle\delta}\big\downarrow & & \big\downarrow{\scriptstyle\delta'} \\ \mathcal{C}_N & \xrightarrow[f^*]{} & f_*\mathcal{C}_M \end{array},$$

where the $\mathcal{C}$*'s are sheaves of continuous functions on the relevant manifolds.*

We wish to state some further properties of R-supermanifolds that will be recalled in the sequel. For a proof, the reader may refer to [**Rt2**].

Proposition 7.1. *Let* $(M,\mathcal{A},\delta)$ *be an* (m,n) *dimensional R-supermanifold, and let* $(U,(x^1,\ldots,x^m,y^1,\ldots,y^n))$ *be a coordinate chart on it. Let us define* $\hat{A}_U$ *as the subsheaf of* $\mathcal{A}_{|U}$ *whose sections do not depend on the odd variables, in the sense that*

$$\hat{A}_U = \left\{ f \in A(U) \mid \frac{\partial f}{\partial y^\alpha} = 0, \quad \alpha = 1,\ldots,n \right\}.$$

There is an isomorphism

$$\hat{\mathcal{A}}_U \otimes_{\mathbf{R}} \bigwedge_{\mathbf{R}} \mathbf{R}^n \to \mathcal{A}_{|U}\,,$$

having identified $\bigwedge_{\mathbf{R}} \mathbf{R}^n$ *with the Grassmann algebra generated by the* y*'s. Moreover, the restriction of* δ *to* $\hat{\mathcal{A}}_U$ *is injective.*

Proposition 7.2. *Let* $(M, \mathcal{A}, \delta)$ *be an* (m, n) *dimensional R-supermanifold, and let* $\mathcal{K} = \operatorname{Ker}\delta$ *be the kernel of the evaluation morphism. Then* $\mathcal{K}$ *is the ideal of* $\mathcal{A}$ *whose sections on an open subset* $U \subset M$ *are the graded functions* $f \in \mathcal{A}(U)$ *such that* $f g_1 \cdots g_n = 0$ *for every choice of* $g_1, \ldots, g_n \in \mathcal{A}_1$.

Proposition 7.3. *Let* $(M, \mathcal{A}, \delta)$ *be an* (m, n) *dimensional R-supermanifold, and, for any* $z \in M$, *let* $T_z M = \operatorname{Der}_{B_L}(\mathcal{A}_z, B_L)$ *be the* B_L*-module whose elements are the graded derivations* $X\colon \mathcal{A}_z \to B_L$. *Then* $T_z M$ *is a free rank* (m, n) *graded* B_L*-module and the elements* $(\frac{\partial}{\partial x^i})_z$, $(\frac{\partial}{\partial y^\alpha})_z$ *defined by*

$$\left(\frac{\partial}{\partial x^i}\right)_z (f) = \widetilde{\frac{\partial f}{\partial x^i}}(z), \qquad \left(\frac{\partial}{\partial y^\alpha}\right)_z (f) = \widetilde{\frac{\partial f}{\partial y^\alpha}}(z) \qquad \text{for all} \quad f \in \mathcal{A}_z,$$

yield a graded basis for $T_z M$. *Finally, there is a canonical isomorphism of graded* B_L*-modules*

$$T_z M \simeq (\mathcal{D}er\,\mathcal{A})_z / (\mathfrak{L}_z \cdot (\mathcal{D}er\,\mathcal{A})_z)\,, \tag{7.2}$$

where $\mathfrak{L}_z$ *is the ideal of germs in* $\mathcal{A}_z$ *which vanish when evaluated, i.e.*

$$\mathfrak{L}_z = \{f \in \mathcal{A}_z \mid \tilde{f}(z) = 0\}.$$

Some comments on Rothstein's axiomatics. It is convenient to restate Rothstein's axiomatics in a slight different manner, more suitable for dealing with the question of topological completeness of the rings of sections of $\mathcal{A}$. Let us consider $(M, \mathcal{A}, \delta)$ as above; that is, M is a Hausdorff, paracompact space; $\mathcal{A}$ is a sheaf on M of graded-commutative B-algebras with identity and $\delta\colon \mathcal{A} \to \mathcal{C}_B$ is an evaluation morphism.

Axiom 3 can obviously be reformulated as follows:

Let $(U, (x^1, \ldots, x^m, y^1, \ldots, y^n))$ *be a coordinate chart. For any* $z \in U$ *the ideal* $\mathfrak{L}_z$ *of* $\mathcal{A}_z$ *is generated by* $\{x^1 - \tilde{x}^1(z), \ldots, x^m - \tilde{x}^m(z), y^1 - \tilde{y}^1(z), \ldots, y^n - \tilde{y}^n(z)\}$.

Axiom 1 allows us to replace this axiom by a weaker requirement; for this we need some preliminary discussion.

Lemma 7.1. *There is an isomorphism of $\mathcal{A}_z/\mathfrak{L}_z$-modules*

$$\begin{aligned} \mathfrak{L}_z/\mathfrak{L}_z^2 &\simeq \mathcal{D}er^*\mathcal{A}_z \otimes_{\mathcal{A}_z} \mathcal{A}_z/\mathfrak{L}_z \\ \bar{f} &\mapsto df \otimes 1 \end{aligned} \tag{7.3}$$

where, as usual, the bar means the class in the quotient.

Proof. It can be shown easily that $df \otimes \bar{g} \mapsto \overline{(f - \tilde{f}(z))g}$ defines a morphism $\mathcal{D}er^*\mathcal{A}_z \otimes_{\mathcal{A}_z} \mathcal{A}_z/\mathfrak{L}_z \to \mathfrak{L}_z/\mathfrak{L}_z^2$ which inverts the previous one. ∎

REMARK 7.1. In the category of graded manifolds — which are, as we shall prove shortly, R-supermanifolds — the right-hand side of Eq. (7.3) represents no more than the cotangent space at z, so that Eq. (7.3) can be thought of as the dual of the isomorphism (7.2). The same happens in the case of G-supermanifolds. ▲

If we denote by $d_z f$ the class in $\mathfrak{L}_z/\mathfrak{L}_z^2$ of the element $f - \tilde{f}(z) \in \mathfrak{L}_z$, Axiom 1 for $(M, \mathcal{A}, \delta)$ implies that — given a coordinate chart $(U, (x^1, \ldots, x^m, y^1, \ldots, y^n))$ — the elements $\{d_z x^i,\ d_z y^\alpha\}$ are a basis for the $\mathcal{A}_z/\mathfrak{L}_z$-module $\mathfrak{L}_z/\mathfrak{L}_z^2$.

Let us suppose furthermore that the rings $\mathcal{A}_z$ are local for every $z \in M$; that is, that $(M, \mathcal{A})$ is a *graded locally ringed space* (Definition II.4.1), as is the case in most interesting examples. Thus, any graded ideal of $\mathcal{A}_z$ is contained in its radical,[6] and hence one can apply Lemma I.1.1 (graded Nakayama's lemma) to obtain that — if $\mathfrak{L}_z$ is finitely generated — the elements $\{x^i - \tilde{x}^i(z),\ y^\alpha - \tilde{y}^\alpha(z)\}$ will be generators of $\mathfrak{L}_z$ if and only if their classes $\{d_z x^i,\ d_z y^\alpha\}$ generate the $\mathcal{A}_z/\mathfrak{L}_z$-module $\mathfrak{L}_z/\mathfrak{L}_z^2$. That is to say, we have proved — under the hypothesis that the rings $\mathcal{A}_z$ are local — the following result.

Lemma 7.2. *Axiom 1 implies Axiom 3 provided that the ideals $\mathfrak{L}_z$ are finitely generated.* ∎

We are thus led to consider the apparently weaker axiom:

Axiom 3′. *For every $z \in M$ the ideal $\mathfrak{L}_z$ is finitely generated.*

[6] This because any graded ideal is contained in a maximal graded ideal. Proof of this statement, which makes use of Zorn's lemma, goes as in the non-graded case [**AtM**].

It is an important fact that Axiom 3′ does not depend on the choice of a coordinate chart. Thus, while in order to check Axiom 3 one has to prove the existence of a Taylor expansion for any coordinate chart, if $(M, \mathcal{A})$ is a graded locally ringed space it is sufficient to show that there is one coordinate chart for which a Taylor expansion exists.

We can summarize this discussion as follows.

Proposition 7.4. *If an R-supermanifold is also a graded locally ringed space, we can replace Axiom 3 by Axiom 3′.* ■

Comparison of Rothstein and supersmooth supermanifolds. None of the classes of supersmooth functions introduced in Definition III.2.1 yields a category of supermanifolds satisfying this axiomatics. In particular, G^∞ supermanifolds do not fulfill Proposition 7.1 (cf. Proposition III.2.2) and in fact they violate Axiom 1. As far as GH^∞ supermanifolds, and the particular case of H^∞ ones, are concerned, neither do they contain the necessary ingredients for producing an R-supermanifold. Indeed, in this case one should choose $B \equiv B_{L'}$, although then the evaluation of a GH^∞ function is B_L-valued, and not $B_{L'}$-valued; *vice versa*, if one sets $B \equiv B_L$, one should regard $\mathcal{GH}_{L'}$ as a sheaf of B_L-algebras. This can be done, for if $p_{L'}: B_L \to B_{L'}$ is the projection of B_L onto $B_{L'}$ obtained by suppressing the extra generators of B_L, the latter can be made into a B_L-module by letting $a \cdot b = p_{L'}(a)\, b$ for all $a \in B_{L'}$ and $b \in B_L$. In this way $\mathcal{GH}_{L'}$ becomes a sheaf of graded B_L-algebras, but the natural evaluation morphism (the identity) is not a morphism of B_L-algebras.

On the other hand, R-supermanifolds turn out to be — whenever we choose $B = B_L$, and an extra axiom is imposed — an extension of G^∞ supermanifolds, in a sense to be specified later. In order to motivate the introduction of a further axiom, let us point out that, contrary to what is claimed in [**Rt2**], it is not true that the image of the evaluation morphism endows M with a structure of a G^∞ supermanifold. Indeed, given an R-supermanifold $(M, \mathcal{A}, \delta)$ over B_L, the sheaf $\mathcal{A}$ is not topologically complete with respect to the even coordinates. The following Example should clarify what we mean.

EXAMPLE 7.1. Let us take $B = \mathbf{R}$, $n = 0$ and $M = \mathbf{R}^m$. If we consider the sheaf $\mathcal{A} = \mathbf{R}[x^1, \dots, x^m]$ of polynomial functions on $\mathbf{R}^m$ and the trivial evaluation morphism $\delta: \mathcal{A} \hookrightarrow \mathcal{C}_{\mathbf{R}}$, $\delta(f) = f$, then $(M, \mathcal{A}, \delta)$ is an R-supermanifold of dimension $(m, 0)$. However $(M, \delta(\mathcal{A})) = (M, \mathbf{R}[x^1, \dots, x^m])$ is certainly not an $(m, 0)$-dimensional G^∞ supermanifold, which in this case would be no more that an m-dimensional smooth manifold. ▲

More generally, if $(M, \mathcal{A}, \delta)$ is an R-supermanifold with $B = B_L$, the sheaf $\delta(\mathcal{A})$ is a subsheaf of the sheaf of G^∞ functions on M, although it may not include all of them. In order to ensure that $(M, \delta(\mathcal{A}))$ is a G^∞ supermanifold, a further axiom must be imposed; that is to say, Rothstein's axiomatics is too general to single out a class of supermanifolds extending ordinary smooth manifolds.

Let us go back to the abstract setting, and consider an R-supermanifold $(M, \mathcal{A}, \delta)$ of dimension (m, n) over a Banach algebra B. Then, if $\| \ \|$ denotes the norm in B, the rings of sections $\mathcal{A}(U)$ of $\mathcal{A}$ on every open subset $U \subset M$ can be topologized by means of the seminorms $p_{L,K}: \mathcal{A}(U) \to \mathbf{R}$ defined by

$$p_{L,K}(f) = \max_{z \in K} \left\| (\widetilde{L(f)})(z) \right\|$$

where L runs over the differential operators of $\mathcal{A}$ on U, and $K \subset U$ is compact. As a consequence of the axioms, the family of seminorms $p_{D_{J,\mu},K}$, where K is a compact subset of a coordinate chart $(W, (x^1, \ldots, x^m, y^1, \ldots, y^n))$ $(W \subset U)$ and $D_{J,\mu} = \left(\frac{\partial}{\partial x}\right)^J \left(\frac{\partial}{\partial y}\right)_\mu$ (see Remark III.1.1 for notation), defines a topology of $\mathcal{A}(U)$, thus endowing it with a structure of locally convex metrizable graded algebra (in fact, Axiom 4 means that $\mathcal{A}(U)$ is Hausdorff).

We are therefore led to introducing the following supplementary axiom.

Axiom 5. (Completeness) *For every open subset $U \subset M$, the space $\mathcal{A}(U)$ is complete with respect to the above topology.*

Axioms 4 and 5, taken together, are equivalent to still another axiom:

Axiom 6. *For every open subset $U \subset M$, the space $\mathcal{A}(U)$ is a graded Fréchet algebra.*

Definition 7.2. *An R^∞-supermanifold over B is an R-supermanifold $(M, \mathcal{A}, \delta)$ over B, additionally satisfying Axiom 5; or, equivalently, it is a triple $(M, \mathcal{A}, \delta)$ fulfilling Axioms 1, 2, 3 and 6.*

If $(M, \mathcal{A}, \delta)$ is an R-supermanifold, and $(U, (x^1, \ldots, x^m, y^1, \ldots, y^n))$ is a coordinate chart, then the algebraic isomorphism $\hat{\mathcal{A}}(U) \otimes \bigwedge_{\mathbf{R}} \mathbf{R}^n \simeq \mathcal{A}(U)$ provided by Proposition 7.1 is a metric isomorphism, when $\hat{\mathcal{A}}(U)$ is endowed with the the induced topology. Thus, $\mathcal{A}(U)$ is complete if and only if $\hat{\mathcal{A}}(U)$ is also complete.

Definition 7.3. *A morphism of R^∞-supermanifolds is a morphism of R-supermanifolds $(f,\psi):(M,\mathcal{A},\delta) \to (N,\mathcal{B},\delta')$ such that $\psi_V:\mathcal{B}(V) \to f_*\mathcal{A}(V)$ is continuous for every open subset $V \subset N$.*

The case $B = B_L$. We will show that whenever the ground algebra B is taken as B_L, R^∞-supermanifolds coincide with the G-supermanifolds previously introduced; in fact, the standard model for R^∞-supermanifolds is simply the standard G-supermanifold over $B_L^{m,n}$.

Proposition 7.5. *The triple $(B_L^{m,n},\mathcal{G},\delta)$, where $(B_L^{m,n},\mathcal{G})$ is the standard G-supermanifold over $B_L^{m,n}$ and $\delta:\mathcal{G} \to \mathcal{C}_L^\infty$ is the usual evaluation morphism, $\delta(f \otimes a) = fa$, is an R^∞-supermanifold.*

Proof. Axiom 1 is Proposition III.4.3. Axiom 2 is obviously fulfilled. On the other hand, since $(B_L^{m,n},\mathcal{G})$ is a graded locally ringed space, in view of Proposition 7.4 it suffices to prove Axiom 3 only for one coordinate chart; e.g., for the natural one. Axiom 3 thus ensues from the Taylor expansion for the functions in $\hat{\mathcal{G}}^\infty$ (Proposition III.2.3). In order to prove Axiom 4, let $U \subset B_L^{m,n}$ be an open set, and let $\sum_{\mu\in\Xi_n} f_\mu \otimes y^\mu \in \mathcal{G}(U)$. If $\delta(D_1 \cdots D_p(\sum_{\mu\in\Xi_n} f_\mu \otimes y^\mu)) = 0$ for arbitrary $D_1,\ldots,D_p \in \mathcal{D}er\mathcal{G}(U)$, then $f_\mu \in \operatorname{Ker}\delta$. Since δ is injective on elements in $\hat{\mathcal{G}}$, it follows that $f_\mu = 0$, i.e. Axiom 4 is satisfied. Axiom 5 is Proposition III.4.5. ∎

The following result is analogous to Lemma 1.1.

Lemma 7.3. *If $(f,\phi):(M,\mathcal{A},\delta) \to (U,\mathcal{G},\delta)$ and $(f,\phi'):(M,\mathcal{A},\delta) \to (U,\mathcal{G},\delta)$ are morphisms of R^∞-supermanifolds, and $\phi(x^i) = \phi'(x^i)$ for $i = 1,\ldots,m$, $\phi(y^\alpha) = \phi'(y^\alpha)$ for $\alpha = 1,\ldots,n$, then $\phi = \phi'$.* ∎

Coordinate charts and automorphisms of $(B_L^{m,n},\mathcal{G})$ regarded as an R^∞-supermanifold are described by the following Lemma.

Lemma 7.4. *Let $U \subset B_L^{m,n}$ be an open subset.*

(1) *A family of sections $(u^1,\ldots,u^m,v^1,\ldots,v^n)$ of $\mathcal{G}$ is a coordinate system for $\mathcal{G}_{|U}$ as an R-supermanifold if and only the evaluations $(\tilde{u}^1,\ldots,\tilde{u}^m,\tilde{v}^1,\ldots,\tilde{v}^n)$ yield a G^∞ coordinate system.*

(2) *Let $\psi = (u^1,\ldots,u^m,v^1,\ldots,v^n)$ be a new coordinate system for $\mathcal{G}_{|U}$ and let $\tilde{\psi}:U \to W \subset B_L^{m,n}$ be the induced homeomorphism $z \mapsto (\tilde{u}^1(z),\ldots,\tilde{u}^m(z),\tilde{v}^1(z),\ldots,\tilde{v}^n(z))$. There exists a unique isomorphism of R^∞-supermanifolds*

$$(f,\phi):(U,\mathcal{G}_{|U},\delta) \to (W,\mathcal{G}_{|W},\delta)$$

such that $\phi(x^i) = u^i$ *for* $i = 1, \ldots, m$, *and* $\phi(y^\alpha) = v^\alpha$ *for* $\alpha = 1, \ldots, n$.

(3) *If* $V \subset B_L^{m,n}$ *is another open subset, every isomorphism* $(f, \chi)\colon (U, \mathcal{G}_{|U}) \cong (V, \mathcal{G}_{|V})$ *of* G^∞ *supermanifolds can be extended (in many ways) to an isomorphism of* R^∞*-supermanifolds* $(f, \phi)\colon (U, \mathcal{G}_{|U}) \cong (V, \mathcal{G}_{|V})$. *Here 'extension' means that the diagram*

$$\begin{array}{ccc} \mathcal{G}_{|V} & \xrightarrow{\phi} & f_*\mathcal{G}_{|U} \\ {\scriptstyle\delta}\downarrow & & \downarrow{\scriptstyle\delta} \\ \mathcal{G}^\infty{}_{|V} & \xrightarrow[\chi]{} & f_*\mathcal{G}^\infty{}_{|U} \end{array}$$

commutes.

Proof. (1) Since $\operatorname{Ker}\delta$ is nilpotent, a matrix of sections of $\mathcal{G}$ is invertible if and only if its evaluation is invertible as well, thus proving the statement.

(2) One can define a ring morphism

$$\phi\colon B_L[x^1, \ldots, x^m] \otimes \bigwedge\langle y^1, \ldots, y^n\rangle \to f_*\mathcal{G}$$

by imposing that $\phi(x^i) = u^i$, $\phi(y^\alpha) = v^\alpha$ for $i = 1, \ldots, m$, $\alpha = 1, \ldots, n$. Since the topology of $\mathcal{G}$ can be described by the seminorms associated with any coordinate chart, ϕ is continuous and therefore induces a morphism between the completions:

$$\phi\colon \mathcal{G} \to f_*\mathcal{G}\,.$$

To see that (f, ϕ) is an isomorphism, we can construct, by the same procedure, a morphism $(f, \psi)\colon (B_L^{m,n}, \mathcal{G}, \delta) \to (U, \mathcal{G}, \delta)$ such that $\psi(u^i) = x^i$, $\psi(v^\alpha) = y^\alpha$ for $i = 1, \ldots, m$, $\alpha = 1, \ldots, n$. In this way, we have two morphisms of R^∞-supermanifolds $(\mathrm{Id}, \mathrm{Id}), (\mathrm{Id}, \phi \circ \psi)\colon (B_L^{m,n}, \mathcal{G}, \delta) \to (B_L^{m,n}, \mathcal{G}, \delta)$ which coincide on a coordinate system, thus finishing the proof by the previous Lemma.

(3) Follows from (1) and (2) since a G^∞ isomorphism transforms G^∞ coordinate systems into G^∞ coordinate systems. ∎

Having introduced the local model of R^∞-supermanifolds, these can be characterized as graded ringed spaces. Moreover, one can show that any R^∞-supermanifold has an underlying G^∞ supermanifold. To this end we need a preliminary result (cf. [**Rt2**]).

Lemma 7.5. *Let* $(M, \mathcal{A}, \delta)$ *be an* (m, n) *dimensional R-supermanifold, and let* (U, φ) *be a local chart for it (i.e.,* $\varphi = (x^1, \ldots, x^m, y^1, \ldots, y^n)$ *is a coordinate* *system on U). For all $f \in \mathcal{A}(U)$, the composition $\tilde{f} \circ \tilde{\varphi}^{-1}$ is a G^∞ function on $\tilde{\varphi}(U) \subset B_L^{m,n}$.*

Proof. Denote $g = \tilde{f} \circ \tilde{\varphi}^{-1}$, and, if $z, w \in U$, denote $a = \tilde{\varphi}(z)$, $b = \tilde{\varphi}(w)$. By Axiom 3, there exist germs $f_A \in \mathcal{A}_z$, with $A = 1, \ldots, m+n$, such that

$$f_z = \tilde{f}(z) + \sum_{i=1}^{m} f_i(x^i - a^i) + \sum_{\alpha=1}^{n} f_\alpha(y^\alpha - a^{m+\alpha}),$$

so that

$$\delta\left(\frac{\partial f}{\partial x^i}\right)(x) = \tilde{f}_i(z), \qquad \delta\left(\frac{\partial f}{\partial y^\alpha}\right)(x) = (-1)^{|f|}\tilde{f}_\alpha(z).$$

Now setting $g_A = \tilde{f}_A \circ \tilde{\varphi}^{-1}$, and applying Axiom 3 again, we can introduce continuous functions g_{AB} such that

$$g(b) = g(a) + \sum_{A=1}^{m+n} g_A(a)\,(b-a)^A + \sum_{A,B=1}^{m+n} g_{AB}(a)\,(b-a)^A\,(b-a)^B\,.$$

It follows that g is of class C^1, and that its Fréchet differential is the $(B_L)_0$-linear operator

$$Dg_a(c) = \sum_{A=1}^{m+n} g_A(a)\,c^A, \quad \text{with} \quad c \in B_L^{m,n}\,. \tag{7.4}$$

By applying the same argument to the partial derivatives $\dfrac{\partial f}{\partial x^i}, \dfrac{\partial f}{\partial y^\alpha}$, we can prove that g is smooth. If g depends only on even variables, then Eq. (7.4) implies, through Proposition II.2.4, that g is G^∞. If g also depends on the odd coordinates y, then, in view of Proposition 7.1, it can be written as $g(x,y) = \sum_{\mu\in\Xi_n} g_\mu(x)y^\mu$, where the g_μ's are G^∞, so that g is also G^∞. ■

Proposition 7.6. *Let $(M, \mathcal{A}, \delta)$ be an (m,n)-dimensional R^∞-supermanifold $(M, \mathcal{A}, \delta)$, with $B = B_L$; then:*

(1) *the pair $(M, \mathcal{A}^\infty)$, with $\mathcal{A}^\infty = \delta(\mathcal{A})$, is a G^∞ supermanifold;*
(2) *$(M, \mathcal{A}, \delta)$ is locally isomorphic, as an R^∞-supermanifold, with the G-supermanifold $(B_L^{m,n}, \mathcal{G}, \delta)$.*

Proof. Let (U, φ) be a coordinate chart for $(M, \mathcal{A}, \delta)$, with $\varphi = (x^1, \ldots, x^m, y^1, \ldots, y^n)$. We should recall that, if $\hat{\mathcal{A}}_\varphi$ is the subsheaf of $\mathcal{A}_{|U}$ whose sections does not depend on the y's, then

$$\mathcal{A}_{|U} \simeq \hat{\mathcal{A}}_\varphi \otimes_{\mathbf{R}} \wedge \mathbf{R}^n\,.$$

We define an injection

$$\hat{T}_\varphi : \hat{\mathcal{A}}_\varphi \hookrightarrow \tilde{\varphi}^{-1}\hat{\mathcal{G}}_{|\tilde{\varphi}(U)}$$

by letting $\hat{T}_\varphi(f) = \tilde{f} \circ \tilde{\varphi}^{-1}$; by Lemma 7.5, $\hat{T}_\varphi(f)$ is a G^∞ function and therefore is a section of $\tilde{\varphi}^{-1}\hat{\mathcal{G}}_{|\tilde{\varphi}(U)}$. Furthermore, $\hat{T}_\varphi$ is a metric isomorphism with its image, so that $\hat{T}_\varphi(\hat{\mathcal{A}}_\varphi)$ is complete. Since $\hat{T}_\varphi(\hat{\mathcal{A}}_\varphi)$ is complete and contains the G^∞ functions which are polynomials in the even coordinates, it contains all the G^∞ functions, that is, $\hat{T}_\varphi$ is an isomorphism. The morphism $\hat{T}_\varphi$ determines a metric isomorphism

$$T_\varphi : \mathcal{A}_{|U} \to \tilde{\varphi}^{-1}\mathcal{G}_{|\tilde{\varphi}(U)}$$

simply by letting $T_\varphi(\sum f_\mu \otimes y^\mu) = \sum \hat{T}_\varphi(f_\mu) \otimes y^\mu$. Now, by means of the diagram

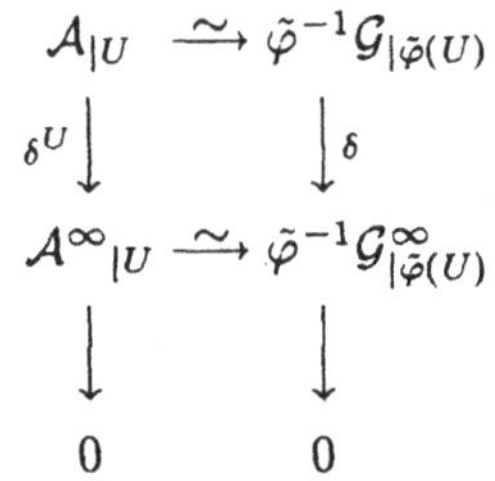

$$\begin{array}{ccc} \mathcal{A}_{|U} & \xrightarrow{\sim} & \tilde{\varphi}^{-1}\mathcal{G}_{|\tilde{\varphi}(U)} \\ \downarrow \delta^U & & \downarrow \delta \\ \mathcal{A}^\infty{}_{|U} & \xrightarrow{\sim} & \tilde{\varphi}^{-1}\mathcal{G}^\infty_{|\tilde{\varphi}(U)} \\ \downarrow & & \downarrow \\ 0 & & 0 \end{array}$$

we simultaneously prove both results. ∎

The second statement of Proposition 7.6 has a converse, so that R$^\infty$-supermanifolds (but not all R-supermanifolds, as erroneously claimed in [**Rt2**]) can be characterized in terms of their local model.

Proposition 7.7. *A triple $(M, \mathcal{A}, \delta')$, where M is a Hausdorff paracompact space, $\mathcal{A}$ is a sheaf on M of graded-commutative B_L-algebras with identity and $\delta' : \mathcal{A} \to \mathcal{C}_B$ is an evaluation morphism, is a (m,n)-dimensional R^∞-supermanifold if and only if it is locally isomorphic with the G-supermanifold $(B_L^{m,n}, \mathcal{G}, \delta)$. This means that for every point $z \in M$ there is an open neighbourhood $U \subset M$ and an isomorphism of graded ringed spaces $(f, \phi) : (U, \mathcal{A}_{|U}) \xrightarrow{\sim} (f(U) \subset B_L^{m,n}, \mathcal{G}_{|f(U)})$ such that $\delta' \circ \phi = f^* \circ \delta$.* ∎

Corollary 7.1. *The category of R^∞-supermanifolds over B_L and the category of G-supermanifolds are equivalent.*

A consequence of this Corollary is that G-supermanifolds can be characterized by Axioms 1, 2, 3′ and 6. Moreover, it should be noticed that the continuity requirement in Definition 7.3 is *a posteriori* redundant, in view of Proposition 1.1.

One should of course check that these axioms are actually independent; the only non-trivial thing to prove is that Axiom 3′ does not follow from Axioms 1, 2 and 6. This has been shown by an example in [**BBHP1,BBHP2**].

Extending G^∞ supermanifolds to G-supermanifolds. A question which arises naturally is whether, given a G^∞ supermanifold M with structure sheaf $\mathcal{A}^\infty$, there exists a G-supermanifold $(M, \mathcal{A}, \delta)$ which extends $(M, \mathcal{A}^\infty)$, in the sense that $\mathcal{A}^\infty = \delta(\mathcal{A})$. This problem has been dealt with in [**Rt2**], of course without any mention to G-supermanifolds; here we wish to report the results obtained there, filling in many details that in [**Rt2**] have been passed by.

If such an extension exists, one has an exact sequence

$$0 \to \mathcal{K} \to \mathcal{A} \to \mathcal{A}^\infty \to 0\,, \tag{7.5}$$

where $\mathcal{K}$ is by definition the kernel of δ. The important fact is that, provided that $L \geq n$, the kernel $\mathcal{K}$ is determined by $\mathcal{A}^\infty$. Then the condition for $(M, \mathcal{A}^\infty)$ to admit an extension by a G-supermanifold can be formulated intrinsically. It is apparent that the problem is trivial if considered locally; namely, any point $p \in M$ has a neighbourhood U such that $(U, \mathcal{A}^\infty{}_{|U})$ admits an extension. Therefore, it comes as no surprise that the obstruction to the existence of a global extension, and the space which classifies the possible extensions when they exist, are cohomological in nature. One indeed has the following result.

Proposition 7.8. *Given a G^∞ supermanifold $(M, \mathcal{A}^\infty)$, let $\mathcal{K} = S^{L+1}(\mathfrak{N}^\infty)$ be the $(L+1)$-th graded symmetric tensor power of $\mathfrak{N}^\infty$ (the nilpotent ideal of $\mathcal{A}^\infty$) over $\mathcal{A}^\infty$ and let us assume that $L \geq n$. There exists a class $c[M, \mathcal{A}^\infty] \in H^2(M, \mathcal{D}er(\mathcal{A}^\infty, \mathcal{K}))$, called the Rothstein class of $(M, \mathcal{A}^\infty)$, which vanishes if and only if there is at least one G-supermanifold which extends $(M, \mathcal{A}^\infty)$. If $c[M, \mathcal{A}^\infty] = 0$, the isomorphism classes of G-supermanifolds extending $(M, \mathcal{A}^\infty)$ are classified by the cohomology group $H^1(M, \mathcal{D}er(\mathcal{A}^\infty, \mathcal{K}))$.* ∎

This Proposition will be proved by demonstrating a series of preliminary results.

Lemma 7.6. *Let $(M, \mathcal{A}, \delta)$ be a G-supermanifold, and let $\mathfrak{N}$ be the nilpotent ideal of $\mathcal{A}$. Then*

$$\operatorname{Ker}\delta \simeq \mathfrak{N}^{L+1}\,. \tag{7.6}$$

Proof. Since (7.6) is an isomorphism of sheaves, we may assume that $(M, \mathcal{A}) \simeq (B_L^{m,n}, \mathcal{G})$. If $p \in B_L^{m,n}$, and $f \in \mathcal{G}_p$, we can write

$$f = \sum_{\mu\in\Xi_L} \sum_{\nu\in\Xi_n} f_{\mu\nu} \otimes \beta_\mu \otimes e^\nu ,$$

where the $f_{\mu\nu}$'s are germs of real-valued C^∞ functions at $\sigma^{m,n}(p)$, $\{e_\alpha, \alpha = 1, \ldots, n\}$ is the canonical basis of $\mathbf{R}^n$, and $\{\beta_\mu, \mu \in \Xi_L\}$ is the canonical basis of B_L. In accordance with our standard notation, we write e^ν for the product $e^{\nu(1)} \cdots e^{\nu(r)}$ in $\bigwedge_{\mathbf{R}} \mathbf{R}^n$ if $\nu = \{\nu(1), \ldots, \nu(r)\}$.

Now, by Proposition 7.2, f is in $\operatorname{Ker}\delta$ if and only if $\sum_{\mu\in\Xi_L} \sum_{\nu\in\Xi_n} f_{\mu\nu}\beta_\mu y^\nu = 0$ for all $y^1, \ldots, y^n \in (B_L)_1$; i.e., if and only if $f_{\mu\nu} = 0$ whenever $d(\mu)+d(\nu) \leq L$. Since $\mathfrak{N}$ is generated by the elements $\{e_\alpha\}$ and $\{\beta_i\}$, with $i = 1, \ldots, L$, this proves the thesis. ∎

We can now prove that the kernel of the morphism $\mathcal{A} \to \mathcal{A}^\infty$ is intrinsic to $\mathcal{A}^\infty$; i.e., it is completely determined by the sheaf $\mathcal{A}^\infty$.

Proposition 7.9. *Under the same hypotheses of Lemma 7.6, let us assume $L \geq n$, and let $\mathfrak{N}^\infty$ be the nilpotent ideal of $\mathcal{A}^\infty = \delta(\mathcal{A})$. Then,* $\operatorname{Ker}\delta$ *is an $\mathcal{A}^\infty$-module, and there exists an isomorphism*

$$\operatorname{Ker}\delta \simeq S^{L+1}(\mathfrak{N}^\infty)$$

where $S^{L+1}(\mathfrak{N}^\infty)$ is the $(L+1)$-th graded symmetric tensor power of $\mathfrak{N}^\infty$ over $\mathcal{A}^\infty$.

Proof. We start by noticing that if $L \geq n$ then Lemma 7.6 implies that $(\operatorname{Ker}\delta)^2 = 0$, and therefore $\operatorname{Ker}\delta$ is an $\mathcal{A}^\infty$-module; indeed, for any $f \in \operatorname{Ker}\delta(U)$, and $g \in \mathcal{A}^\infty(U)$, set $f \cdot g = fh$, where $h \in \mathcal{A}(U)$ is such that $\delta(h) = g$. Since $(\operatorname{Ker}\delta)^2 = 0$, the choice of h in $\delta^{-1}(g)$ is immaterial.

Now, let $\mathcal{K}$ be the $(L+1)$-th graded symmetric tensor power of $\mathfrak{N}^\infty$ over $\mathcal{A}^\infty$, and let us define a morphism of $\mathcal{A}^\infty$-modules $\lambda \colon \mathcal{K} \to \operatorname{Ker}\delta$ by letting

$$f_1 \odot \cdots \odot f_{L+1} \mapsto \hat{f}_1 \cdots \hat{f}_{L+1} ,$$

where each $\hat{f}_i$ is a section of $\mathcal{A}$ such that $\delta(\hat{f}_i) = f_i$ (again, the choice of such $\hat{f}_i$'s is immaterial because $(\operatorname{Ker}\delta)^2 = 0$). The surjectivity of λ follows from the fact that $\delta(\mathfrak{N}) = \mathfrak{N}^\infty$. We prove that λ is injective by exhibiting a left inverse for it. We can again assume that $(M, \mathcal{A}) = (B_L^{m,n}, \mathcal{G})$. If $f \in (\operatorname{Ker}\delta)_p$, with $p \in M$, because of Lemma 7.6 we can write

$$f = \sum_{d(\mu)+d(\nu)>L} f_{\mu\nu} \otimes \beta_\mu \otimes e^\nu ,$$

where the $f_{\mu\nu}$'s are again germs of real-valued C^∞ functions at $\sigma^{m,n}(p)$. Since $\delta(\mathfrak{N}) = \mathfrak{N}^\infty$, and $d(\mu) + d(\nu) > L$, we have $\delta(\beta_\mu \otimes e^\nu) = \sum_j a_{j1}^{\mu\nu} \cdots a_{j(L+1)}^{\mu\nu}$, with the a's germs in $\mathfrak{N}_p^\infty$. Then the map

$$f \mapsto \sum_{d(\mu)+d(\nu)>L} \sum_j f_{\mu\nu}\, a_{j1}^{\mu\nu} \odot \cdots \odot a_{j(L+1)}^{\mu\nu}$$

is well defined and inverts λ. ∎

Given a G^∞ supermanifold $(M, \mathcal{A}^\infty)$, we now construct its local extensions to G-supermanifolds. Let $\mathfrak{A} = \{(U_j, \psi_j)\}$ be a G^∞ atlas, and let us consider for each j the sheaf $\mathcal{A}_j$ on U_j defined by $\mathcal{A}_j = \psi_j^{-1}\mathcal{G}_{|\psi_j(U_j)}$, where $\mathcal{G}$ is the structure sheaf of the standard G-supermanifold over $B_L^{m,n}$. From Proposition 7.9 we obtain an exact sequence

$$0 \to S^{L+1}(\mathfrak{N}^\infty{}_{|\psi_j(U_j)}) \to \mathcal{G}_{|\psi_j(U_j)} \xrightarrow{\delta} \mathcal{G}^\infty{}_{|\psi_j(U_j)} \to 0 ,$$

$\mathfrak{N}^\infty$ being the nilpotent subsheaf of $\mathcal{G}^\infty$, and hence another exact sequence

$$0 \to \mathcal{K}_{|U_j} \xrightarrow{\lambda_j} \mathcal{A}_j \xrightarrow{\delta_j} \mathcal{A}^\infty{}_{|U_j} \to 0 .$$

For each j, the triple $(U_j, \mathcal{A}_j, \delta_j)$ is apparently a G-supermanifold, whereas, by the proof of the above Proposition, λ_j is described by $\lambda_j(f_1 \odot \cdots \odot f_{L+1}) = f_1^j \cdots f_{L+1}^j$, where the f_i's are sections of $\mathfrak{N}^\infty{}_{|U_j}$ and f_i^j is any section of $\mathcal{A}_j$ such that $\delta_j(f_i^j) = f_i$.

Lemma 7.7. *There exist isomorphisms $\psi_{jh}\colon \mathcal{A}_{h|U_{jh}} \to \mathcal{A}_{j|U_{jh}}$ such that the following diagram commutes:*

$$\begin{array}{ccccc} \mathcal{K}_{|U_{jh}} & \xrightarrow{\lambda_h} & \mathcal{A}_{h|U_{jh}} & \xrightarrow{\delta_h} & \mathcal{A}^\infty{}_{|U_{jh}} \\ & {\scriptstyle\lambda_j}\searrow & \downarrow{\scriptstyle\psi_{jh}} & \swarrow{\scriptstyle\delta_j} & \\ & & \mathcal{A}_{j|U_{jh}} & & \end{array} \qquad (7.7)$$

where U_{jh} denotes the intersection $U_j \cap U_h$.

Proof. The existence of ψ_{jh} commuting with δ_h, δ_j is an easy consequence of Lemma 7.4. The proof of $\lambda_j = \psi_{jh}\lambda_h$ follows from the very definition of λ_j, λ_h, for if $f_1, \dots, f_{L+1}$ are sections of $\mathfrak{N}^\infty{}_{|U_{jh}}$ and $f_1^h, \dots, f_{L+1}^h$ are sections of $\mathcal{A}_{h|U_{jh}}$ such that $\delta_h(f_i^h) = f_i$, $i = 1, \dots, L+1$, in such a way that $\lambda_h(f_1 \odot \cdots \odot f_{L+1}) = f_1^h \cdots f_{L+1}^h$, then one has $(\psi_{jh}\lambda_h)(f_1 \odot \cdots \odot f_{L+1}) = \psi_{jh}(f_1^h) \cdots \psi_{jh}(f_{L+1}^h) = \lambda_j(f_1 \odot \cdots \odot f_{L+1})$ because $\delta_j(\psi_{jh}(f_i^h)) = f_i$ for $i = 1, \dots, L+1$. ∎

We now construct the cohomology class $c[M, \mathcal{A}^\infty]$ mentioned in Proposition 7.8. As customary, we shall write $U_{j_1,\dots,j_N}$ for the intersection $U_{j_1} \cap \cdots \cap U_{j_N}$. We define morphisms $\xi_{jhk}: \mathcal{A}_{j|U_{jhk}} \to \mathcal{A}_{j|U_{jhk}}$ by letting $\xi_{jhk} = \psi_{jh} \circ \psi_{hk} \circ \psi_{kj}$. The condition for the sheaves $\mathcal{A}_j$ to glue is apparently that the morphisms ξ_{jhk} should be the identity morphisms (cf. Section III.4). In view of the commutativity of (7.7), the morphisms $\psi_{jhk} = \xi_{jhk} - \mathrm{Id}$ take values in $\mathcal{K}_{|U_{jhk}}$, and vanish on $\mathcal{K}_{|U_{jhk}}$, thus giving rise to morphisms

$$\Delta_{jhk}: \mathcal{A}^\infty{}_{|U_{jhk}} \to \mathcal{K}_{|U_{jhk}}$$

which fulfill

$$\psi_{jhk} = \lambda_j \circ \Delta_{jhk} \circ \delta_j \,. \tag{7.8}$$

Since $\mathcal{K}$ is a square zero ideal, the morphisms Δ_{jhk} satisfy a Leibniz rule, i.e. they are elements in $\mathcal{D}er(\mathcal{A}^\infty, \mathcal{K})(U_{jhk})$.

Lemma 7.8.

(1) *The collection of the Δ_{jhk}'s is a 2-cocycle for the Čech cohomology of M with coefficients in the sheaf $\mathcal{D}er(\mathcal{A}^\infty, \mathcal{K})$, so that a cohomology class $c[\{\psi_{jh}\}] \in H^2(M, \mathcal{D}er(\mathcal{A}^\infty, \mathcal{K}))$ is defined.*

(2) *The class $c[\{\psi_{jh}\}]$ is independent of the isomorphisms $\psi_{jh}: \mathcal{A}_{h|U_{jh}} \to \mathcal{A}_{j|U_{jh}}$ fulfilling diagram (7.7), thus defining a class $c[\mathfrak{A}]$ that depends only on the atlas $\mathfrak{A}$.*

(3) *The class $c[\mathfrak{A}]$ is in fact independent of the atlas, thus yielding a class $c[M, \mathcal{A}^\infty]$ depending only on $(M, \mathcal{A}^\infty)$.*

Proof. (1) Let us fix indexes j, h, k, l, and write $\tau_p = \psi_{pj|U_{jhkl}}$, where p takes all values j, h, k. Now, the morphisms $\tau_p\tau_q^{-1}$ satisfy diagram (7.7), and we have, as above, that $\tau_p\tau_q^{-1} = \psi_{pq} + \lambda_p \circ \Delta_{pq} \circ \delta_q$ for some $\Delta_{pq} \in \mathcal{D}er(\mathcal{A}^\infty, \mathcal{K})(U_{jhkl})$. Then,

$$\Delta_{pqr|U_{jhkl}} = \Delta_{pq} + \Delta_{qr} + \Delta_{rp} \,,$$

from which we easily conclude.

(2) If $\bar{\psi}_{jh}: \mathcal{A}_{h|U_{jh}} \to \mathcal{A}_{j|U_{jh}}$ are isomorphisms fulfilling (7.7), similar arguments show that $\bar{\psi}_{jh} = \psi_{jh} + \lambda_j \circ \Delta_{jh} \circ \delta_h$ for some $\Delta_{jh} \in \mathcal{D}er(\mathcal{A}^\infty, \mathcal{K})(U_{jh})$, thus proving that

$$\bar{\Delta}_{jhk} = \Delta_{jhk} + \Delta_{hj} + \Delta_{jk} + \Delta_{kh} ,$$

$\{\bar{\Delta}_{jhk}\}$ being the 2-cocycle constructed from the $\bar{\psi}_{jh}$'s, which proves the statement.

(3) Since $c[\mathfrak{A}]$ is invariant under refinement of the atlas, and since different atlases always have a common refinement, whenever we consider different atlases $\mathfrak{A}$ and $\mathfrak{A}'$ we may assume that they have the same open sets, i.e. $\mathfrak{A} = \{(U_j, \psi_j)\}$ and $\mathfrak{A}' = \{(U_j, \psi_j')\}$. Then, as in the proof of Lemma 7.7, we have isomorphisms $\tau_j: \mathcal{A}_j \to \mathcal{A}_j'$ yielding commutative diagrams

$$\begin{array}{ccccc} \mathcal{K}_{|U_j} & \xrightarrow{\lambda_j} & \mathcal{A}_j & \xrightarrow{\delta_j} & \mathcal{A}^\infty{}_{|U_j} \\ & {\scriptstyle \lambda_j'}\searrow & \downarrow{\scriptstyle \tau_j} & \swarrow{\scriptstyle \delta_j'} & \\ & & \mathcal{A}_j' & & \end{array}$$

Now, since the isomorphisms $\psi_{jh}' = \tau_j^{-1} \psi_{jh} \tau_h$ on U_{jh} verify diagram (7.7), we can construct the 2-cocycle with them. However, direct computation shows that $\psi_{jhk}' = \tau_j(\lambda_j \circ \Delta_{jhk} \circ \delta_j)\tau_j^{-1}$, so that $\Delta_{jhk}' = \Delta_{jhk}$, thus finishing the proof. ∎

We now show that the class $c[M, \mathcal{A}^\infty] \in H^2(M, \mathcal{D}er(\mathcal{A}^\infty, \mathcal{K}))$ vanishes if and only if there is at least one G-supermanifold extending $(M, \mathcal{A}^\infty)$. Indeed, if $c[M, \mathcal{A}^\infty] = 0$ we have — possibly after refining the atlas —

$$\Delta_{jhk} = \Delta_{jh} + \Delta_{hk} + \Delta_{kj} ;$$

if we set

$$\zeta_{jh} = \psi_{jh} - \lambda_j \circ \Delta_{jh} \circ \delta_h ,$$

a direct calculation shows that $\zeta_{jh} \circ \zeta_{hk} \circ \zeta_{kj} = \mathrm{Id}$, which means that the ζ_{jh}'s glue the sheaves $\mathcal{A}_j$, yielding the desired G-supermanifold. Conversely, if there exists such an extension $(M, \mathcal{A})$ we have — again possibly after refining the atlas — isomorphisms $\gamma_j: \mathcal{A}_{|U_j} \xrightarrow{\sim} \mathcal{A}_j$ such that the maps $\gamma_{jh} = \gamma_j \gamma_h^{-1}: \mathcal{A}_{h|U_{jh}} \to \mathcal{A}_{j|U_{jh}}$ verify diagram (7.7). It follows that $\psi_{jh} = \gamma_{jh} + \lambda_j \circ \Delta_{jh} \circ \delta_h$ for some derivation $\Delta_{jh} \in \mathcal{D}er(\mathcal{A}^\infty, \mathcal{K})(U_{jh})$. Direct computation shows that $\Delta_{jhk} = \Delta_{jh} + \Delta_{hk} + \Delta_{kj}$, that is, $c[M, \mathcal{A}^\infty] = 0$.

The last part of Proposition 7.8 claims that for a given G-supermanifold $(M, \mathcal{A}, \delta)$, the G-supermanifolds extending $(M, \mathcal{A}^\infty)$ are classified by the cohomology group $H^1(M, \mathcal{D}er(\mathcal{A}^\infty, \mathcal{K}))$. The only non trivial thing to show is how to construct a G-supermanifold extending $(M, \mathcal{A}^\infty)$ from a cohomology class. Indeed, if $\{U_j\}$ is an open cover of M and $\Delta_{jh} \in \mathcal{D}er(\mathcal{A}^\infty, \mathcal{K})(U_{jh})$ is a 1-cocycle, the isomorphisms $\psi_{jh}: \mathcal{A}_{|U_{jh}} \xrightarrow{\sim} \mathcal{A}_{|U_{jh}}$ defined by $\psi_{jh} = \mathrm{Id} + \lambda \circ \Delta_{jh} \circ \delta$ verify $\psi_{jh}\psi_{hk}\psi_{kj} = 1$, allowing us to glue the sheaves $\mathcal{A}_{|U_j}$. Thus we obtain a new G-supermanifold $(M, \mathcal{A}', \delta')$, locally isomorphic with $(M, \mathcal{A}, \delta)$, which is also an extension of $(M, \mathcal{A}^\infty)$. One can check directly that equivalent cocycles yield isomorphic G-supermanifolds. This eventually concludes the proof of Proposition 7.8.

Let us notice that for G^∞ DeWitt supermanifolds the sheaf $\mathcal{D}er(\mathcal{A}^\infty, \mathcal{K})$ is acyclic, as we shall see in next Chapter, and therefore these supermanifolds admit unique extensions to G-supermanifolds. We can thus anticipate the following result: *The standard G-supermanifold over $B_L^{m,n}$ is the unique (up to isomorphism) G-supermanifold which extends the canonical G^∞ supermanifold over $B_L^{m,n}$.*

Supermanifolds over arbitrary ground algebras. For the sake of simplicity, we have limited our discussion of Rothstein's axiomatics mostly to the case $B = B_L$. The general case has been dealt with in [**BBHP1,BBHP2**], where several results presented in this book have been extended to that setting.

Graded manifolds as R-supermanifolds. We conclude this Section by showing that graded manifolds fit into Rothstein's axiomatics; indeed, whenever the choice $B = \mathbf{R}$ is made, Axioms 1, 2, 3′ and 6 yield the category of graded manifolds. The only defining property of graded manifolds that is not straightforward to prove is local triviality, which is assured by Proposition 7.1 together with the completeness requirement given by Axiom 6.

Chapter V
Cohomology of supermanifolds

Ogni parte ha inclinazione a ricongiungersi
al suo tutto per fuggire dalla sua imperfezione.
LEONARDO DA VINCI

The aim of this Chapter is to unfold a basic cohomological theory for supermanifolds, which will be exploited in the next Chapter to study the structure of superbundles; in particular to build a theory of characteristic classes. This cohomology theory does not embody only trivial extensions of results valid for differentiable manifolds. For instance, the natural analogue of the de Rham theorem does not hold in general and, similarly, in the case of complex supermanifolds there is, generally speaking, no analogue of the Dolbeault theorem. These features are consequences of the fact that the structure sheaf of a supermanifold does not need to be cohomologically trivial. Related to this is also the fact that the cohomology of the complex of global graded differential forms on a G-supermanifold $(M, \mathcal{A})$ (i.e. the 'super de Rham cohomology' of $(M, \mathcal{A})$) depends on the G-supermanifold structure of $(M, \mathcal{A})$, so that homeomorphic and even smoothly diffeomorphic G-supermanifolds may have a different super de Rham cohomology; that is, super de Rham cohomology is a fine invariant of the supermanifold structure.

Most results of this Chapter were first presented in the papers [**BB2,3**].

1. de Rham cohomology of graded manifolds

Graded manifolds are not very interesting as far as their cohomology is concerned. In the real case, the structure sheaf of a graded manifold $(X, \mathcal{A})$ is fine (cf. Lemma III.1.1), and therefore $\mathcal{A}$, and all sheaves $\Omega^k_{\mathcal{A}}$ of graded differential forms, are acyclic (this follows from Corollaries 3.3 and 3.4 of Chapter II). This implies that the cohomology of the complex $\Omega^\bullet_{\mathcal{A}}(X)$ coincides with the de Rham cohomology of X. In the complex analytic case, a similar argument allows one to prove a Dolbeault-type theorem. Here we do not give the details of this second result, since it is completely analogous to the Dolbeault theorem for complex analytic DeWitt supermanifolds (cf. Section 3).

The complex of sheaves $\mathcal{A}^\bullet$ is exact, and, moreover, it is a resolution of the constant sheaf $\mathbf{R}$ on X; i.e., the sequence of sheaves of $\mathbf{R}$-modules

$$0 \longrightarrow \mathbf{R} \longrightarrow \mathcal{A} \xrightarrow{d} \Omega^1_{\mathcal{A}} \xrightarrow{d} \Omega^2_{\mathcal{A}} \xrightarrow{d} \dots \tag{1.1}$$

is exact. This 'graded Poincaré Lemma' is most easily proved by working in local coordinates and proceeding on the analogy of the usual Poincaré Lemma (see also Proposition 2.1). By defining the de Rham cohomology of $(X, \mathcal{A})$, denoted by $H^\bullet_{DR}(X, \mathcal{A})$, as the cohomology of the complex of graded vector spaces $\mathcal{A}^\bullet(X)$, from (1.1), and using the ordinary de Rham theorem, we obtain the following result (cf. [Kos] Theorem 4.7).

Proposition 1.1. *There is a canonical isomorphism*

$$H^k_{DR}(X, \mathcal{A}) \simeq H^k_{DR}(X) \qquad \text{for all} \quad k \geq 0.$$

■

Here $H^\bullet_{DR}(X)$ denotes the usual de Rham cohomology of X.

2. Cohomology of graded differential forms

In this and the following Section we study certain topics related to the cohomology of G-supermanifolds. It should be stressed that all the results presented here are still valid, *mutatis mutandis*, in other categories of supermanifolds (e.g. H^∞ and GH^∞ supermanifolds).

Let $(M, \mathcal{A})$ be a G-supermanifold. The sheaves $\Omega^k \otimes_{\mathbf{R}} B_L$ of smooth B_L-valued differential forms on M provide a resolution of the constant sheaf B_L on M, in the sense that the differential complex of sheaves of graded-commutative

B_L-algebras $\Omega^\bullet \otimes_{\mathbf{R}} B_L$ (with $\Omega^0 \otimes_{\mathbf{R}} B_L \equiv \mathcal{C}_L^\infty$) is a resolution of the constant sheaf B_L, i.e. the sequence

$$0 \to B_L \to \mathcal{C}_L^\infty \xrightarrow{d} \Omega^1 \otimes_{\mathbf{R}} B_L \xrightarrow{d} \Omega^2 \otimes_{\mathbf{R}} B_L \to \dots \tag{2.1}$$

is exact. The cohomology associated with this complex via the global section functor $\Gamma(\cdot, M)$, i.e. the cohomology of the complex $\Omega^\bullet(M) \otimes_{\mathbf{R}} B_L$, is denoted by $H^\bullet_{DR}(M, B_L)$, and is called the *B_L-valued de Rham cohomology* of M (more precisely, of the differentiable manifold underlying M). Since B_L is a finite-dimensional real vector space, the universal coefficient theorem [Go] entails the (otherwise obvious) isomorphism

$$H^\bullet_{DR}(M, B_L) \simeq H^\bullet_{DR}(M) \otimes_{\mathbf{R}} B_L \,. \tag{2.2}$$

By virtue of the de Rham theorem (Proposition II.3.1), Eq. (2.2) can be equivalently written as

$$H^\bullet_{DR}(M, B_L) \simeq H^\bullet(M, B_L) \,. \tag{2.3}$$

By $H^\bullet(M, \cdot)$ we designate interchangeably the Čech or sheaf cohomology functor, which coincide since the base space is paracompact.

In order to gain information not on the topological or smooth structure of M, but rather on its G-supermanifold structure, we therefore need to define a new cohomology, obtained via a resolution of B_L different from the differential complex (2.1). To this end, we consider the sheaves $\Omega^k_{\mathcal{A}}$ of graded differential forms. The following result is a generalization of the usual Poincaré lemma (cf. [Bru]).

Proposition 2.1. *Given a G-supermanifold $(M, \mathcal{A})$, the differential complex of sheaves of graded B_L-algebras on M*

$$0 \to B_L \to \mathcal{A} \xrightarrow{d} \Omega^1_{\mathcal{A}} \xrightarrow{d} \Omega^2_{\mathcal{A}} \to \dots \tag{2.4}$$

is a resolution of B_L.

Proof. Since the claim to be proved is a local matter, we may assume that $(M, \mathcal{A}) = (B_L^{m,n}, \mathcal{G})$; moreover, it is enough to show that, if U is an open ball around the origin in $B_L^{m,n}$ then any closed graded differential k-form $\lambda \in \Omega^k_{\mathcal{G}}(U)$ is exact; i.e., there exists a graded differential $(k-1)$-form $\eta \in \Omega^{k-1}_{\mathcal{G}}(U)$ such that $\lambda = d\eta$. Given coordinates $(z^1, \dots, z^{m+n})$ in U, let $\omega = dz^{A_k} \wedge \dots \wedge$

$dz^{A_1}\omega_{A_1\dots A_k} \in \Omega^k_{\mathcal{H}^\infty}(U)$ be an H^∞ graded differential k-form on U $(k > 0)$; let us set

$$\hat{K}\omega(z) = (-1)^k\, k\; dz^{A_{k-1}} \wedge \cdots \wedge dz^{A_1} z^B \int_0^1 t^{k-1}\omega_{BA_1\dots A_{k-1}}(tz)\,dt\,.$$

Proposition III.4.2 yields an isomorphism $\Omega^k_{\mathcal{G}}(U) \simeq \Omega^k_{\mathcal{H}^\infty}(U) \otimes_{\mathbf{R}} B_L$; it is therefore possible to introduce a homotopy operator $K\colon \Omega^k_{\mathcal{G}}(U) \to \Omega^{k-1}_{\mathcal{G}}(U)$, defined by

$$K(\omega \otimes a) = \hat{K}\omega \otimes a\,.$$

One can indeed verify easily that $dK\lambda + Kd\lambda = \lambda$ for any section $\lambda \in \mathcal{G}^k(U)$, so that, if $d\lambda = 0$, then $\lambda = d(K\lambda)$. The case $k = 0$ has been left out. However, if $f \in \mathcal{G}(U)$, by writing f as $f = \sum_i f_i \otimes a_i$ with $f_i \in \mathcal{H}^\infty(U)$ and $a_i \in B_L$, the condition $df = 0$ implies directly that f is a constant in B_L. ■

Definition 2.1. *Given a G-supermanifold $(M, \mathcal{A})$, the cohomology of the complex*

$$\mathcal{A}(M) \xrightarrow{d} \Omega^1_{\mathcal{A}}(M) \xrightarrow{d} \Omega^2_{\mathcal{A}}(M) \to \dots\,, \tag{2.5}$$

denoted by $H^\bullet_{SDR}(M, \mathcal{A})$, is called the super de Rham cohomology of $(M, \mathcal{A})$.

The operation of taking the SDR cohomology of a G-supermanifold is functorial. Indeed, given a G-morphism $(f, \phi)\colon (M, \mathcal{A}) \to (N, \mathcal{B})$, it is easily proved that the morphism $\Omega^\bullet_{\mathcal{B}}(N) \to f_*\Omega^\bullet_{\mathcal{A}}(M)$ induced by ϕ commutes with the exterior differential, and therefore yields a morphism of graded B_L-modules $\phi^\sharp\colon H^\bullet_{SDR}(N, \mathcal{B}) \to H^\bullet_{SDR}(M, \mathcal{A})$. It should be noticed that the functor $H^\bullet_{SDR}(\cdot)$ does not fulfill the Eilenberg-Steenrod [Spa] axiomatics for cohomology (if it did, it would coincide with the B_L-valued de Rham cohomology functor) since it does not satisfy the excision axiom. Moreover, the functor $H^\bullet_{SDR}(\cdot)$ does not give rise to topological invariants; indeed, in Example 2.2 we shall show two homeomorphic supermanifolds having different SDR cohomology. On the other hand, it is easily verified that the graded B_L-modules $H^k_{SDR}(M, \mathcal{A})$ are invariants associated with the G-supermanifold structure of M. Indeed, if $(f, \phi)\colon (M, \mathcal{A}) \to (N, \mathcal{B})$ is a G-isomorphism, it is easily proved that $\phi^\sharp\colon H^\bullet_{SDR}(N, \mathcal{B}) \to H^\bullet_{SDR}(M, \mathcal{A})$ is an isomorphism.

The most natural thing to do to gain insight into the geometric significance of the groups $H^k_{SDR}(M, \mathcal{A})$ — which, as a matter of fact, are graded B_L-modules — is to compare them with the cohomology groups $H^k(M, B_L)$, which have a natural structure of graded B_L-modules as well. The morphisms $\Omega^k_{\mathcal{A}}(M) \to$

$\Omega^k(M) \otimes_{\mathbf{R}} B_L$ induced by the morphism $\delta: \mathcal{A} \to \mathcal{C}_L^\infty$ give rise to a morphism of differential complexes, which induces in cohomology a morphism of graded B_L-modules

$$\varrho^k: H^k_{SDR}(M, \mathcal{A}) \to H^k_{DR}(M, B_L) \qquad \forall k \geq 0. \tag{2.6}$$

In degree zero, ϱ^0 is an isomorphism, in that one has manifestly

$$H^0_{SDR}(M, \mathcal{A}) \simeq (B_L)^C \simeq H^0_{DR}(M, B_L),$$

where C is the number of connected components of M, which we assume to be finite. In degree higher than zero, we have, as a straightforward application of the abstract de Rham theorem (Proposition II.2.4), the following result.

Proposition 2.2. *Let $(M, \mathcal{A})$ be a G-supermanifold, and fix an integer $q \geq 1$. If $H^k(M, \Omega^p_{\mathcal{A}}) = 0$ for $0 \leq p \leq q-1$ and $1 \leq k \leq q$, there are isomorphisms*

$$H^k_{SDR}(M, \mathcal{A}) \simeq H^k(M, B_L) \qquad \text{for} \qquad 0 \leq k \leq q.$$

∎

From Eq. (2.3), still working under the hypotheses of Proposition 2.2, we obtain isomorphisms

$$H^k_{SDR}(M, \mathcal{A}) \simeq H^k_{DR}(M, B_L) \qquad \text{for} \qquad 0 \leq k \leq q.$$

Proposition 2.2 provides a useful tool for investigating the cohomological properties of the structure sheaf of a G-supermanifold. For instance, it suffices to exhibit a G-supermanifold $(M, \mathcal{A})$ such that $H^1_{SDR}(M, \mathcal{A}) \neq H^1_{DR}(M, B_L)$ to deduce that, in general, the sheaf $\mathcal{A}$ cannot be expected to be acyclic (we recall that a sheaf $\mathcal{F}$ on a topological space X is acyclic if $H^k(M, \mathcal{F}) = 0$ for all $k > 0$).

EXAMPLE 2.1.[1] Consider Example 2.1 of Chapter III; since $L' = L$, the pair $(M, \mathcal{G}^\infty)$ is already a G-supermanifold. Thus, from Eq. (III.2.11) we have

$$\mathcal{Z}^1(M) = \Omega^1_{\mathcal{A}}(M) \simeq [\mathcal{C}^\infty(\mathbf{R}) \otimes_{\mathbf{R}} \mathfrak{N}_L] \oplus \mathbf{R}_L$$

$$\mathcal{B}^1(M) = \mathcal{C}^\infty(\mathbf{R}) \otimes_{\mathbf{R}} \mathfrak{N}_L;$$

[1]This example already appeared in [**Ra**].

here $\mathbf{R}_L$ is $\mathbf{R}$ with the B_L-module structure induced by the body map $\sigma\colon B_L \to \mathbf{R}$ (cf. Section I.1), and $C^\infty(\mathbf{R})$ is the vector space of smooth real-valued functions on $\mathbf{R}$. Since $H^1_{SDR}(M,\mathcal{A}) \equiv \mathcal{Z}^1(M)/\mathcal{B}^1(M)$ (as a quotient of submodules of $\Omega^1_{\mathcal{A}}(M)$), we obtain

$$H^1_{SDR}(M,\mathcal{A}) \simeq \mathbf{R}_L\,.$$

On the other hand, the B_L-valued de Rham cohomology of M is easily calculated and turns out to be

$$H^1_{DR}(M) \otimes B_L \simeq B_L\,,$$

so that by virtue of Proposition 2.2 we can infer that $H^1(M,\mathcal{A}) \neq 0$. Indeed, a simple direct computation yields

$$H^0(M,\mathcal{A}) \simeq \mathbf{R} \oplus [C^\infty(\mathbf{R}) \otimes \mathfrak{N}_L], \qquad H^1(M,\mathcal{A}) \simeq C^\infty(\mathbf{R}) \otimes B_L\,. \tag{2.7}$$

▲

EXAMPLE 2.2. In Examples 2.2 and 2.3 of Chapter III two GH^∞ supermanifold structures were given to the topological space $T^2 \times \mathbf{R}^2$ (of course, at that stage it was not clear whether the two supermanifold structures were actually inequivalent). By tensoring the structure sheaves by B_L, we obtain two G-supermanifolds, that we denote by $(M_1,\mathcal{A}_1)$ (that obtained from Example 2.2) and $(M_2,\mathcal{A}_2)$ (from Example 2.3). Direct computations show that

$$H^1_{DR}(M_1,B_L) = H^1_{DR}(M_2,B_L) = B_L \oplus B_L,$$

$$H^1_{SDR}(M_1,\mathcal{A}_1) = B_L \otimes_{B_{L'}} B_L,$$

$$H^1_{SDR}(M_2,\mathcal{A}_2) = B_L.$$

From this we learn that the two G-supermanifold structures, and therefore the original GH^∞ structures, are inequivalent, and that the structure sheaf of either G-supermanifold is not acyclic. ▲

The non-acyciclity of their structure sheaf is not a peculiarity of G-supermanifolds, in that all supersmooth (i.e. G^∞ or H^∞ or GH^∞) supermanifolds, and obviously also Rothstein supermanifolds, share this property. In particular, the structure sheaf of a supermanifold is generically not fine, and this entails that it has no supersmooth partition of unity, contrary to differentiable manifolds, but in analogy with complex manifolds. This cohomological affinity between supermanifolds and complex manifolds will be a kind of *leitmotiv* in the developments to follow.

On the other hand, we have seen that the structure sheaf of a (real) graded manifold *is* acyclic. This — as a consequence of the results established in Section IV.6 — suggests that the structure sheaf of any (real) supermanifold of the DeWitt type is acyclic, as we shall actually prove in the next Section.

3. Cohomology of DeWitt supermanifolds

We wish to prove that the structure sheaf of a real DeWitt supermanifold is acyclic; in the complex analytic case, a Dolbeault-type theorem holds.

Even though DeWitt supermanifolds were defined in terms of the coarse (DeWitt) topology, the cohomology of a DeWitt supermanifold $(M, \mathcal{A})$ will be studied by considering in M the fine topology; this is advantageous because in this way M is paracompact. Thus, we continue to confuse the sheaf and Čech cohomologies with coefficients in sheaves on M.

Let us start by considering the real case (for the time being, we defer to say 'real'). We need the following Lemma, which is obtained from a result given in [**Bre**] (Exercise IV.18) by strengthening certain hypotheses (this makes its statement simpler and more directly applicable to our setting).

Lemma 3.1. *Let X and Y be topological spaces, with Y locally euclidean, and $\mathcal{F}$ a sheaf of abelian groups on X; let us assume that all groups $H^k(X, \mathcal{F})$ are finitely generated. Then for all $n \geq 0$ there is an exact sequence of abelian groups*

$$0 \to \bigoplus_{j+k=n} H^j(X, \mathcal{F}) \otimes_{\mathbf{Z}} H^k(Y, \mathbf{Z}) \to H^n(X \times Y, \pi^{-1}\mathcal{F}) \to$$

$$\to \bigoplus_{j+k=n+1} \operatorname{Tor}\left[H^j(X, \mathcal{F}), H^k(Y, \mathbf{Z})\right] \to 0$$

*where $\operatorname{Tor}[\cdot, \cdot]$ denotes the torsion product [**Go,HiS**] and $\pi: X \times Y \to X$ is the canonical projection.* ∎

We can now prove our first basic result.

Proposition 3.1. *The G-supermanifold $(B_L^{m,n}, \mathcal{G})$ is cohomologically trivial:*

$$H^k(B_L^{m,n}, \mathcal{G}) = 0 \qquad \forall k > 0. \tag{3.1}$$

Proof. In view of the definitions of the sheaves $\mathcal{GH}$ and $\mathcal{G}$ (see Sections III.2 and III.4), one has an isomorphism (all tensor products are over $\mathbb{R}$)

$$\mathcal{G} \simeq (\sigma^{m,n})^{-1}(\mathcal{C}^{\infty}_{\mathbb{R}^m} \otimes \bigwedge \mathbb{R}^n \otimes B_L).$$

Therefore, applying Lemma 3.1 with the following identifications:

$$X = \mathbb{R}^m, \qquad Y = \mathfrak{N}_L^{m,n}, \qquad \mathcal{F} = \mathcal{C}^{\infty}_{\mathbb{R}^m} \otimes \bigwedge \mathbb{R}^n \otimes B_L,$$

we obtain (since $H^k(\mathfrak{N}_L^{m,n}, \mathbb{Z}) = 0$ for $k > 0$ and $H^0(\mathfrak{N}_L^{m,n}, \mathbb{Z}) = \mathbb{Z}$),

$$H^k(B_L^{m,n}, \mathcal{G}) \simeq H^k(\mathbb{R}^m, \mathcal{C}^{\infty}_{\mathbb{R}^m} \otimes \bigwedge \otimes B_L).$$

Now, since the sheaf of rings $\mathcal{C}^{\infty}_{\mathbb{R}^m}$ is fine, the sheaf $\mathcal{C}^{\infty}_{\mathbb{R}^m} \otimes \bigwedge \mathbb{R}^n \otimes B_L$ of $\mathcal{C}^{\infty}_{\mathbb{R}^m}$-modules is soft, and therefore is acyclic (cf. Corollaries 3.3 and 3.4 of Chapter II), which yields the sought result. ∎

Coarse partitions of unity. DeWitt supermanifolds do not admit partitions of unity in a strict sense, that is to say, there cannot exist partitions of unity subordinated to any locally finite cover, since the structure sheaf of a DeWitt supermanifold is not soft, and therefore not even fine. However, any DeWitt supermanifold has a particular kind of partition of unity, that we call a *coarse partition of unity* (we recall from Section IV.6 that a cover of a DeWitt supermanifold is said to be coarse if its sets are open in the DeWitt topology).

Lemma 3.2. *Let $(M, \mathcal{A})$ be a DeWitt G-supermanifold, with body M_B and projection $\Phi: M \to M_B$. For any locally finite coarse cover $\mathfrak{U} = \{U_j\}$ of M there exists a family $\{g_j\}$ of global sections of $\mathcal{A}$ such that*

(1) $\operatorname{Supp} g_j \subset U_j$;
(2) $\sum_j g_j = 1$.

Proof. This result is proved in the same manner as Lemma III.1.1. With no loss of generality we can assume that the sets U_j are coordinate neighbourhoods, and then $\mathcal{A}(U_j) \simeq \mathcal{H}^{\infty}(\hat{U}_j) \otimes_{\mathbb{R}} B_L$, where $\hat{U}_j$ is the image of U_j in $B_L^{m,n}$ through the coordinate map. Denoting by W the union of the $\hat{U}_j$'s, it is obviously possible to define functions $\hat{r}_j \in \mathcal{H}^{\infty}(W)$ whose supports lie in $\hat{U}_j$, and are such that $\sum_j \hat{r}_j = 1$; one simply defines

$$\hat{r}_j(x^1, \ldots, x^m, y^1, \ldots, y^n) = Z_0(t_j)(x^1, \ldots, x^m),$$

where Z_0 is the Z-expansion, and $\{t_j\}$ is a smooth partition of unity of the sheaf of C^∞ functions on $\sigma^{m,n}(W) \subset \mathbf{R}^m$ subordinated to the cover $\{\sigma^{m,n}(\hat{U}_j)\}$. The functions $\hat{\tau}_j$ do not sum up to 1, but this can be realized by normalizing them.

Now, the quantities $\hat{\tau}_j \otimes 1$ can be regarded as sections τ_j in $\mathcal{A}(U_j)$ and extended by zero outside U_j, thus yielding global sections of $\mathcal{A}$. Letting $h = \sum_j \tau_j$, we have $\sigma(\delta^M(h)) = 1$, so that h is invertible, and we may set $g_j = h^{-1}\tau_j$. The sections g_j satisfy the required properties by construction. ∎

Corollary 3.1. *Let $\mathfrak{U}$ be a locally finite coarse cover of M. Then*

$$\check{H}^k(\mathfrak{U}, \mathcal{F}) = 0, \qquad k > 0,$$

for any sheaf $\mathcal{F}$ of $\mathcal{A}$-modules. ∎

If we consider in M the coarse topology (let us denote the resulting space by M_{DW}) the sheaf $\mathcal{A}$ is apparently fine; however, this does not allow us to conclude that the sheaf cohomology of $\mathcal{A}$ is trivial, since M_{DW} is not paracompact. In any case, one can conclude that the Čech cohomology $\check{H}^\bullet(M_{DW}, \mathcal{A})$ (or the cohomology $\check{H}^\bullet(M_{DW}, \mathcal{F})$, where $\mathcal{F}$ is any $\mathcal{A}$-module) *is* trivial, since the direct limit over the covers involved in the definition of the Čech cohomology can be taken on coarse covers.

Cohomology of DeWitt G-supermanifolds. We can now state the main result of this section.

Proposition 3.2. *The structure sheaf $\mathcal{A}$ of a DeWitt G-supermanifold $(M, \mathcal{A})$ is acyclic.*

Proof. Any $p \in M_B$ has a system of neighbourhoods $\mathfrak{W}$ such that for all $W \in \mathfrak{W}$ the supermanifold $(\Phi^{-1}(W), \mathcal{A}_{|\Phi^{-1}(W)})$ is isomorphic to $(B_L^{m,n}, \mathcal{G})$; therefore, $\mathcal{A}_{|\Phi^{-1}(W)}$ is acyclic. We are then in the hypotheses of Proposition II.2.5, and hence:

$$H^k(M, \mathcal{A}) \simeq H^k(M_B, \Phi_*\mathcal{A}), \qquad k \geq 0\,.$$

The sheaf $\Phi_*\mathcal{A}$ is fine by Lemma 3.2, and hence acyclic, so that we achieve the thesis. ∎

The reader will notice that the same procedure that brought to Proposition 3.2 can be applied to the structure sheaves of an H^∞ or GH^∞ DeWitt supermanifolds, which are therefore acyclic as well.

Corollary 3.2. *Any locally free $\mathcal{A}$-module $\mathcal{F}$ is acyclic.*

Proof. Let us at first assume that $\mathcal{F}$ trivializes on a coarse cover. Then, since $\Phi_*\mathcal{F}$ is a $\Phi_*\mathcal{A}$-module, the same proof of the previous Proposition applies. Now we must prove that $\mathcal{F}$ actually trivializes on a coarse cover. Without any loss of generality we may assume that $(M,\mathcal{A}) = (B_L^{m,n},\mathcal{G})$, and that $\mathcal{F}$ trivializes on subsets of $B_L^{m,n}$ which are diffeomorphic to open balls. Let U be one of these subsets; then $\mathcal{F}(U) \simeq \mathcal{G}^{p|q}(U)$. In view of the definition of the sheaf $\mathcal{G}$, if V is any other set of this kind such that $\Phi^{-1}\Phi(U) = \Phi^{-1}\Phi(V) = W$, then $\mathcal{F}(U) \simeq \mathcal{F}(V)$, so that one has $\mathcal{F}_{|W} = \mathcal{G}^{p|q}{}_{|W}$. ∎

For instance, the sheaf of derivations $\mathcal{D}er\mathcal{A}$ and sheaves $\Omega_{\mathcal{A}}^k$ of graded differential forms on $(M,\mathcal{A})$ are acyclic.

SDR cohomology of DeWitt supermanifolds. The previous results have an immediate consequence in connection with the super de Rham cohomology of DeWitt supermanifolds.

Proposition 3.3. [2] *The super de Rham cohomology of a DeWitt supermanifold $(M,\mathcal{A})$ is isomorphic with the B_L-valued de Rham cohomology of the body manifold M_B:*

$$H^\bullet_{SDR}(M) \simeq H^\bullet_{DR}(M_B) \otimes_{\mathbf{R}} B_L\,. \tag{3.2}$$

Proof. We have already seen that the sheaves of graded differential forms $\Omega_{\mathcal{A}}^k$ are acyciclic, $H^k(M,\Omega_{\mathcal{A}}^p) = 0$ for all $k > 0$ and $p \geq 0$. Accordingly, Proposition 1.2 implies

$$H^\bullet_{SDR}(M) \simeq H^\bullet(M, B_L)\,. \tag{3.3}$$

On the other hand, M is a fibration over M_B with a contractible fibre, so that $H^\bullet_{DR}(M) \simeq H^\bullet_{DR}(M_B)$, and Eq. (3.3) is equivalent to Eq. (3.2).[3] ∎

Dolbeault theorem. Let $(M,\mathcal{B})$ be an (m,n)-dimensional complex G-supermanifold. We recall that $\Omega_{\mathcal{B}}^p$ is the sheaf of holomorphic graded p-forms on $(M,\mathcal{B})$, while $\Omega_{\mathcal{I}}^{p,q}$ is the sheaf of graded differential forms of type (p,q). Here $\mathcal{I}$ is the complexification of the sheaf $\mathcal{A}$, i.e. $\mathcal{I} = \mathcal{A} \otimes_{\mathbf{R}} \mathbf{C}$.

---

[2] This result was already stated in [**Ra**].

[3] In [**BB2**] we gave a slightly different proof, which does not involve the sheaf cohomology of $\mathcal{A}$, but requires spectral sequence techniques.

Lemma 3.3. *The complex* $\Omega_{\mathcal{I}}^{p,0} \xrightarrow{\bar\partial} \Omega_{\mathcal{I}}^{p,1} \xrightarrow{\bar\partial} \dots$ *is a resolution of* $\Omega_{\mathcal{B}}^{p}$, *i.e. the sequence of sheaves of graded* C_L*-modules*

$$0 \to \Omega_{\mathcal{B}}^{p} \to \Omega_{\mathcal{I}}^{p,\bullet} \tag{3.4}$$

is exact.

Proof. This is the transposition to the supermanifold setting of the so-called $\bar\partial$-Poincaré or Grothendieck or Dolbeault Lemma, and is proved by mimicking the proof valid in the case of complex manifolds (see e.g. [GrH]), in the same way as the ordinary Poincaré Lemma has been generalized to Proposition 2.1. ■

The sheaves $\Omega_{\mathcal{I}}^{p,q}$ are acyclic by Corollary 3.2, so that the resolution (3.4) of the sheaf of holomorphic graded p-forms on $(M,\mathcal{B})$, by the abstract de Rham theorem, computes the cohomology of M with coefficients in $\Omega_{\mathcal{B}}^{p}$.

The cohomology of the complex

$$\Omega_{\mathcal{I}}^{p,0}(M) \xrightarrow{\bar\partial} \Omega_{\mathcal{I}}^{p,1}(M) \xrightarrow{\bar\partial} \dots$$

is denoted by $H_{\bar\partial}^{p,\bullet}(M,\mathcal{B})$, and is called the *Dolbeault cohomology* of $(M,\mathcal{B})$. More precisely, we let

$$H_{\bar\partial}^{p,q}(M,\mathcal{B}) = \frac{\operatorname{Ker}\bar\partial\colon \Omega_{\mathcal{I}}^{p,q}(M) \to \Omega_{\mathcal{I}}^{p,q+1}(M)}{\operatorname{Im}\bar\partial\colon \Omega_{\mathcal{I}}^{p,q-1}(M) \to \Omega_{\mathcal{I}}^{p,q}(M)}.$$

The previous discussion leads to a Dolbeault-type theorem, valid for DeWitt supermanifolds. For a non-DeWitt supermanifold, the non-acyclicity of the structure sheaf is, generally speaking, an obstruction to the validity of such a theorem.

Proposition 3.4. *Let* $(M,\mathcal{B})$ *be a complex DeWitt G-supermanifold. There are isomorphisms of graded* C_L*-modules*

$$H_{\bar\partial}^{p,q}(M,\mathcal{B}) \simeq H^{q}(M,\Omega_{\mathcal{B}}^{p}).$$

■

Cohomology of G^∞ DeWitt supermanifolds. Proposition 3.2, which states the acyclicity of the structure sheaf of a DeWitt G-supermanifold, can be shown to hold true also in the case of the sheaf $\mathcal{A}^\infty$ of G^∞ functions on a DeWitt supermanifold.

Proposition 3.5. *The structure sheaf of a G^∞ DeWitt supermanifold is acyclic.*

Proof. Working as in Lemma 3.2, one can construct a coarse G^∞ partition of unity on M, so that the sheaf $\Phi_*\mathcal{A}^\infty$ is fine, and therefore acyclic. Let us now consider for a while the G^∞ DeWitt supermanifold $(B_L^{m,n}, \mathcal{G}^\infty)$. Lemma 3.1 implies

$$H^k(B_L^{m,n}, (\sigma^{m,n})^{-1}(\sigma^{m,n})_*\mathcal{G}^\infty) \simeq H^k(\mathbb{R}^m, (\sigma^{m,n})_*\mathcal{G}^\infty) = 0$$

for all $k > 0$. Since $(\sigma^{m,n})^{-1}(\sigma^{m,n})_*\mathcal{G}^\infty \simeq \mathcal{G}^\infty$ by the very definition of the sheaf $\mathcal{G}^\infty$, the result is proved for the supermanifold $(B_L^{m,n}, \mathcal{G}^\infty)$. The result for a generic G^∞ DeWitt supermanifold now follows from Proposition II.2.5. ∎

4. Again on the structure of DeWitt supermanifolds

We are now in possession of the tools needed to complete the investigation of the relationship between the various categories of DeWitt supermanifolds that we began in Section IV.6. The result we aim at establishing is the following: any H^∞ or G^∞ or G-supermanifold structure on a DeWitt supermanifold determines compatible structures of the two other types (we shall clarify shortly what we mean by 'compatible'). Thus, the sets of isomorphism classes of the following objects

(1) H^∞ DeWitt supermanifolds;

(2) GH^∞ DeWitt supermanifolds;

(3) G^∞ DeWitt supermanifolds;

(4) DeWitt G-supermanifolds;

(5) graded manifolds,

all having the same body manifold X, and the same odd dimension n, are in a one-to-one correspondence. Moreover, anyone of these objects corresponds to a rank n vector bundle over X, and *vice versa*.

We have already established in Section IV.6 the relationship between H^∞ DeWitt supermanifolds and graded manifolds. To complete our analysis, we need the following result.

Proposition 4.1. *Any G^∞ DeWitt supermanifold $(M, \mathcal{A}^\infty)$ carries one and only one compatible G-supermanifold structure.* ■

This amounts to saying that there is a sheaf $\mathcal{A}$ of graded B_L-algebras on M, and a B_L-algebra morphism $\delta: \mathcal{A} \to \mathcal{C}_L^M$ such that $(M, \mathcal{A}, \delta)$ is a DeWitt G-supermanifold, and $\operatorname{Im}\delta = \mathcal{A}^\infty$. Moreover, such a G-supermanifold structure is unique up to isomorphisms.

In accordance with the discussion of Section IV.7, a possible proof for Proposition 4.1 consists in showing that

$$H^k(M, \mathcal{D}er(\mathcal{A}^\infty, \mathcal{K})) = 0 \qquad \text{for} \quad k = 1, 2. \tag{4.1}$$

We recall that the sheaf $\mathcal{K}$ can be regarded as the $(L+1)$-st graded symmetric power of $\mathfrak{N}^\infty$ over $\mathcal{A}^\infty$, where $\mathfrak{N}^\infty$ is the sheaf of nilpotents of $\mathcal{A}^\infty$. Eq. (4.1) is proved by a sequence of partial results. We start with a key result which we take from [**Rt2**].

Lemma 4.1. *Let $(B_L^{m,n}, \mathcal{G})$ be the standard G-supermanifold over $B_L^{m,n}$. One has an isomorphism of sheaves of graded B_L-modules $\mathcal{D}er(\mathcal{G}^\infty, \mathcal{K}) \simeq \mathcal{D}er(\mathcal{G}, \mathcal{K})$.*

Proof. The map $\delta: \mathcal{G} \to \mathcal{G}^\infty$ induces a morphism

$$\begin{aligned} \mathcal{D}er(\mathcal{G}^\infty, \mathcal{K}) &\to \mathcal{D}er(\mathcal{G}, \mathcal{K}) \\ D &\mapsto \bar{D} \end{aligned} \tag{4.2}$$

given by $\bar{D}(f) = D(\delta(f))$. Since δ is surjective (cf. Proposition III.4.1), the morphism (4.2) is injective. To prove its surjectivity, consider coordinates $(x^1, \dots, x^m, y^1, \dots, y^n)$, and for any $\bar{D} \in \mathcal{D}er(\mathcal{G}, \mathcal{K})(U)$, with $U \subset B_L^{m,n}$, let $\bar{D} = \sum_{i=1}^m D^i \frac{\partial}{\partial x^i} + \sum_{\alpha=1}^n D^\alpha \frac{\partial}{\partial y^\alpha}$, with D^i, $D^\alpha \in \mathcal{K}(U)$. Since in this case $\mathcal{K} \simeq \mathfrak{N}^{L+1}$, where $\mathfrak{N}$ is the nilpotent ideal of $\mathcal{A}$ (cf. Lemma IV.7.6), and since $\mathfrak{N}$ is locally generated by the elements $\{\beta_i,\, i = 1, \dots, L\}$ of the canonical basis of $\mathbf{R}^L$, and by the odd coordinates y^α, we have

$$\frac{\partial}{\partial x^i}(\mathfrak{N}^{L+1}) \subset \mathfrak{N}^{L+1}, \qquad \frac{\partial}{\partial y^\alpha}(\mathfrak{N}^{L+1}) \subset \mathfrak{N}^L,$$

and therefore $\bar{D}(\mathcal{K}) \subset \mathfrak{N}^{L+1} \cdot \mathfrak{N}^L = 0$, so that $\bar{D}$ lies in the image of the morphism (4.2); indeed, one can define $D(f) = \bar{D}(g)$, where g is any section in $\mathcal{G}(U)$ which is mapped to $f \in \mathcal{G}^\infty(U)$ by δ. ■

Lemma 4.2. *The sheaf $\mathcal{K}$ over $B_L^{m,n}$ is acyclic.*

Proof. One writes the long cohomology exact sequence associated with the sequence (IV.7.5) and applies Propositions 3.1 and 3.5. ∎

Corollary 4.1. *The sheaf $\mathcal{D}er(\mathcal{G}^\infty, \mathcal{K})$ over $B_L^{m,n}$ is acyclic.*

Proof. From Lemma 4.1 we obtain $\mathcal{D}er(\mathcal{G}^\infty, \mathcal{K}) \simeq \mathcal{K} \otimes_{\mathcal{G}} \mathcal{D}er\mathcal{G} \simeq \mathcal{K}^{m|n}$, the second isomorphism being due to the fact that $Der\mathcal{G}$ is free of rank (m,n); Lemma 4.2 allows to conclude. ∎

Corollary 4.2. *The sheaf $\mathcal{D}er(\mathcal{A}^\infty, \mathcal{K})$ over a G^∞ DeWitt supermanifold $(M, \mathcal{A}^\infty)$ is acyclic.*

Proof. In view of Corollary 4.1, any $p \in M_B$ has a system of neighbourhoods whose counterimages are acyclic for the sheaf $\mathcal{D}er(\mathcal{A}^\infty, \mathcal{K})$. By Proposition II.2.5 we obtain $H^k(M, Der(\mathcal{A}^\infty, \mathcal{K}) \simeq H^k(M_B, \Phi_* Der(\mathcal{A}^\infty, \mathcal{K}))$ for all $k > 0$. But $\Phi_* Der(\mathcal{A}^\infty, \mathcal{K})$ is a module over the fine sheaf $\Phi_* \mathcal{A}^\infty$, and therefore, by Corollaries 3.3 and 3.4 of Chapter II, it is acyclic. ∎

Corollary 4.2 implies Eq.(4.1), and therefore provides a proof of Proposition 4.1.

Now we examine various relationships that occur between DeWitt supermanifolds of different categories.

1. A GH^∞ or H^∞ DeWitt supermanifold produces a DeWitt G-supermanifold simply by tensoring its structure sheaf by B_L.

2. A G^∞ DeWitt supermanifold yields a DeWitt G-supermanifold through the extension procedure discussed in Section IV.7, which is always possible as shown in Proposition 4.1.

3. A DeWitt G-supermanifold $(M, \mathcal{A})$ produces an H^∞ (and therefore GH^∞ and G^∞) supermanifold as follows. Let $\mathbf{R}_L$ be the real field $\mathbf{R}$ regarded as a B_L-module by means of the body map $\sigma: B_L \to \mathbf{R}$, and let

$$\mathcal{H}_M = \mathcal{A} \otimes_{B_L} \mathbf{R}_L. \tag{4.3}$$

In order to prove that $(M, \mathcal{H}_M)$ is an H^∞ supermanifold, let us investigate the effects of the recipe (4.3) in the case $(M, \mathcal{A}) = (B_L^{m,n}, \mathcal{G})$. Since $\mathcal{G} \simeq \mathcal{H}^\infty \otimes_{\mathbf{R}} B_L$, we have $\mathcal{G} \otimes_{B_L} \mathbf{R}_L \simeq \mathcal{H}^\infty$. Therefore, the space $(M, \mathcal{H}_M)$ is locally isomorphic with the space $(B_L^{m,n}, \mathcal{H}^\infty)$; that is to say, $(M, \mathcal{H}_M)$ is an H^∞ supermanifold.

The situation can be described pictorially by the following 'diagram':

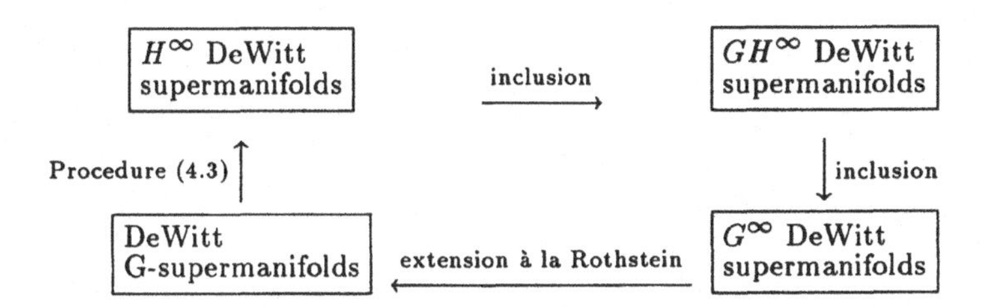

Any time we make a loop in this diagram we get back (up to isomorphism) to the manifold we started from. In this sense, the various supermanifold structures that can be imposed on a DeWitt supermanifold are compatible.

More formally, we have obtained the following result.

Proposition 4.2. *The sets of isomorphism classes of*

(1) H^∞ *DeWitt supermanifolds;*
(2) GH^∞ *DeWitt supermanifolds;*
(3) G^∞ *DeWitt supermanifolds;*
(4) *DeWitt G-supermanifolds;*

are isomorphic. ∎

Chapter VI
Geometry of supervector bundles

E o esplendor das mapas, caminho abstracto para a imaginação concreta
Letras e riscos irregulares abrindo para a maravilha
F. PESSOA

Our purpose in this Chapter is to study the main features of the theory of vector bundles in the category of G-supermanifolds.

Connections on supervector bundles are introduced in Section 1; a distinguished feature of superbundles, which stresses once more the similarity between supermanifolds and complex manifolds, is that a superbundle may not admit connections on it. Thus, one can define a cohomological invariant of the bundle (its *Atiyah class*) which vanishes if and only if the bundle admits connections.

Section 2 is devoted to superline bundles, and their cohomological classification, while in Section 3 a theory of Chern classes of complex supervector bundles is presented. The interesting property here is that supervector bundles have both 'even' and 'odd' Chern classes, in consideration of the fact that the monoid of supervector bundles over a fixed G-supermanifold is naturally graded.

Subsequently, in Section 4 we discuss how Chern classes can be represented in terms of curvature forms.

1. Connections

Supervector bundles or principal superfibre bundles over supermanifolds do not necessarily carry connections; since supermanifolds may not admit partitions of unity, the usual proofs of the existence of connections do not apply. Indeed, superbundles are in this respect akin to holomorphic bundles on complex manifolds [**Ati,Ksz**].

The problem of the existence of connections on superbundles, apart from its own interest from a purely geometric viewpoint, is relevant to string theory and field theory over topologically non-trivial supermanifolds; the use of non-trivial superspaces is important e.g. for the solution of the anomaly problem for supersymmetric gauge theory [**BoPT2,BruL**].

In this Section, proceeding largely by analogy with complex manifolds, we analyze this problem. It turns out that one can attach to any superbundle Ξ a cohomology class $b(\Xi)$ whose vanishing is equivalent to the existence of a connection on Ξ. Another important feature of the theory is that, whenever the base supermanifold is DeWitt, a superbundle carries connections.

Let $\Xi \equiv ((\xi, \mathcal{A}_\xi), \pi)$ be an SVB of rank (r, s) on a supermanifold $(M, \mathcal{A}, \delta)$; thus, the sheaf $\mathcal{E}$ of G-sections of the projection $\pi : (\xi, \mathcal{A}_\xi) \to (M, \mathcal{A})$ is a locally free $\mathcal{A}$-module of rank (r, s).

Definition 1.1. *A connection ∇ on Ξ is an even morphism of sheaves of graded B_L-modules*

$$\nabla : \mathcal{E} \to \mathcal{H}om(\mathcal{D}er\mathcal{A}, \mathcal{E}) \equiv \mathcal{E} \otimes_{\mathcal{A}} \Omega^1_{\mathcal{A}}, \tag{1.1}$$

satisfying the Leibniz rule

$$\nabla(sf) = \nabla(s)f + s \otimes df \quad \forall f \in \mathcal{A}(U),\ s \in \mathcal{E}(U), \quad \text{and } \forall \text{ open } U \subset M\,.$$

Here $\Omega^1_{\mathcal{A}}$ is the sheaf of graded differential 1-forms on $(M, \mathcal{A})$ (cf. Section IV.4). If Ξ is a trivial bundle, and an isomorphism $\mathcal{E} \simeq \mathcal{A}^{r|s}$ has been fixed, there is a canonical 'flat' connection on Ξ, given by

$$\nabla(\sum e_A\, s^A) = \sum e_A \otimes ds^A\,,$$

where $\{e_A\}$ is the canonical basis of $\mathcal{A}^{r|s}$.

It is convenient to introduce the sheaf $\mathcal{J}(\Xi) = \mathcal{E} \oplus (\mathcal{E} \otimes_{\mathcal{A}} \Omega^1_{\mathcal{A}})$, equipped with the structure of graded $\mathcal{A}$-module induced by

$$(s \oplus \alpha)f = sf \oplus (\alpha f + s \otimes df)$$

for all $f \in \mathcal{A}(U)$, $s \in \mathcal{E}(U)$, $\alpha \in (\mathcal{E} \otimes_{\mathcal{A}} \Omega^1_{\mathcal{A}})(U)$, and for all open $U \subset M$. $\mathcal{J}(\Xi)$ is apparently the first jet extension of the sheaf $\mathcal{E}$ (cf. [**Ksz**] for the definition of jet extension in the ordinary case, and [HeM1], where connections are also considered, for the case of graded manifolds).

We consider the exact sequence of graded $\mathcal{A}$-modules

$$0 \rightarrow \mathcal{E} \otimes \Omega^1_{\mathcal{A}} \rightarrow \mathcal{J}(\Xi) \rightarrow \mathcal{E} \rightarrow 0, \tag{1.2}$$

which need not be split,[1] due to the non-trivial $\mathcal{A}$-module structure of $\mathcal{J}(\Xi)$.

Proposition 1.1. *The sequence (1.2) is split if and only if there exists a connection on* Ξ.

Proof. Given a connection ∇ on Ξ, the map $\lambda : \mathcal{E} \rightarrow \mathcal{J}(\Xi)$ given by $\lambda(s) = s \oplus \nabla(s)$ is a splitting of (1.2). Conversely, denoting by $\pi_2 : \mathcal{J}(\Xi) \rightarrow \mathcal{E} \otimes_{\mathcal{A}} \Omega^1_{\mathcal{A}}$ the projection, a splitting λ of (1.2) determines the connection $\nabla = \pi_2 \circ \lambda$. ■

The sequence (1.2) determines an element $b(\Xi) \in H^1(M, \mathcal{H}om(\mathcal{E}, \mathcal{E} \otimes_{\mathcal{A}} \Omega^1_{\mathcal{A}}))$, that we call the *Atiyah class* of Ξ, in the following standard way [Hirz]. We apply the functor $\mathcal{H}om(\mathcal{E}, \bullet)$ to the exact sequence (1.2); since $\mathcal{E}$ is locally free, we obtain another exact sequence,[2]

$$0 \rightarrow \mathcal{H}om(\mathcal{E}, \mathcal{E} \otimes_{\mathcal{A}} \Omega^1_{\mathcal{A}}) \rightarrow \mathcal{H}om(\mathcal{E}, \mathcal{J}(\Xi)) \rightarrow \mathcal{H}om(\mathcal{E}, \mathcal{E}) \rightarrow 0. \tag{1.3}$$

The induced cohomology sequence contains the segment

$$\begin{aligned} H^0(M, \mathcal{H}om(\mathcal{E}, \mathcal{E} \otimes_{\mathcal{A}} \Omega^1_{\mathcal{A}})) \rightarrow H^0(M, \mathcal{H}om(\mathcal{E}, \mathcal{J}(\Xi))) \rightarrow \\ \rightarrow H^0(M, \mathcal{H}om(\mathcal{E}, \mathcal{E})) \rightarrow H^1(M, \mathcal{H}om(\mathcal{E}, \mathcal{E} \otimes_{\mathcal{A}} \Omega^1_{\mathcal{A}})). \end{aligned}$$

The identity morphism Id: $\mathcal{E} \rightarrow \mathcal{E}$ is of course an element in $H^0(M, \mathcal{H}om(\mathcal{E}, \mathcal{E}))$; its image in $H^1(M, \mathcal{H}om(\mathcal{E}, \mathcal{E} \otimes_{\mathcal{A}} \Omega^1_{\mathcal{A}}))$ is by definition the Atiyah class $b(\Xi)$

[1] We recall that an exact sequence (say, of modules) $0 \rightarrow M \rightarrow N \xrightarrow{p} Q \rightarrow 0$ is *split* if $N \simeq M \oplus Q$. A *splitting* of the exact sequence is a morphism $i : Q \rightarrow N$ such that $p \circ i = \mathrm{Id}$; the existence of at least one of such a morphism is apparently equivalent to the splitness of the sequence. Let us also notice that in ordinary differential geometry all exact sequences of smooth vector bundles do split, due to the existence of smooth partitions of unity [Hus].

[2] In view of Proposition I.2.2, the sequence (1.3) can also be written

$$0 \rightarrow (\mathcal{E} \otimes_{\mathcal{A}} \Omega^1_{\mathcal{A}}) \otimes \mathcal{E}^* \rightarrow \mathcal{J}(\Xi) \otimes \mathcal{E}^* \rightarrow \mathcal{E} \otimes \mathcal{E}^* \rightarrow 0$$

and is therefore obtained from (1.2) by tensoring with $\mathcal{E}^*$.

of Ξ. The vanishing of $b(\Xi)$ is equivalent to the existence of an element $\lambda \in H^0(M, \mathcal{H}om(\mathcal{E}, \mathcal{J}(\Xi)))$ whose image is I, which is no more than a splitting of the sequence (1.2).

A cocycle representing $b(\Xi)$ can be obtained in terms of a local trivialization of Ξ. Indeed, let ∇_j be the flat connection on $\Xi_{|U_j}$ determined by a fixed trivialization of Ξ relative to a cover $\{U_j\}$ of M. The 1-cocycle

$$\{b_{jk} \equiv \nabla_k - \nabla_j\} \tag{1.4}$$

is a representative of $b(\Xi)$.

It is possible to express $b(\Xi)$ in terms of the transition morphisms of Ξ; these can be regarded as automorphisms of the sheaf $\mathcal{A}^{r|s}{}_{|U_j \cap U_k}$:

$$g_{jk}\colon \mathcal{A}^{r|s}{}_{|U_j \cap U_k} \to \mathcal{A}^{r|s}{}_{|U_j \cap U_k}$$

(cf. Section IV.3). A trivialization of Ξ given by an open cover $\{U_j\}$ with sections $s_j \in \mathcal{E}_{|U_j}$ determines transition morphisms such that $\hat{s}_j = g_{jk}\hat{s}_k$,[3] where $\hat{s}_j$ and $\hat{s}_k$ are the sections s_j and s_k restricted to $U_j \cap U_k$ and represented in $\mathcal{A}^{r|s}{}_{|U_j \cap U_k}$. Inserting this into Eq. (1.4) we obtain

$$b_{jk} = -dg_{jk}\, g_{jk}^{-1}\,. \tag{1.5}$$

Since in general the structure sheaf $\mathcal{A}$ of a supermanifold is not acyclic, the sheaf $\mathcal{H}om(\mathcal{E}, \mathcal{E} \otimes_{\mathcal{A}} \Omega^1_{\mathcal{A}}))$ has non-trivial cohomology as well, so that the Atiyah class of an SVB need not vanish; therefore, in contrast to smooth bundles, and in analogy with holomorphic bundles, a superbundle does not necessarily admit connections.

EXAMPLE 1.1. We construct a non-trivial SVB which admits connections, even though the structure sheaf of its base supermanifold is not acyclic. We consider the GH^∞ supermanifold described in Example III.2.1; by tensoring its structure sheaf by B_L, we obtain a G-supermanifold $(M, \mathcal{A})$. We notice parenthetically that the graded tangent bundle to $(M, \mathcal{A})$ is a trivial rank (1,0) SVB; i.e., it is a trivial superline bundle (cf. Section IV.3 and next Section).

We consider the rank $(1, 0)$ SVB Ξ defined by the transition morphisms

$$g_{12}|V_1 = \mathrm{Id}, \qquad g_{12}|V_2 = -\,\mathrm{Id}\,;$$

[3] Juxtaposition here denotes matrix multiplication.

V_1 and V_2 are the connected components of $(U_1 \cap U_2) \times \mathbf{R}$ (the sets U_1, U_2 were defined in Example III.2.1). Topologically, the total space of Ξ is a Möbius band times a Euclidean space. Ξ is not trivial, while its Atiyah class vanishes as a consequence of Eq. (1.5), so that it carries a connection. ▲

In the next Section, when the cohomological classification of superline bundles will become available, we shall demonstrate the existence of SVB's which do not admit connections.

On the other hand, in the case of DeWitt supermanifolds we have the following result, which relies on their cohomological triviality.

Proposition 1.2. *The Atiyah class of any SVB over a DeWitt G-supermanifold $(M, \mathcal{A})$ vanishes.*

Proof. The sheaf $\mathcal{H}om(\mathcal{E}, \mathcal{E} \otimes_{\mathcal{A}} \Omega^1_{\mathcal{A}}))$ is a locally free $\mathcal{A}$-module, so that it is acyclic by Corollary V.3.2. ■

Curvature. Having fixed a connection ∇ on the SVB $\Xi = ((\xi, \mathcal{A}_\xi), \pi)$, the morphism (1.1) can be extended to morphisms (denoted by the same symbol)

$$\nabla : \mathcal{E} \otimes_{\mathcal{A}} \Omega^p_{\mathcal{A}} \to \mathcal{E} \otimes_{\mathcal{A}} \Omega^{p+1}_{\mathcal{A}}, \quad p \geq 0. \tag{1.6}$$

A simple direct computation shows that the morphism

$$\nabla^2 : \mathcal{E} \to \mathcal{E} \otimes_{\mathcal{A}} \Omega^2_{\mathcal{A}}$$

is $\mathcal{A}$-linear, and therefore determines an element $R \in \operatorname{Hom}(\mathcal{E}, \mathcal{E} \otimes_{\mathcal{A}} \Omega^2_{\mathcal{A}})$, that is, a global section of the sheaf $\mathcal{H}om(\mathcal{E}, \mathcal{E}) \otimes_{\mathcal{A}} \Omega^2_{\mathcal{A}}$, i.e. a graded differential 2-form with values in $\mathcal{H}om(\mathcal{E}, \mathcal{E})$, which is the *curvature* of the connection ∇. As usual, this obeys the *Bianchi identity*:

$$\nabla R = 0. \tag{1.7}$$

Connection and curvature forms. If we introduce a cover $\{U_j\}$ of M over which Ξ trivializes, and $\{e_1^{(j)}, \dots, e_{r+s}^{(j)}\}$ is a homogeneous basis of $\mathcal{E}(U_j)$, we can represent a connection ∇ over Ξ in terms of a collection $\{\nabla^{(j)}\}$ of matrix-valued graded differential 1-forms, each defined on the open set U_j (local connection forms); the curvature R can be similarly represented by a collection $\{R^{(j)}\}$ of matrix-valued graded differential 2-forms (local curvature forms). To this end, we set

$$\nabla(e_A^{(j)}) = \sum_{B=1}^{r+s} e_B^{(j)} \otimes \nabla_{AB}^{(j)}, \qquad R(e_A^{(j)}) = \sum_{B=1}^{r+s} e_B^{(j)} \otimes R_{AB}^{(j)},$$

with the index A running from 1 to $r+s$. In terms of these forms, the definition of curvature reads

$$R_{AB}^{(j)} = d\nabla_{AB}^{(j)} + \sum_{C=1}^{r+s} \nabla_{AC}^{(j)} \wedge \nabla_{CB}^{(j)}$$

(this is the so-called *Cartan structural equation*), while the Bianchi identity reads

$$dR_{AB}^{(j)} + \sum_{C=1}^{r+s} \left(\nabla_{AC}^{(j)} \wedge R_{CB}^{(j)} - R_{AC}^{(j)} \wedge \nabla_{CB}^{(j)}\right) = 0\,.$$

On the overlap $U_j \cap U_k$ of two trivializing patches there are two different local connection (or curvature) forms, and these are intertwined by the usual relations

$$\nabla^{(j)} = g_{kj}^{-1}\,\nabla^{(k)}\,g_{kj} + g_{kj}^{-1}\,dg_{kj} \tag{1.8}$$

$$R^{(j)} = g_{kj}^{-1}\,R^{(k)}\,g_{kj}\,. \tag{1.9}$$

2. Superline bundles

In this and in the following Sections, we deal with a theory of characteristic classes for *complex* supervector bundles (CSVB'S) which parallels the usual theory of Chern classes for smooth complex vector bundles. Complex supervector bundles are defined exactly in the same way as 'real' SVB's (cf. Section IV.3), but using the complexification $\mathcal{I} = \mathcal{A} \otimes_{\mathbf{R}} \mathbf{C}$ of the structure sheaf of a G-supermanifold $(M, \mathcal{A})$ rather than $\mathcal{A}$ itself. Thus, a rank (r,s) CSVB over $(M, \mathcal{A})$ has a standard fibre whose underlying topological space is $C_L^{r|s}$, while its sheaf of sections is a rank (r,s) locally free graded $\mathcal{I}$-module. Notice that the evaluation map $\delta: \mathcal{A} \to \mathcal{A}^\infty$ extends naturally to a morphism $\delta: \mathcal{I} \to \mathcal{I}^\infty$, where $\mathcal{I}^\infty$ is the complexification of $\mathcal{A}^\infty$.

We consider first the case of *complex superline bundles* (CSLB's), i.e. CSVB's of rank either (1,0) or (0,1). In both cases a CSLB is specified by the assignment of its transition morphisms relative to a cover $\mathfrak{U} = \{U_j\}$ of M; each transition morphism g_{jk} is a section in $\mathcal{I}_0^\star(U_j \cap U_k)$, where $\mathcal{I}_0^\star$ denotes the subsheaf of $\mathcal{I}_0$ whose sections are invertible (the symbol * we use to denote invertible subsheaves should not be confused with the symbol * denoting dual module). The transition morphisms satisfy the multiplicative cocycle condition

$$g_{jk}\, g_{kh}\, g_{hj} = \mathrm{Id}\,,$$

while, on the other hand, two CSLB's are isomorphic if and only if their transition morphisms differ by a coboundary, in the sense that

$$g'_{jk} = \lambda_j\, g_{jk}\, \lambda_k^{-1}\,,$$

where $\{\lambda_j\}$ is a 0-cocycle of $\mathcal{I}_0^\star$ relative to the cover $\mathfrak{U}$.

Thus, the isomorphism classes of CSLB's — having fixed at the outset whether we are dealing with the rank (1,0) or (0,1) case — are in a one-to-one correspondence with the elements of the cohomology group $H^1(M, \mathcal{I}_0^\star)$, where $\mathcal{I}_0^\star$ is considered as a sheaf of abelian groups with respect to its multiplicative structure. This allows us to introduce, as in the ordinary case, an integral cohomology class which, in a sense to be elucidated later, classifies the CSLB's over $(M, \mathcal{A})$.

Obstruction class and super Picard group. We start by defining an exponential map $\exp\colon C_L \to C_L^\star$ by letting

$$\exp z = \sum_{k=0}^{\infty} \frac{(2\pi i z)^k}{k!} \tag{2.1}$$

where for all $z \in C_L$ the series converges in the vector space C_L (here i is the imaginary unit). Hence, there is an exact sequence of abelian groups

$$0 \to \mathbf{Z} \to C_L \xrightarrow{\exp} C_L^\star \to 1\,. \tag{2.2}$$

Applying all this pointwise to C_L-valued G^∞ functions, we obtain an exact sequence

$$0 \to \mathbf{Z} \to \mathcal{I}_0^\infty \xrightarrow{\exp} \mathcal{I}_0^{\infty\star} \to 1\,, \tag{2.3}$$

where we have considered only the even part of the sheaf $\mathcal{I}_0^\infty$ for convenience. We also define an exponential map $\exp\colon \mathcal{I} \to \mathcal{I}^\star$ by the same prescription (2.1).

Complexifying the exact sequence (IV.7.5), and taking the even parts, we obtain an exact sequence

$$0 \longrightarrow \mathcal{D} \longrightarrow \mathcal{I}_0 \longrightarrow \mathcal{I}_0^\infty \longrightarrow 0$$

where $\mathcal{D}$ is a square zero ideal. It follows that on $\mathcal{D}$, the morphism exp reduces to $f \mapsto 1 + 2\pi i f$.

With the aim of extending the exact sequence (2.2) to the sheaf $\mathcal{I}_0$, we consider the commutative diagram

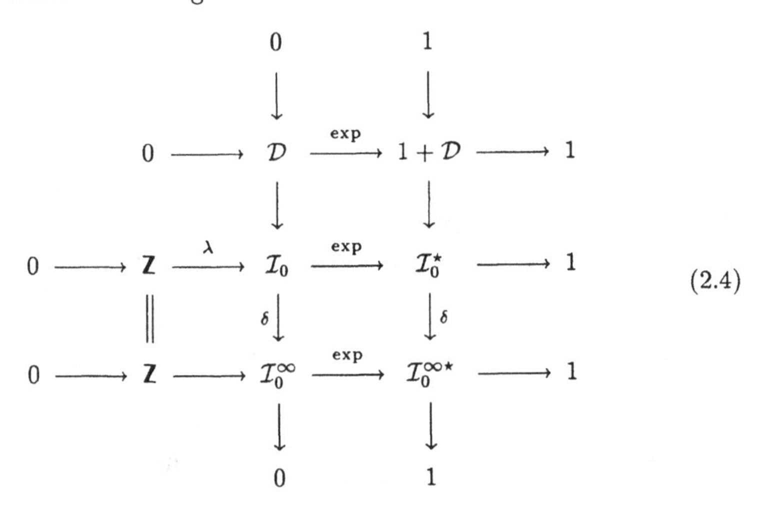

(2.4)

where the abelian groups in the column on the right are taken with their multiplicative structure, and the exactness of the middle row sequence has yet to be proven.

Lemma 2.1. *The sequence*

$$0 \longrightarrow \mathbf{Z} \stackrel{\lambda}{\longrightarrow} \mathcal{I}_0 \stackrel{\exp}{\longrightarrow} \mathcal{I}_0^\star \longrightarrow 1 \tag{2.5}$$

is exact.

Proof. It is obvious that λ is injective, and that $\operatorname{Im}\lambda \subset \operatorname{Ker}\exp$. To show that $\operatorname{Ker}\exp \subset \operatorname{Im}\lambda$ we resort to diagram (2.4). If — for a suitable open set $U \subset M$ — we have $\exp f = 0$, then $\delta(f) = z \in \mathbf{Z}$. Setting $f = z + k$, we have $k \in \mathcal{D}(U)$. Then $\exp(f) = 1$ implies $k = 0$, i.e. $f \in \mathbf{Z}$.

To show that exp is surjective, let us consider $f \in \mathcal{I}_0^\star(U)$. There is a $g \in \mathcal{I}_0(U)$ such that $\delta(\exp(g)) = \delta(f)$, so that $f - \exp(g) \in \mathcal{D}(U)$. Since $\exp(g)$ is invertible, we may set $f - \exp(g) = \exp(g)\, 2\pi i\, h$ with $h \in \mathcal{D}(U)$, so that $f = \exp(g + h)$. ∎

We now consider the exact cohomology sequence induced by (2.5); it contains the segment

$$H^1(M,\mathbf{Z}) \to H^1(M,\mathcal{I}_0) \to H^1(M,\mathcal{I}_0^\star) \xrightarrow{\partial} H^2(M,\mathbf{Z}) \to H^2(M,\mathcal{I}_0)\,. \tag{2.6}$$

Let Λ be a CSLB over $(M, \mathcal{A})$; we denote by the same symbol the class it determines in $H^1(M, \mathcal{I}_0^\star)$.

Definition 2.1. *The element $\partial(\Lambda) \in H^2(M, \mathbf{Z})$ is the obstruction class of the CSLB Λ.*

In the case of smooth complex line bundles over smooth manifolds, since the relevant structure sheaf is acyclic, the obstruction map ∂ is an isomorphism, that is to say, two line bundles are isomorphic if and only if they have the same obstruction class. In the present case this is no longer true; since the sheaf $\mathcal{I}_0$ has, in general, non-trivial cohomology, the morphism ∂ has both a kernel and a cokernel. However, this same reasoning proves the following result.

Proposition 2.1. *Two CSLB's over a DeWitt supermanifold are isomorphic if and only if they have the same obstruction class.* ∎

Thus, in general CSLB's behave like holomorphic line bundles on complex manifolds; indeed, we may define a super Picard group

$$\mathrm{SPic}^0(M,\mathcal{A}) = \frac{H^1(M,\mathcal{I}_0)}{\operatorname{Im} H^1(M,\mathbf{Z})}$$

which classifies the complex superline bundles whose obstruction class vanishes. Obviously, $\mathrm{SPic}^0(M, \mathcal{A}) = 0$ if $(M, \mathcal{A})$ is DeWitt.

It should be noticed that the super Picard group is neither a topological nor a differentiable invariant, but depends (obviously up to isomorphism) on the G-supermanifold structure. This fact is illustrated once more by Example II.2.1; in that case we certainly have $\mathrm{SPic}^0(M, \mathcal{A}) \neq 0$ (cf. next Example). On the other hand, the underlying smooth manifold $S^1 \times \mathbf{R}$ admits a DeWitt G-supermanifold structure in an obvious way, and the super Picard group of this supermanifold vanishes.

We can now prove the existence of supervector bundles which do not admit connections. To this end we need a preliminary result.

Lemma 2.2. *Let $(M, \mathcal{A})$ be a (1,0) dimensional G-supermanifold. A CSLB Λ over $(M, \mathcal{A})$ admits connections if and only if it can be given constant transition morphisms.*

Proof. The "if" part of this claim follows from Eq. (1.5). To show the converse we notice that the vanishing of the Atiyah class of Λ can be written, again according to Eq. (1.5), in the form

$$d \log g_{jk} = \tau_k - \tau_j$$

with $\{\tau_j\}$ a 0-cochain for the Čech cohomology of $\mathcal{H}om(\mathcal{E}, \mathcal{E} \otimes \Omega^1_{\mathcal{A}})$ with respect to a suitable cover of M. Since $\dim(M, \mathcal{A}) = (1,0)$, we have $d\tau_j = d\tau_k = 0$, and the cover can be chosen so as to give $\tau_j = d\lambda_j$ for all j's. The transition morphisms $g'_{jk} = \exp(\lambda_j)\, g_{jk} \exp(-\lambda_k)$ are equivalent to the g_{jk}'s and are constant. ∎

EXAMPLE 2.1. We consider again the G-supermanifold built over the GH^∞ supermanifold of Example III.2.1 as the base supermanifold. By the previous Lemma, we can prove that there are CSLB's on (M, A) which do not have connections simply by showing that there are CSLB's on $(M, \mathcal{A})$ which cannot be given constant transition morphisms. Since a CSLB with constant transition morphisms determines an element of $H^1(M, (C_L)_0^\star)$, this amounts to saying that it is not possible to find a surjective morphism $H^1(M, (C_L)_0^\star) \to H^1(M, \mathcal{I}_0^\star)$. In our example, $H^1(M, (C_L)_0^\star) \simeq (C_L)_0^\star$; the group $H^1(M, \mathcal{I}_0^\star)$ is computed by considering the exact sheaf sequence (2.5), which induces the exact cohomology sequence

$$\begin{aligned} 0 \to H^0(M, \mathbf{Z}) \to H^0(M, \mathcal{I}_0) \to H^0(M, \mathcal{I}_0^\star) \to \\ \to H^1(M, \mathbf{Z}) \to H^1(M, \mathcal{I}_0) \to H^1(M, \mathcal{I}_0^\star) \to 0\,; \end{aligned}$$

recalling Eq. (II.2.14) we obtain

$$\begin{aligned} 0 \to \mathbf{Z} \to \mathbf{C} \oplus [C^\infty(\mathbf{R}) \otimes \mathfrak{P}_L] \to \mathbf{C}^\star \oplus [C^\infty(\mathbf{R}) \otimes \mathfrak{P}_L] \to \\ \to \mathbf{Z} \to C^\infty(\mathbf{R}) \otimes \mathfrak{P}_L \to H^1(M, \mathcal{I}_0^\star) \to 0 \end{aligned}$$

where $\mathfrak{P}_L$ is the nilpotent ideal of $(C_L)_0$. From this we obtain by direct computation

$$H^1(M, \mathcal{I}_0^\star) \simeq C^\infty(\mathbf{R};\, S^1) \oplus (C^\infty(\mathbf{R}) \otimes \mathfrak{P}_L)\,,$$

where $C^\infty(\mathbf{R}; S^1)$ is the group of smooth maps from the real line to S^1. Thus, $H^1(M,(C_L)_0^\star)$ is finite-dimensional over $\mathbf{R}$, while $H^1(M,\mathcal{I}_0^\star)$ is infinite-dimensional, so that a surjection from the first space onto the second cannot exist. ▲

Underlying G^∞ bundles. Any CSLB Λ on a G-supermanifold $(M,\mathcal{A})$ has an underlying G^∞ superline bundle (cf. Section IV.3), which we denote by $\delta(\Lambda)$. If $\{g_{jk}\}$ is a set of transition morphisms, then $\delta(\Lambda)$ can be given transition functions $\{\delta(g_{jk})\}$; moreover, the morphism $H^1(M,\mathcal{I}_0^\star) \to H^1(M,\mathcal{I}_0^{\infty\star})$ induced by $\delta\colon \mathcal{I}_0^\star \to \mathcal{I}_0^{\infty\star}$ maps the isomorphism class of Λ to the isomorphism class of $\delta(\Lambda)$. An obstruction class can be attached to $\delta(\Lambda)$ by means of the exponential sheaf sequence (2.3); the cohomology diagram obtained from

$$\begin{array}{ccccccccc} 0 & \longrightarrow & \mathbf{Z} & \xrightarrow{\lambda} & \mathcal{I}_0 & \xrightarrow{\exp} & \mathcal{I}_0^\star & \longrightarrow & 1 \\ & & \| & & \downarrow{\scriptstyle\delta} & & \downarrow{\scriptstyle\delta} & & \\ 0 & \longrightarrow & \mathbf{Z} & \longrightarrow & \mathcal{I}_0^\infty & \xrightarrow{\exp} & \mathcal{I}_0^{\infty\star} & \longrightarrow & 1 \end{array}$$

shows that the obstruction classes of Λ and $\delta(\Lambda)$ can be identified.

Associated smooth bundles. Given a complex superline bundle Λ over M, we can associate with it a smooth line bundles over the smooth manifold underlying M, that with a slight abuse of language, we again call M. Let us consider the sheaf morphism $\rho\colon \mathcal{I}_0 \to \mathcal{C}_M$ defined by the composition

$$\mathcal{I}_0 \xrightarrow{\delta} \mathcal{I}_0^\infty \xrightarrow{\sigma} \mathcal{C}_M\,, \tag{2.7}$$

where σ is the body map, and $\mathcal{C}_M$ is now the sheaf of germs of smooth C-valued functions on M. There is a commutative diagram

$$\begin{array}{ccccccccc} 0 & \longrightarrow & \mathbf{Z} & \longrightarrow & \mathcal{I}_0 & \longrightarrow & \mathcal{I}_0^\star & \longrightarrow & 0 \\ & & \| & & \downarrow{\scriptstyle\rho} & & \downarrow{\scriptstyle\rho} & & \\ 0 & \longrightarrow & \mathbf{Z} & \longrightarrow & \mathcal{C}_M & \longrightarrow & \mathcal{C}_M^\star & \longrightarrow & 0 \end{array}$$

which induces the commutative cohomology diagram

$$\begin{array}{ccccccc} H^1(M,\mathbf{Z}) & \longrightarrow & H^1(M,\mathcal{I}_0) & \longrightarrow & H^1(M,\mathcal{I}_0^\star) & \longrightarrow & H^2(M,\mathbf{Z}) \\ & & \downarrow{\scriptstyle\rho} & & \downarrow{\scriptstyle\rho} & & \downarrow{\scriptstyle\mathrm{Id}} \\ & & 0 & \longrightarrow & H^1(M,\mathcal{C}_M^\star) & \longrightarrow & H^2(M,\mathbf{Z}) \end{array}$$

(one has $H^1(M, \mathcal{C}_M) = 0$ since $\mathcal{C}_M$ is fine). According to this diagram, $\rho(\Lambda)$ is a smooth line bundle over M with the same obstruction class as Λ; moreover, the transition functions of $\rho(\Lambda)$ are obtained from those of Λ by evaluating with δ and taking the body. Since smooth line bundles are classified by their obstruction class, while superline bundles are not, non-isomorphic superline bundle may have isomorphic associated smooth line bundles. Consider for instance a non-trivial CSLB Λ over the supermanifold of Example III.2.1 (cf. Example 2.1): since all smooth complex line bundles over $S^1 \times \mathbf{R}$ are trivial, Λ and the trivial CSLB over $(M, \mathcal{A})$ have the same associated smooth bundle. The spaces of superline bundles, whose associated smooth line bundles are isomorphic, are obviously isomorphic with $\mathrm{SPic}^0(M, \mathcal{A})$.

Holomorphic superline bundles. Holomorphic supervector bundles over complex G-supermanifolds are defined along the same lines as supervector bundles over real G-supermanifolds (see Section IV.3). In particular, holomorphic superline bundles (HSLB's) over a complex G-supermanifold $(M, \mathcal{B})$ are in correspondence to rank $(1,0)$ or $(0,1)$ locally free $\mathcal{B}$-modules, so that their isomorphism classes can be identified with elements in $H^1(M, \mathcal{B}_0^\star)$.

Superline bundles over DeWitt supermanifolds. If (M, A) is a (real) DeWitt G-supermanifold, the sheaf $\mathcal{I}_0$ is acyclic; then the obstruction morphism $\partial \colon H^1(M, \mathcal{I}_0^\star) \to H^2(M, \mathbf{Z})$ is bijective, and the CSLB's over $(M, \mathcal{A})$ are classified by their obstruction class. Moreover, M is homotopic to its body M_B, so that $H^2(M, \mathbf{Z}) \simeq H^2(M_B, \mathbf{Z})$; we therefore expect an isomorphism $H^1(M, \mathcal{I}_0^\star) \simeq H^1(M_B, \mathcal{C}_{M_B}^\star)$ to hold. Indeed, it suffices to consider the exact sequence of sheaves over M_B

$$1 \to \mathcal{F} \to \Phi_* \mathcal{I}_0^* \to \mathcal{C}_{M_B}^\star \to 1 \tag{2.8}$$

where $\mathcal{F}$ is the subgroup in $\Phi_* \mathcal{I}_0$ generated over $\mathcal{C}_{M_B}^\star$ by elements in $1 + (\Phi_* \mathcal{I}_0)^2$. The sheaf $\mathcal{F}$ is acyclic because it is a $\mathcal{C}_{M_B}$-module via the exponential map, so that $H^1(M_B, \Phi_* \mathcal{I}_0^\star) \simeq H^1(M_B, \mathcal{C}_{M_B}^\star)$. On the other hand, any point $p \in M_B$ has a system of neighbourhoods $\{U\}$ such that $H^k(U, \mathbf{Z}) = 0$ for $k > 0$ which, in view of the exact sequence (2.5) and of the acyclicity of $\mathcal{I}_0$, implies $H^k(U, \Phi_* \mathcal{I}_0^\star) = 0$ for $k > 0$. The second condition of Proposition II.2.5 is therefore fulfilled, so that $H^1(M_B, \Phi_* \mathcal{I}_0^\star) \simeq H^1(M, \mathcal{I}_0^\star)$, which gives the required isomorphism $H^1(M, \mathcal{I}_0^\star) \simeq H^1(M_B, \mathcal{C}_{M_B}^\star)$. Thus we have proved the following result:

Proposition 2.2. *Let $(M, \mathcal{A})$ be a DeWitt G-supermanifold with body M_B. There is a one-to-one correspondence between CSLB's over $(M, \mathcal{A})$ and smooth complex line bundles over M_B.*

If $(M, \mathcal{B})$ is a complex DeWitt supermanifold, the body M_B of M is a complex manifold; if $\mathcal{O}$ denotes the sheaf of germs of holomorphic functions on M_B, reasoning as in the real case one obtains a commutative diagram

$$\begin{array}{ccccccc} H^1(M,\mathbb{Z}) & \longrightarrow & H^1(M,\mathcal{B}_0) & \longrightarrow & H^1(M,\mathcal{B}_0^\star) & \longrightarrow & H^2(M,\mathbb{Z}) \\ \downarrow & & \downarrow & & \downarrow & & \downarrow \\ H^1(M_B,\mathbb{Z}) & \longrightarrow & H^1(M_B,\mathcal{O}) & \longrightarrow & H^1(M_B,\mathcal{O}^*) & \longrightarrow & H^2(M_B,\mathbb{Z}) \end{array} \quad . \tag{2.9}$$

In this case the body manifold has a Picard group as well,

$$\mathrm{Pic}^0(M_B) = \frac{H^1(M_B,\mathcal{O})}{\mathrm{Im}\, H^1(M_B,\mathbb{Z})},$$

and the diagram

$$\begin{array}{ccccccc} H^1(M,\mathbb{Z}) & \longrightarrow & H^1(M,\mathcal{B}_0) & \longrightarrow & \mathrm{SPic}^0(M,\mathcal{B}) & \longrightarrow & 0 \\ {\scriptstyle\Phi}\downarrow & & {\scriptstyle\rho}\downarrow & & \downarrow{\scriptstyle\varpi} & & \\ H^1(M_B,\mathbb{Z}) & \longrightarrow & H^1(M_B,\mathcal{O}) & \longrightarrow & \mathrm{Pic}^0(M_B) & \longrightarrow & 0 \end{array}$$

defines a morphism of abelian groups $\varpi\colon \mathrm{SPic}^0(M,\mathcal{B}) \to \mathrm{Pic}^0(M_B)$, which in general is neither surjective nor injective. Nonetheless, whenever $(M,\mathcal{B})$ is split (cf. Section IV.6), the fact that $\mathcal{B} \simeq \Phi^{-1} \bigwedge \xi \otimes C_L$ for some holomorphic vector bundle ξ on M_B entails that ϖ is surjective.

EXAMPLE 2.2. Let M_B be a complex torus. After assuming $L = 2$, $L' = 0$, we endow the space $M = M_B \times \mathbb{C}^2$ with the trivial structure of (1,1) complex DeWitt G-supermanifold $(M,\mathcal{B})$, in such a way that $\mathcal{B} \simeq \Phi^{-1}\mathcal{O} \otimes \bigwedge\mathbb{C} \otimes C_L$. Direct computation shows that $\mathrm{SPic}^0(M,\mathcal{B}) = \mathrm{Pic}^0(M_B) \times \mathfrak{P}_L$, so that $\mathrm{Ker}\,\varpi = \mathfrak{P}_L$. This simple situation shows that the morphism ϖ is not injective, even though $(M,\mathcal{B})$ is split. ▲

More generally, in the case when the complex DeWitt supermanifold $(M,\mathcal{B})$ is split, and its body M_B is compact Kähler, its super Picard group is related to the Picard group of M_B in a simple way. Indeed, recalling that the Picard group of an ordinary compact complex Kähler manifold $(X, \mathcal{O}_X)$ is a complex manifold of the same dimension as the complex vector space $H^1(X, \mathcal{O}_X)$ [**GrH**], we obtain the following result.

Proposition 2.3. $\mathrm{SPic}^0(M, \mathcal{B})$ *is a complex DeWitt supermanifold of dimension* $(p, 0)$, *where* $p = \sum_{k=0}^{n} \dim H^1(M_B, \bigwedge^k \xi)$. *The body of* $\mathrm{SPic}^0(M, \mathcal{B})$ *is diffeomorphic with the manifold* $\mathrm{Pic}^0(M_B) \times \bigoplus_{k=1}^{n} H^1(M_B, \bigwedge^k \xi)$. ∎

If $m = n = 1$, M_B is compact, and ξ is a spin structure over M_B, $(M, \mathcal{A})$ is said to be a split super Riemann surface [**Hod**]. In this case, using Serre duality [**Ser**], we obtain

Corollary 2.1. $\dim \mathrm{SPic}^0(M, \mathcal{B}) = (g + q, 0)$, *where* g *is the genus of* M_B, *and* $q = \dim H^0(M_B, \xi)$. ∎

REMARK 2.1. It should be noticed that the dimension of this super Picard varieties disagrees with some results in the literature (cf. e.g. [**GN**]), where for instance the dimension computed in Corollary 2.1 would be (g, q). However, it seems natural to give the super Picard group a supermanifold structure in the same way as the ordinary Picard group is endowed with a complex manifold structure; thus, if for instance we consider a case where $H^1(M_B, \mathbf{Z}) = 0$, the super Picard group $\mathrm{SPic}^0(M, \mathcal{B})$ reduces to $H^1(M, \mathcal{B}_0)$, which is the even part of the graded C_L-module $H^1(M, \mathcal{B})$. If the latter is free — which is always the case when $(M, \mathcal{B})$ is split [**Hod**] — $\mathrm{SPic}^0(M, \mathcal{B})$ has a natural structure of purely even supermanifold, which is compatible with the one described in Proposition 2.3. ▲

3. Characteristic classes

We now proceed to the construction of characteristic classes for supervector bundles; given a G-supermanifold $(M, \mathcal{A})$, and a CSVB Ξ on it, we shall associate with Ξ both *even* and *odd* Chern classes. All these classes will be elements in the cohomology ring $H^\bullet(M, \mathbf{Z})$. The Chern classes we are going to build are meant to be generalizations of the obstruction class of a CSLB, as defined in the previous Section; since the latter coincides with the obstruction class of the underlying G^∞ line bundle, it is quite natural to attach characteristic classes not directly to the CSVB Ξ, but rather to its underlying G^∞ vector bundle. In a sense, we shall associate characteristic classes with the equivalence class of

CSVB's having isomorphic underlying G^∞ bundles. In fact, one cannot expect that integer valued cohomology classes are able to discriminate between CSVB's with the same underlying G^∞ bundle.

Therefore — since this will make our job much easier — in this Section *all SVB's are intended to be G^∞ vector bundles*, and all morphisms are G^∞.

The approach we intend to follow is the constructive one, based on the introduction of the universal bundles via projectivization of the vector bundle, which was devised in the ordinary case by Grothendieck [Gro3].

On G^∞ vector bundles. The definition of G^∞ vector bundles was given in Chapter IV (Definition IV.3.2). Here we wish to define the concepts of *subbundle* and *quotient bundle* for the category of complex G^∞ vector bundles. Let M be a G^∞ supermanifold, and let $p: E \to M$ be a rank (r, s) complex G^∞ vector bundle on it. We say that a collection $\{F_z \subset E_z\}_{z \in M}$ of free rank (h, k) graded submodules of the fibres of E (with $h \leq r$, $k \leq s$) define a subbundle $q: F \to M$ of E if — denoting by F the union $\bigcup_{z\in M} F_z$ — there is a cover $\{U_j\}$ of M and a local trivialization

$$\varphi_j: E_{|U_j} \to U_j \times C_L^{r|s}$$

such that the restriction

$$\varphi_{j|F_{U_j}}: F_{U_j} \to U_j \times C_L^{r|s}$$

takes values in $U_j \times C_L^{h|k}$; here $F_{U_j} = \bigcup_{z\in U_j} F_z$. With this assumption one can indeed equip F with the structure of a rank (h, k) complex G^∞ vector bundle.

Associated with the trivialization $\{\varphi_j\}$ there are transition functions g_{jk} displaying the block structure

$$g_{jk} = \begin{pmatrix} h_{jk} & l_{jk} \\ 0 & w_{jk} \end{pmatrix};$$

the maps h_{jk} are the transition functions of the bundle F. On the other hand, we can define another complex G^∞ vector bundle on M, called the quotient bundle $Q = E/F$, which is the bundle whose fibre at $z \in M$ is the free graded C_L-module $Q_z = E_z/F_z$, with the G^∞ maps w_{jk} as transition functions, relative to the cover $\{U_j\}$.

The fact that F is a subbundle of E, and that Q is their quotient, will be usually stated by saying that the sequence $0 \to F \to E \to Q \to 0$ is exact. It is easily verified that, if $0 \to \Xi' \to \Xi \to \Xi'' \to 0$ is an exact sequence of SVB's in the category of G-supermanifolds (which amounts to saying the corresponding sequence of modules is exact), then their underlying G^∞ bundles give rise to an exact sequence as well.

Projective superspaces. Let r and s be nonnegative integers. We recall that $GL_{C_L}[r|s]$ (henceforth simply denoted by $GL[r|s]$) is the group of even automorphisms of $C_L^{r|s}$, whose elements can be regarded as matrices displaying the block form (I.3.1). After fixing another pair of nonnegative integers h, k with $h \leq r$ and $k \leq s$, we define $GL(h,k;r,s)$ as the subgroup of $GL[r|s]$ whose elements are matrices with the form

$$\begin{pmatrix} A & B & C & D \\ 0 & E & 0 & F \\ G & H & L & P \\ 0 & Q & 0 & R \end{pmatrix},$$

where the blocks have the following dimensions, both horizontal and vertical: h, $r-h$, k, $s-k$. Quite evidently, $GL(h,k;r,s)$ is an H^∞ DeWitt supermanifold with body $Gl(h;r) \times Gl(k;s)$, where $Gl(h;r)$ is the subgroup of matrices in $Gl(r;\mathbf{C})$ (ordinary Lie group) with the form

$$\begin{pmatrix} A & B \\ 0 & C \end{pmatrix}.$$

If we perform the (algebraic) quotient of groups $GL[r|s]/GL(h,k;r,s)$, the resulting space, denoted by $G_{h,k}(r,s)$, can be endowed with a structure of H^∞ DeWitt supermanifold, of even dimension $h(r-h)+k(s-k)$, odd dimension $k(r-h)+h(s-k)$, and body $G_h(r) \times G_k(s)$, where $G_h(r)$ is the Grassmann manifold of h-planes in $\mathbf{C}^r$. The supermanifold $G_{h,k}(r,s)$ parametrizes the rank (h,k) free graded submodules of $C_L^{r|s}$.

Now, let W be a rank (r,s) free graded C_L-module, and let us define

$\mathbf{P}_{1,0}(W)$ = space of rank (1,0) free graded sub-C_L- modules of W

$\mathbf{P}_{0,1}(W)$ = space of rank (0,1) free graded sub-C_L-modules of W.

From the previous discussion it follows also that $\mathbf{P}_{1,0}(W)$ and $\mathbf{P}_{0,1}(W)$ are both DeWitt supermanifolds, with dimensions $(r-1,s)$ and $(s-1,r)$ respectively, and bodies isomorphic with the complex projective spaces $\mathbf{CP}^{r-1}$ and $\mathbf{CP}^{s-1}$. It follows that $\mathbf{P}_{1,0}(W)$ (resp. $\mathbf{P}_{0,1}(W)$) has the same integer cohomology as $\mathbf{CP}^{r-1}$ (resp. $\mathbf{CP}^{s-1}$).

Universal bundles. On $\mathbf{P}_{1,0}(W)$ we may define a tautological bundle S_0, which is the rank (1,0) subbundle of $\mathbf{P}_{1,0}(W) \times W$ formed by the pairs (u, v) such that $v \in u$; analogously, one defines a rank (0,1) tautological bundle S_1 on $\mathbf{P}_{0,1}(W)$, which is a subbundle of $\mathbf{P}_{0,1}(W) \times W$. Now, let $\bar{W}$ be the body of W, i.e. the vector space $\bar{W} = W \otimes_{C_L} \mathbf{C}_L$, where $\mathbf{C}_L$ is $\mathbf{C}$ with the C_L- module structure given by the body map $\sigma: C_L \to \mathbf{C}$; the space $\bar{W}$ is graded, $\bar{W} = \bar{W}_0 \oplus \bar{W}_1$. Denoting by $\bar{S}_i$, $i = 0, 1$, the tautological bundles of the projective spaces $\mathbf{P}(\bar{W}_i)$, the body of S_i (in the sense of DeWitt supermanifolds) is simply $\bar{S}_i$, whence one has commutative diagrams

$$\begin{array}{ccccccccc} 0 & \longrightarrow & S_i & \longrightarrow & \mathbf{P}_{1-i,i}(W) \times W & \longrightarrow & Q_i & \longrightarrow & 0 \\ & & \downarrow & & \downarrow & & \downarrow & & \\ 0 & \longrightarrow & \bar{S}_i & \longrightarrow & \mathbf{P}(\bar{W}_i) \times \bar{W}_i & \longrightarrow & \bar{Q}_i & \longrightarrow & 0 \end{array} \qquad i = 0, 1 \quad (3.1)$$

where Q_i and $\bar{Q}_i$ are by definition the quotient (super)bundles. The following theorem is a straightforward consequence of (3.1) and of classical results concerning the cohomology of projective bundles [MiS].

Proposition 3.1. *The integer cohomology of* $\mathbf{P}_{1,0}(W)$ *is freely generated over* $\mathbf{Z}$ *by* $\{1, x, x^2, \ldots, x^{r-1}\}$, *where* x *is the obstruction class of* S_0. *Analogously, the integer cohomology of* $\mathbf{P}_{0,1}(W)$ *is freely generated over* $\mathbf{Z}$ *by* $\{1, t, t^2, \ldots, t^{s-1}\}$, *where* t *is the obstruction class of* S_1. ■

Let us define the (ordinary) Lie group

$$PGL[r|s] = \frac{GL[r|s]}{(C_L)_0^\star I},$$

together with the canonical projection $\lambda: GL[r|s] \to PGL[r|s]$; as usual, the space $PGL[r|s]$ can be given a structure of H^∞ supermanifold.[4] $PGL[r|s]$ acts in a natural way on $\mathbf{P}_{1,0}(W)$ and $\mathbf{P}_{0,1}(W)$. Given a CSVB $p: E \to M$, whose transition functions relative to a fixed cover are g_{ij}, we define its even and odd projectivizations as follows: $\mathbf{P}_{1,0}(E)$ (resp. $\mathbf{P}_{0,1}(E)$) is the bundle on

[4] As a matter of fact, $PGL[r|s]$ is a Lie supergroup, cf. Chapter VII.

M whose standard fibre over $x \in M$ is $\mathbf{P}_{1,0}(E_x)$ (resp. $\mathbf{P}_{0,1}(E_x)$) and whose transition functions are the maps $\lambda \circ g_{ij}$. We shall denote by $\pi_i: \mathbf{P}_{1-i,i}(E) \to M$, $i = 0, 1$, the bundle projections. The operation of taking the projectivizations is functorial, in the sense that if $f: M \to N$ is a G^∞ morphism, and E is an CSVB over N, there are G^∞ maps $F_i: \mathbf{P}_{1-i,i}(f^{-1}E) \to \mathbf{P}_{1-i,i}(E)$ such that the following diagram commutes:

$$\begin{array}{ccc} \mathbf{P}_{1-i,i}(f^{-1}E) & \xrightarrow{F_i} & \mathbf{P}_{1-i,i}(E) \\ \downarrow & & \downarrow \\ M & \xrightarrow[f]{} & N \end{array} \qquad i = 0, 1.$$

$\mathbf{P}_{1,0}(E)$ and $\mathbf{P}_{0,1}(E)$ carry tautological bundles defined in the obvious way; $S_0(E) \to \mathbf{P}_{1,0}(E)$ has rank (1,0), while $S_1(E) \to \mathbf{P}_{0,1}(E)$ has rank (0,1). There are two tautological exact sequences,

$$0 \to S_i(E) \to \pi_i^{-1}E \to Q_i(E) \to 0, \qquad i = 0, 1.$$

The assignment of the tautological bundles is functorial as well, i.e. there are commutative diagrams

$$\begin{array}{ccccccccc} 0 & \longrightarrow & S_i(f^{-1}E) & \longrightarrow & \pi_i^{-1}(f^{-1}E) & \longrightarrow & Q_i(f^{-1}E) & \longrightarrow & 0 \\ & & \downarrow & & \downarrow & & \downarrow & & \\ 0 & \longrightarrow & S_i(E) & \longrightarrow & \pi_i^{-1}E & \longrightarrow & Q_i(E) & \longrightarrow & 0 \end{array} \qquad i = 0, 1.$$

In order to obtain information about the integer cohomology of the projectivizations of E, we must use the Leray-Hirsch theorem [**Hus**]. We need it in the following weaker form than the one given in [**Hus**]. The cohomology groups involved in the statement can be regarded as sheaf cohomology groups with coefficients in the constant sheaf with stalk K.

Proposition 3.2. (Leray-Hirsch) *Let $p: Q \to M$ be a locally trivial topological bundle, with standard fibre F, and let K be a principal ideal domain.*[5] *Assume there are cohomology classes $\{a_1 \dots a_q\}$ in $H^\bullet(Q, K)$ that when restricted to the*

[5] We recall that a *principal ideal domain* is a commutative ring K with no zero divisors such that every ideal is of the form bK for some $b \in K$.

fibres of Q generate freely over K the cohomology of the fibres with coefficients in K. Then $H(Q,K)$ is a free $H(M,K)$-module generated by $\{a_1 \dots a_q\}$. ■

If we consider the bundles $\mathbf{P}_{1-i,i}(E)$ over M, the hypotheses of the Leray-Hirsch theorem are fulfilled as a consequence of Proposition 3.1, so that we have

Proposition 3.3. *The following isomorphisms of $\mathbf{Z}$-modules hold:*

$$H(\mathbf{P}_{1-i,i}(E),\mathbf{Z}) \simeq H(M,\mathbf{Z}) \otimes_{\mathbf{Z}} H(\mathbf{P}_{1-i,i}(C_L^{r|s}),\mathbf{Z}), \quad i = 0,1.$$

■

Characteristic classes of smooth bundles. Before introducing characteristic classes for supervector bundles, we would like to recall the general features of the Chern classes of smooth vector bundles over differentiable manifolds. Chern classes can be characterized axiomatically; the literature on this topic is vast, see e.g. [**Hirz,Hus,MiS,Vai**]. We follow in particular [**Vai**].

Let X be a differentiable manifold.

Axiom 1. *For each isomorphism class ξ of complex vector bundles of rank r over X, the h-th Chern class of ξ is an element $c_h(\xi)$ in $H^{2h}(X,\mathbf{Z})$ for $h = 1,\dots,r$, while $c_0(\xi) = 1$.*

Let us define the *total Chern class* $c(\xi) = \sum_{h=0}^{r} c_h(\xi)$.

Axiom 2. (Normalization) *If ξ is an isomorphism class of line bundles, then $c_1(\xi)$ is minus the obstruction class of ξ.*

Axiom 3. (Functoriality) *For any smooth map $f: X \to Y$ into a differentiable manifold Y, and for any vector bundle ξ over Y, one has $c(f^{-1}\xi) = f^\sharp(c(\xi))$.*

Axiom 4. (Whitney product formula) *For all vector bundles ξ, η over X one has $c(\xi \oplus \eta) = c(\xi) \smile c(\eta)$, where $\smile$ denotes the cup product in the ring $H^\bullet(X,\mathbf{Z})$.*

For a definition of the cup product the reader may refer to [**Go,Bre**].

Characteristic classes of supervector bundles. Given a rank (r,s) CSVB $p: E \to M$, we can straightforwardly introduce its *even* and *odd Chern classes* as follows: if x and t are, respectively, the obstruction classes of the even and odd tautological bundles of the projectivizations of E, we let (with reference to Proposition 3.1)

$$x^r = -\sum_{j=1}^{r} C_j^0(E)\, x^{r-j}, \qquad t^s = -\sum_{k=1}^{s} C_k^1(E)\, t^{s-k}, \tag{3.2}$$

so that $C_j^0(E)$ and $C_k^1(E)$ are elements in $H^{2j}(M,\mathbb{Z})$ and $H^{2k}(M,\mathbb{Z})$, respectively. Correspondingly, there are two total Chern classes:

$$C^0(E) = \sum_{j=0}^{r} C_j^0(E), \qquad C^1(E) = \sum_{k=0}^{s} C_k^1(E). \tag{3.3}$$

According to this definition, a rank (r,s) CSVB has r even and s odd Chern classes.

We wish now to prove that the Chern classes of a CSVB satisfy analogous properties to those verified by the Chern classes of a smooth bundle over a differentiable manifold. The normalization and functoriality properties are readily proved.

Proposition 3.4. *If E has rank (1,0), then*

$$C^0(E) = 1 - \delta(E); \qquad C^1(E) = 1\,, \tag{3.4}$$

while, if E has rank (0,1),

$$C^1(E) = 1 - \delta(E); \qquad C^0(E) = 1\,. \tag{3.5}$$

Proof. If rank $E = (1,0)$, then E has only an even projectivization; moreover, $S_0(E) \simeq E$, so that (3.4) follows. A similar argument applies to the rank (0,1) case. ∎

Proposition 3.5. *If $f\colon M \to N$ is a G^∞ morphism, and E is a CSVB over N, then*

$$C^i(f^{-1}E) = f^\sharp C^i(E), \qquad i = 0,1.$$

Proof. This property follows from the functoriality of the projectivized and tautological bundles. ∎

In order to prove a Whitney product formula, we need some further constructions; in particular, we must show that a rank (r,s) G^∞ vector bundle on M determines two smooth bundles $\bar{E}_0$ and $\bar{E}_1$ on M, with rank r and s, respectively. Indeed, the body map, regarded as a sheaf morphism $\mathcal{I}^\infty \to \mathcal{C}_M$ (we recall that $\mathcal{I}^\infty$ is the complexification of $\mathcal{A}^\infty$), endows $\mathcal{C}_M$ with a structure of $\mathcal{I}^\infty$-module; if $\mathcal{E}$ is the sheaf of sections of E, then $\mathcal{E} \otimes_{\mathcal{I}^\infty} \mathcal{C}_M$ is a rank $r+s$ smooth complex vector bundle which splits as a direct sum $\bar{E}_0 \oplus \bar{E}_1$, as required. The same result can be obtained by applying the body map to the transition functions of E, thus obtaining matrix-valued maps with a block-diagonal structure; the diagonal blocks are the transition functions of E_0 and E_1.

This construction entails the existence of vector bundle maps $E \to \bar{E}_i$; these can be lifted to maps between the projectivized bundles $\mathbf{P}_{1-i,i}(E) \to \mathbf{P}(\bar{E}_i)$ and between the tautological bundles, so that one obtains commutative diagrams of morphisms of smooth vector bundles

$$\begin{array}{ccccccccc} 0 & \longrightarrow & S_i(E) & \longrightarrow & \pi_i^{-1}E & \longrightarrow & Q_i(E) & \longrightarrow & 0 \\ & & \downarrow & & \downarrow & & \downarrow & & \\ 0 & \longrightarrow & S(\bar{E}_i) & \longrightarrow & \bar{p}_i^{-1}\bar{E}_i & \longrightarrow & Q(\bar{E}_i) & \longrightarrow & 0 \end{array} \qquad i=0,1$$

where $\bar{p}_i$ is the bundle projection $\bar{E}_i \to M$. The commutativity of these diagrams implies that, for fixed i, $S_i(E)$ and $S(\bar{E}_i)$ have the same obstruction class. This in turn implies

Lemma 3.1. $C^0(E) = c(\bar{E}_0), \quad C^1(E) = c(\bar{E}_1)$. ∎

It is now possible to prove Whitney's formula.

Proposition 3.6. *If $0 \to E \to F \to G \to 0$ is an exact sequence of CSVB's, then*

$$C^i(F) = C^i(E) \smile C^i(G), \qquad i = 0,1. \tag{3.6}$$

Proof. By tensoring the given exact sequence with $\mathcal{C}_M$ one obtains an exact sequence of smooth vector bundles over M

$$0 \to \bar{E} \to \bar{F} \to \bar{G} \to 0$$

which splits (as all sequences of smooth vector bundles do, cf. [**Hus**]), thus yielding isomorphisms $\bar{F}_i \simeq \bar{E}_i \oplus \bar{G}_i$. The ordinary Whitney formula then yields $c(\bar{F}_i) = c(\bar{E}_i) \smile c(\bar{G}_i)$, which, together with Lemma 3.1, implies the thesis. ∎

It should be noticed that we have stated the Whitney product formula in terms of exact sequences of CSVB's rather than in terms of direct sums of CSVB's, since, due to the non-acyciclity of the structure sheaf of the base supermanifolds, not all exact sequences of CSVB's split.

We conclude this section by introducing the Chern character of a CSVB. For a given rank (r, s) CSVB E over M, through the formal factorizations [Hirz]

$$\sum_{j=0}^{r} C_j^0(E)\, x^j = \prod_{j=1}^{r}(1+\gamma_j x), \qquad \sum_{k=0}^{s} C_k^1(E)\, t^k = \prod_{k=1}^{s}(1+\delta_k t)$$

we define the even and odd Chern characters of E

$$Ch^0(E) = \sum_{j=1}^{r} e^{\gamma_j}, \qquad Ch^1(E) = \sum_{k=1}^{s} e^{\delta_k},$$

and the total Chern character

$$Ch(E) = Ch^0(E) - Ch^1(E)\,. \tag{3.7}$$

Of course $Ch(E) \in H(M, \mathbb{Z})$, and there is a decomposition

$$Ch(E) = \sum_{i=0}^{\infty} Ch_i(E), \qquad Ch_i(E) \in H^{2i}(M, \mathbb{Z});$$

in particular, one has $Ch_0(E) = r - s$.

The choice of the minus sign in Eq. (3.7) is related to the possibility of representing, under suitable conditions, the Chern character in terms of curvature forms, and eventually stems from the minus sign involved in the definition of graded trace (cf. Section I.3). Let $\Xi = \Xi_0 \oplus \Xi_1$ be a rank (r, s) CSVB (now we mean a CSVB in the sense of Section IV.3, and not its underlying G^∞ bundle), and let $\Pi\Xi = \Xi_1 \oplus \Xi_0$; here Π is the parity change functor (we may think of it as acting on the module of sections, cf. Section IV.3). Then we have trivially $Ch(\Pi\Xi) = -Ch(\Xi)$, and therefore $Ch(\Xi \oplus \Pi\Xi) = 0$.

The analogue of the Whitney product formula for Chern characters reads as follows: if $0 \to E \to F \to G \to 0$ is an exact sequence of CSVB's, then

$$Ch^i(F) = Ch^i(E) + Ch^i(G), \qquad i = 0, 1\,. \tag{3.8}$$

Uniqueness of Chern classes. It is possible to see that, as in the case of ordinary complex vector bundles, the normalization, functiorial, and additivity properties characterize uniquely the Chern classes of CSVB's; more precisely, any family of maps $\{d_k\}$ from the monoid of CSVB's over a supermanifold M into the cohomology groups $H^{2k}(M, \mathbf{Z})$ which satisfy Proposition 4.4, 4.5, and 4.6, necessarily coincide with the Chern classes.

In order to prove this fact we first need a rather technical result, expressed by the following statement.

Proposition 3.7. *Let $E \to M$ a rank (r, s) CSVB. There exists a G^∞ morphism $f: N \to M$ such that*

(1) *$f^\sharp: H^\bullet(M, \mathbf{Z}) \to H^\bullet(N, \mathbf{Z})$ is injective;*
(2) *there is on N a chain of sub-CSVB's*

$$0 = F_0 \subset F_1 \subset \cdots \subset F_{r+s} = f^{-1}(E)$$

such that all quotients F_j/F_{j-1} have either rank $(1,0)$ or $(0,1)$.

Proof. This result is proved by double induction on the rank of E. If $\operatorname{rank} E = (1,0)$ or $(0,1)$ the result is trivial. Suppose now that $\operatorname{rank} E = (r+1, s)$ and consider the even projectivization of E, $\pi_0 : \mathbf{P}_{1,0}(E) \to M$; the cohomology map $\pi_0^* : H(M, \mathbf{Z}) \to H(\mathbf{P}_{1,0}(E), \mathbf{Z})$ is injective by Leray-Hirsch. The pullback bundle $\pi_0^{-1}E \to \mathbf{P}_{1,0}(E)$ has a tautological superline subbundle $S_0(E) \to \mathbf{P}_{1,0}(E)$, and the quotient superbundle $Q_0(E)$ has rank (r, s). By the induction hypothesis, there is a $\mathcal{G}$-map $g : N \to \mathbf{P}_{1,0}(E)$ satisfying the properties in the statement of this Lemma. Then the composition $f = \pi_0 \circ g : N \to M$ yields the required map. The induction on the odd rank is proved in the same way. ∎

Now, for any $k \in \mathbf{N}$ let d_k be a law that with any CSVB E over M associates an element in $H^{2k}(M, \mathbf{Z})$, and let us assume that $d_0(E) = 1$ and $d_k(E) = 0$ if $\operatorname{rank} E = (r, s)$ and $k > r + s$. Let $d(E) = \sum_{k=1}^{r+s} d_k(E)$.

Proposition 3.8. *If the maps d_k satisfy the following properties:*

(1) *if L is a CSLB, then $d(L) = 1 - \delta(L)$, where $\delta(L)$ is the obstruction class of L;*
(2) *d is functorial, in the sense that if $f: N \to M$ is a G^∞ map, and E is a CSVB on M then $d(f^{-1}E) = f^\sharp d(E)$;*
(3) *d is additive, in the sense that if $0 \to E' \to E \to E'' \to 0$ is an exact sequence of CVSB's, then $d(E) = d(E') \smile d(E'')$;*

then $d(E) = C(E)$ for all E's.

Proof. By (1) $d_1(L) = C_1(L)$ for every CSLB L, so that additivity implies that $d(E) = C(E)$ whenever E admits a chain of sub-CSVB's such that all quotients are CSLB's. Finally, for every CSVB $E \to M$ there is a G^∞ morphism $f\colon N \to M$ such that $f^{-1}E$ admits such a chain, and $f^\sharp\colon H^\bullet(M, \mathbb{Z}) \to H^\bullet(N, \mathbb{Z})$ is injective. The functoriality property allows to conclude. ■

4. Characteristic classes in terms of curvature forms

A classical result in bundle theory, usually known as the *Chern-Weil theorem*, states that the characteristic classes of a complex vector bundle ξ over a differentiable manifold X can be realized as cohomology classes in $H^\bullet_{DR}(M)$ in terms of the curvature of a connection on ξ. In this Section we consider the extension of this result to the case of G-supermanifolds. It turns out that, while the result is readily proved in the case of superline bundles, there seem to be obstructions to its extension to higher rank SVB's, unless the base supermanifold is DeWitt.

Let us first consider a complex superline bundle Ξ over a G-supermanifold $(M, \mathcal{A})$; assuming that Ξ has a vanishing Atiyah class, let ∇ be a connection on it. A glance at Eq. (1.9) shows that in the case of CSLB's the curvature R of ∇ induces a globally defined graded differential 2-form on $(M, \mathcal{A})$, that we denote by R again. The Bianchi identity states that R is closed, and therefore a cohomology class $[R] \in H^2_{SDR}(M, \mathcal{I})$ is singled out. (Here [ ] denotes a cohomology class in $H^\bullet_{SDR}(M, \mathcal{I})$, where $\mathcal{I}$ is the complexification of the structure sheaf $\mathcal{A}$ of M, cf. Section VI.2). The important fact is that $[R]$ is independent of the connection; indeed, if ∇' is another connection, the difference $\eta = \nabla - \nabla'$ is a globally defined graded differential 1-form on $(M, \mathcal{A})$, so that $R - R' = d\eta$.

The Chern class $C_1(\Xi)$ of Ξ lies in $H^2(M, \mathbb{Z})$; in order to compare it with $[R]$, we need to map both cohomology classes into $H^2(M, C_L)$. Let $\{g_{jk}\}$ be transition morphisms of Ξ with respect to a suitable cover $\{U_j\}$ of M; then, by its very definition, $C_1(\Xi)$ is represented by the Čech 2-cocycle

$$\frac{1}{2\pi i}\log(g_{jk} + g_{kh} + g_{hj})\,. \tag{4.1}$$

We can of course regard this as a cocycle for the sheaf C_L on M. On the other hand, from Eq. (1.8), and recalling the abstract de Rham theorem (Proposition

II.2.4), we see that the morphism $H^2_{SDR}(M,\mathcal{I}) \to H^2(M,C_L)$ maps $[R]$ exactly into the Čech cocycle (4.1), so that we obtain the following representation theorem.

Proposition 4.1. *Let Ξ be a CSLB with vanishing Atiyah class, let $C_1(\Xi)$ be its first Chern class, regarded as an element in $H^2(M,C_L)$, and denote by λ: $H^2_{SDR}(M,\mathcal{I}) \to H^2(M,C_L)$ the morphism ensuing from the abstract de Rham theorem. Then,*

$$C_1(\Xi) = \frac{i}{2\pi}\lambda([R]),$$

where R is the curvature form of any connection on Ξ. ■

Elementary invariant polynomials. In order to generalize Proposition 4.1 to higher rank SVB's we need some algebraic preliminaries, related to the study of Ad-invariant polynomials on the general linear graded Lie algebra. We shall use some elements of the corresponding theory in the ordinary case, for which the reader may refer to [**GrH**]. Let r, s be two fixed nonnegative integers; for the sake of simplicity, we denote by $\mathfrak{G}$ the graded Lie C_L-algebra $M_{C_L}[r|s]$ formed by the $(r+s)\times(r+s)$ matrices with entries in C_L, graded in the usual way. The elementary invariant polynomials on $\mathfrak{G}$ are defined by the analogy with the usual theory (cf. [**GrH**]); however, we do not know whether these functions generate all the invariant polynomials on $\mathfrak{G}$, as it happens in the ordinary case.

The adjoint action of $GL[r|s]$ over $\mathfrak{G}$ is defined as usual by $\mathrm{Ad}_H X = HXH^{-1}$, for $X \in \mathfrak{G}$ and $H \in GL[r|s]$.[6] The N-th elementary invariant polynomial is a mapping $P^N: \mathfrak{G} \to C_L$, which we first define on $\mathfrak{G}_0$ by means of the equation

$$P^N(X) = \frac{1}{N!}\left[\frac{d^N}{dt^N}\,\mathrm{Ber}(I+tX)\right]_{t=0}. \tag{4.2}$$

The Ad-invariance of these polynomials, namely, the property $P^N(\mathrm{Ad}_H X) = P^N(X)$ for all $H \in GL[r|s]$, is assured by Eq. (I.3.6).

A more explicit representation of these polynomials can be obtained as follows. Let $n = r+s$, and consider n complex variables $\lambda_1,\ldots,\lambda_n$; let $\tau_1,\ldots,\tau_n$

[6] A more general definition of adjoint representation, for a generic Lie supergroup, will be given in the next Chapter.

be the polynomials

$$\tau_N(\lambda_1,\ldots,\lambda_n)=\sum_{i=1}^{n}(\lambda_i)^N,\qquad N=1,\ldots,n,$$

and let f_N, $N=1,\ldots,n$ be the polynomials defined by the conditions

$$\sigma_N(\lambda_1,\ldots,\lambda_n)=f_N(\tau_1(\lambda_1,\ldots,\lambda_n),\ldots,\tau_N(\lambda_1,\ldots,\lambda_n)),\qquad N=1,\ldots,n,$$

where the σ_N's are the symmetric elementary functions of $\lambda_1,\ldots,\lambda_n$.

Proposition 4.2. *For all $X\in\mathfrak{G}_0$, and all $N=1,\ldots,r+s$ the following identity holds:*

$$P^N(X)=f_N(\operatorname{Str}X,\operatorname{Str}X^2,\ldots,\operatorname{Str}X^N).\tag{4.3}$$

Proof. For small enough values of a real parameter t there exists a smooth function $Y(t)$ of t with values in $\mathfrak{G}_0$ such that $I+tX=\exp Y(t)$, so that (cf. Section I.3)

$$\operatorname{Ber}(I+tX)=\exp\operatorname{Str}Y(t).$$

We must therefore compute the quantities $\left[\frac{d^N}{dt^N}\exp\operatorname{Str}Y(t)\right]_{t=0}$; a direct calculation shows a result of the type

$$\operatorname{Str}Y^{(N)}(0)+\alpha_1\operatorname{Str}Y^{(N-1)}(0)\operatorname{Str}Y'(0)+\cdots+\alpha_{k(N)}\left(\operatorname{Str}Y'(0)\right)^N,$$

where $Y^{(i)}$ denotes the i-th derivative of Y, $k(N)$ is a suitable integer, and the coefficients α_i are real numbers which are apparently independent of the value of s. We can therefore assume $s=0$, so that the claim reduces to the the classical well-known result. ∎

With the aid of this more explicit representation we can compute the first few polynomials:

$$P^1(X)=\operatorname{Str}X,\qquad P^2(X)=\frac{1}{2}\left[(\operatorname{Str}X)^2-\operatorname{Str}X^2\right],$$

$$P^3(X)=\frac{1}{6}(\operatorname{Str}X)^3-\frac{1}{2}(\operatorname{Str}X^2)(\operatorname{Str}X)+\frac{1}{3}\operatorname{Str}X^3.$$

Furthermore, equation (4.3) can be regarded as a *definition* of the N-th elementary invariant polynomial for the whole graded Lie algebra $\mathfrak{G}$ (and not only its even part). Its Ad-invariance is now assured by Eq. (I.3.4). Obviously, these polynomials are naturally defined on any subalgebra of $\mathfrak{G}$.

Finally, we need to introduce the *polarization* $\tilde{P}^N$ of the elementary invariant polynomial P^N, which is a graded symmetric C_L-multilinear morphism

$$\tilde{P}^N : \underbrace{\mathfrak{G} \times \cdots \times \mathfrak{G}}_{N} \to C_L$$

satisfying the properties

$$\tilde{P}^N(X, \ldots, X) = P^N(X)$$

$$\tilde{P}^N(\mathrm{Ad}_H X_1, \ldots, \mathrm{Ad}_H X_N) = \tilde{P}^N(X_1, \ldots, X_N)$$

for all $H \in GL[r|s]$ and $X, X_1, \ldots, X_N \in \mathfrak{G}$. It is not hard to verify that the polarization is indeed uniquely defined by the first of these properties, together with graded symmetry.

Differential calculus of forms with values in a module. We wish to consider invariant polynomials of the curvature form of a connection, so that we need to introduce some elements of the differential calculus for graded differential forms with values in a module.

Given a G-supermanifold $(M, \mathcal{A})$, a differential calculus for graded differential forms with values in a locally free $\mathcal{A}$-module $\mathcal{F}$ makes a sense only when a graded derivation law on $\mathcal{F}$ is fixed (see [HeM1]). However, graded forms with values in a free module can be differentiated with respect to the trivial derivation law, which in practice means that one can avoid using the general theory of module-valued graded differential forms. Indeed, if $\mathcal{F} = \mathcal{A} \otimes_{B_L} F$, where F is a free B_L-module, the module of $\mathcal{F}$-valued (that is, F-valued) graded differential k-forms is simply $\Omega^k_{\mathcal{A}}(U) \otimes_{B_L} F$, and we can define an exterior differential by letting

$$d(\omega \otimes u) = d\omega \otimes u$$

for every graded differential k-form $\omega \in \Omega^k_{\mathcal{A}}(U)$ and every vector $u \in F$; we also let

$$D(f \otimes u) = D(f) \otimes u$$

for any section $f \in \mathcal{A}(U)$ and any derivation $D \in \mathcal{D}er\, \mathcal{A}(U)$. Furthermore, if F is a graded B_L-algebra, one can define a wedge product between F-valued graded differential forms, simply by extending Eq. (IV.4.1). Finally, if $\mathfrak{F}$ is a graded Lie B_L-algebra, one can define a bracket between sections of $\Omega^k_{\mathcal{A}} \otimes_{B_L} \mathfrak{F}$ and $\Omega^h_{\mathcal{A}} \otimes_{B_L} \mathfrak{F}$, which yields a section of $\Omega^{k+h}_{\mathcal{A}} \otimes_{B_L} \mathfrak{F}$ according to the rule

$$[\eta,\tau](D_1,\dots,D_{k+h}) = \\ \frac{1}{(k+h)!} \sum_{\sigma\in\mathfrak{S}_{k+h}} (-1)^{\Delta_3(\sigma,D,\tau)} \left[\eta(D_{\sigma(1)},\dots,D_{\sigma(k)}), \tau(D_{\sigma(k+1)},\dots,D_{\sigma(k+h)})\right],$$

where the D's are homogeneous sections of $\operatorname{Der}\mathcal{A}$, $\mathfrak{S}$ is the permutation group; furthermore, $\Delta_3(\sigma,D,\tau) = |\sigma| + \Delta_2(\sigma,D,\tau)$, where $|\sigma|$ is the parity of the permutation σ, and the symbol Δ_2 has the same meaning as in Eq. (I.2.10).

These operations fulfill the following properties.

Proposition 4.3. *Let φ and ψ be sections of $\Omega^k_{\mathcal{A}} \otimes_{B_L} \mathfrak{F}$ and $\Omega^h_{\mathcal{A}} \otimes_{B_L} \mathfrak{F}$ respectively, where $\mathfrak{F}$ is a graded Lie B_L-algebra. The following identities hold:*

$$d([\varphi,\psi]) = [d\varphi,\psi] + (-1)^k [\varphi, d\psi];$$
$$[\varphi,\psi] = -(-1)^{kh+|\varphi|\,|\psi|} [\psi,\varphi];$$
$$[\varphi,[\varphi,\varphi]] = 0.$$

∎

Let us now return to the case where the module where the graded differential forms take values is $\mathfrak{G} \equiv M_{C_L}[r|s]$, which is both an associative graded B_L-algebra and a graded Lie B_L-algebra. We denote by $\mathfrak{G}^N$ the N-th graded tensor power of $\mathfrak{G}$ over C_L (cf. Section I.2), and, given a G-supermanifold $(M,\mathcal{A})$, let us consider the sheaf $\mathcal{L}^{k,N} = \Omega^k_{\mathcal{A}} \otimes_{B_L} \mathfrak{G}^N$ of graded differential k-forms on $(M,\mathcal{A})$ with values in $\mathfrak{G}^N$. We can apply the differential calculus so far developed to the sections of these sheaves.

We also consider the graded C_L-module $\mathbf{W}^N(\mathfrak{G})$ whose elements are the graded module morphisms $P\colon \mathfrak{G}^N \to C_L$ which are graded symmetric and adjoint-invariant, i.e.

$$P(Z_1\otimes\dots Z_i\otimes Z_{i+1}\otimes\cdots\otimes Z_N) = (-1)^{|Z_i||Z_{i+1}|} P(Z_1\otimes\dots Z_{i+1}\otimes Z_i\otimes\cdots\otimes Z_N) \tag{4.4}$$

for all homogeneous $Z_1,\dots,Z_N \in \mathfrak{G}$, and

$$P(\operatorname{Ad}_H Z_1 \otimes\cdots\otimes \operatorname{Ad}_H Z_N) = P(Z_1\otimes\cdots\otimes Z_N) \tag{4.5}$$

for all $H \in GL[r|s]$. The latter condition implies that

$$\sum_{i=1}^{N}(-1)^{|Z|(|Z_1|+\cdots+|Z_N|)}\, P(Z_1 \otimes \cdots \otimes [Z, Z_i] \otimes \cdots \otimes Z_N) = 0, \qquad (4.6)$$

where Z is another homogeneous element in $\mathfrak{G}$.

We define $\mathbf{W}(\mathfrak{G}) = \bigoplus_{N\in\mathbf{N}} \mathbf{W}^N(\mathfrak{G})$ and make it into a graded C_L-algebra by defining the following product: if $P \in \mathbf{W}^N(\mathfrak{G})$ and $Q \in \mathbf{W}^h(\mathfrak{G})$, then PQ is the element in $\mathbf{W}^{N+N'}(\mathfrak{G})$ which acts on a tensor product of homogeneous elements in $\mathfrak{G}$ according to the law

$$PQ(Z_1 \otimes \cdots \otimes Z_{p+q}) = \sum_{\sigma\in\mathfrak{S}_{N+N'}} \frac{(-1)^{|\sigma|+\Delta_1(\sigma,Z)}}{N!N'!} P(Z_{\sigma(1)}\otimes\cdots\otimes Z_{\sigma(N)})\otimes Q(Z_{\sigma(N+1)}\otimes\cdots\otimes Z_{\sigma(N+N')})$$

where the symbol Δ_1 here has the same meaning as in Eq. (I.2.8).

Now, if U is an open set in some G-supermanifold $(M, \mathcal{A})$, and φ is a $\mathfrak{G}^N$-valued graded differential k-form on U, i.e. $\varphi \in \mathcal{L}^{k,N}(U)$, by composition with an element $P \in \mathbf{W}^N(\mathfrak{G})$ we obtain a C_L-valued graded differential k-form on U, say $P(\varphi)$. One easily shows that

$$dP(\varphi) = P(d\varphi)\,.$$

Furthermore, property (4.6) implies that, given homogeneous sections $\psi_1 \in \mathcal{L}^{k_1,1}(U), \ldots, \psi_N \in \mathcal{L}^{k_N,1}(U)$, and $\varphi \in \mathcal{L}^{1,1}(U)$, then

$$\sum_{i=1}^{N}(-1)^{\sum_{j=1}^{i} k_j+|\varphi|\,|\psi_j|}\, P(\psi_1 \wedge \cdots \wedge [\psi_j, \varphi] \wedge \cdots \wedge \psi_N) = 0\,. \qquad (4.7)$$

Invariant polynomials of curvature. Now, let Ξ be a rank (r, s) CSVB over $(M, \mathcal{A})$. Assuming that Ξ has a vanishing Atiyah class, let ∇ be a connection on it, whose local curvature forms relative to a certain trivializing cover $\{U_j\}$ are denoted by $R^{(j)}$. If $P \in \mathbf{W}^N(\mathfrak{G})$, we set for brevity $P(X, \ldots, X) = P(X)$; then, in view of the Ad-invariance property (4.5), we have

$$P(R^{(j)}) = P(R^{(k)}) \quad \text{on} \quad U_j \cap U_k\,,$$

thus defining a global graded differential $2N$-form $P(R)$ on (M, A).

Proposition 4.4. *The graded differential $2N$-form $P(R)$ is closed, $dP(R) = 0$. Moreover, the super de Rham cohomology class $[P(R)] \in H^{2N}_{SDR}(M, \mathcal{I})$ does not depend on the connection.*

Proof. By using the Bianchi identity and the identity (4.7), we obtain:

$$dP(R) = dP(R, \ldots, R) = \\ N\, P(dR, R, \ldots, R) = N\, P([R, \nabla], R, \ldots, R) = 0\,.$$

To prove the second claim, we show that, given two connections ∇_0, ∇_1, the difference $P(R_1) - P(R_0)$ is an exact form. For a real parameter t, we introduce the connection $\nabla_t = \nabla_0 + t\eta$, where η is the graded differential 1-form $\nabla_1 - \nabla_0$; the curvature R_t of ∇_t satisfies the condition

$$\frac{d}{dt} R_t = d\eta + [\nabla_t,\, \eta]\,.$$

This equation yields

$$\frac{d}{dt} P(R_t) = N\, P(\frac{d}{dt} R_t, R_t, \ldots, R_t) = \\ N\, P(d\eta, R_t, \ldots, R_t) + N\, P([\nabla_t,\, \eta], R_t, \ldots, R_t) = \\ N\, dP(\eta, R_t, \ldots, R_t)\,.$$

By integrating over t between 0 and 1, we eventually obtain:

$$P(R_1) - P(R_0) = N\, d \int_0^1 P(\eta, R_t, \ldots, R_t)\,.$$

∎

The Chern-Weil theorem. In order to prove the representation theorem for a CSVB of arbitrary rank, say Ξ of rank (r, s), let us assume that Ξ has vanishing Atiyah class, and let ∇ be a connection on it, with curvature form R. For $k = 1, \ldots, r + s$, let

$$d_k(\Xi) = \left[P^k \left(\frac{i}{2\pi} R \right) \right]\,. \tag{4.8}$$

As a consequence of Proposition 4.4, the class $d_k(\Xi)$ depends only on the bundle Ξ.

We wish to use Proposition 3.8 to show that the classes $d_k(\Xi)$ coincide (up to action of a morphism) with the Chern classes of Ξ. To this end, we first notice that Proposition 4.1 can be restated in the form $C_1(\Xi) = \lambda(d_1(\Xi))$. Secondly, we check functoriality: if Ξ is a CSVB on $(N, \mathcal{B})$, and $F = (f, \phi): (M, \mathcal{A}) \to (N, \mathcal{B})$ is a G-morphism, then

$$d_k(F^{-1}\Xi) = f^\sharp d_k(\Xi).$$

Finally, we have to verify additivity. Using the same notations as above, we set $d(\Xi) = \sum_{k=0}^{r+s} d_k(\Xi)$, with $d_0 = 1$. At this point, we need to assume that the base supermanifold is DeWitt.

Proposition 4.5. *Let $0 \to \Xi' \to \Xi \to \Xi'' \to 0$ be an exact sequence of CSVB's over a DeWitt G-supermanifold $(M, \mathcal{A})$. Then,*

$$d(\Xi) = d(\Xi') \smile d(\Xi'') .$$

Proof. Since all exact sequences of CSVB's over DeWitt G-supermanifolds split, so that $\Xi \simeq \Xi' \oplus \Xi''$, and all CSVB's over DeWitt G-supermanifolds admit connections, we can choose connections on Ξ' and Ξ'' and put on Ξ the direct sum connection. Then the matrix of the curvature forms has a block diagonal structure; inserting this into Eq. (4.2) we obtain the result. ∎

Thus, resorting to Proposition 3.8 applied to the subcategory of CSVB's over DeWitt G-supermanifolds, we eventually obtain the Chern-Weil theorem for CSVB's over DeWitt G-supermanifolds.

Proposition 4.6. *Let Ξ be a rank (r, s) CSVB over a DeWitt G-supermanifold, and, for all $k = 1, \ldots, r+s$, let $C_k(\Xi)$ be its k-th Chern class regarded as an element in $H^{2k}(M, C_L)$. Finally, let $\lambda: H^{2k}_{SDR}(M, \mathcal{I}) \to H^{2k}(M, C_L)$ be the morphism ensuing from the abstract de Rham theorem. Then,*

$$C_k(\Xi) = \lambda\left([P^k(\frac{i}{2\pi}R)]\right)$$

where R is the curvature form of any connection on Ξ. ∎

This result can be also stated in terms of the Chern character of Ξ:

$$Ch_k(\Xi) = \lambda\left([\mathrm{Str}(\frac{i}{2\pi}R)^k]\right) .$$

Chapter VII

Lie supergroups and principal superfibre bundles

Y ¿qué importa errar lo menos
quien ha acertado lo más ?
P. CALDERÓN DE LA BARCA

This last Chapter is devoted to developing the rudiments of a theory of Lie supergroups within the category of G-supermanifolds, together with the basic definitions related to principal superfibre bundles and associated superbundles. Since a G-supermanifold structure is not determined by the underlying topological space, the group axioms must be expressed in categorial terms.[1] This is what happens in the theory of algebraic groups, whose guidelines will be followed here. Most of our material will be taken, with the necessary modifications, from [**Wat,Hum**], where the theory of algebraic groups is developed, from [**Pen**], containing the theory of Lie supergroups and their representations in the context of algebraic graded manifolds (superschemes), and from [**Lop**], which is devoted to graded Lie groups (i.e., Lie groups in the framework of the Berezin-Leïtes-Kostant graded manifold theory, as earlier considered in [**Kos**]).

After supplying the definition of a G-Lie supergroup, we show how the notion of left-invariant derivation permits one, in analogy with the ordinary case, to attach a Lie superalgebra to any G-Lie supergroup. Then, we define the concepts of action of a G-Lie supergroup on a G-supermanifold, and of the quotient of such an action. We can thus introduce principal superfibre bundles and associated superfibre bundles.

[1] Even though we shall follow a more direct approach, this fact could be stated by saying that Lie supergroups are *group objects* in the category of G-supermanifolds.

1. Lie supergroups

Let us at first consider an ordinary Lie group H; one then has the multiplication morphism $m: H \times H \to H$, $m(x, y) = x \cdot y$, the inversion morphism $s: H \to H$, $s(x) = x^{-1}$ and the unit element $e: \{e\} \to H$ (here $\{e\}$ is the unit element regarded as a Lie group with a single element). The associativity property, $(x \cdot y) \cdot z = x \cdot (y \cdot z)$, the unit property, $x \cdot e = e \cdot x = x$ for every $x \in H$, and the inverse property, $x \cdot x^{-1} = x^{-1} \cdot x = e$, can be stated in terms of commutative diagrams:

$$\begin{array}{ccc} H \times H \times H & \xrightarrow{Id \times m} & H \times H \\ {\scriptstyle m\times \mathrm{Id}}\downarrow & & \downarrow{\scriptstyle m} \\ H \times H & \xrightarrow{m} & H \end{array} \qquad \text{(associativity)},$$

$$\begin{array}{ccccc} \{e\} \times H & \xrightarrow{e\times \mathrm{Id}} & H \times H & \xleftarrow{\mathrm{Id}\times e} & H \times \{e\} \\ & \searrow & \downarrow & \swarrow & \\ & & H & & \end{array} \qquad \text{(unit property)},$$

and

$$\begin{array}{ccccc} H & \xrightarrow{(s,\mathrm{Id})} & H \times H & \xleftarrow{(\mathrm{Id},s)} & H \\ \downarrow & & {\scriptstyle m}\downarrow & & \downarrow \\ \{e\} & \xrightarrow{e} & H & \xleftarrow{e} & \{e\} \end{array} \qquad \text{(inverse property)}.$$

This way of describing a Lie group structure may appear to be unnecessarily complicated, and possibly produces an unpleasant impression of formality, but nevertheless it is the only description which leads directly to the introduction of the notion of Lie supergroup (at least for finite-dimensional ground algebras such as B_L). Indeed, as is usual for G-supermanifolds, not all the information about a G-Lie supergroup is contained in the underlying topological space, which in this case is an ordinary Lie group; thus, supergroup properties must be stated in terms of morphisms of graded ringed spaces. The diagrams that the reader will encounter will be best understood by rephrasing them in terms of the corresponding properties of ordinary Lie groups.

In order to avoid the cumbersome notation where G-supermanifolds and G-morphisms are pairs of objects, we shall denote G-supermanifolds with hats, $\widehat{M} = (M, \mathcal{A})$, and write only $\widehat{M}$ (except when an explicit reference to the sheaf is necessary). The structure sheaf of a G-supermanifold $\widehat{M}$ will be denoted by $\mathcal{A}_{\widehat{M}}$. A G-supermanifold morphism $(f,\phi):(M,\mathcal{A}_M) \to (N,\mathcal{A}_N)$ will be simply denoted by $\widehat{f}:\widehat{M} \to \widehat{N}$, using the notation $\widehat{f}^*$ for the sheaf morphism $\widehat{f}^* = \phi: \mathcal{A}_{\widehat{N}} \to f_*\mathcal{A}_{\widehat{M}}$ (of course $\widehat{f}^* \neq f^*$).

Let us denote by $\widehat{z} = (z, B_L)$ a single point endowed with its trivial $(0,0)$ dimensional G-supermanifold structure. For any G-supermanifold $\widehat{M}$, there are natural isomorphisms $\widehat{z} \times \widehat{M} \simeq \widehat{M} \times \widehat{z} \simeq \widehat{M}$, so that these spaces will be identified in the sequel.

Definition 1.1. *A G-supermanifold $\widehat{H}$ is said to be a G-Lie supergroup if there exist morphisms of G-supermanifolds*

$$\begin{array}{ll} \widehat{m}: \widehat{H} \times \widehat{H} \to \widehat{H} & \textit{(multiplication)} \\ \widehat{e}: \widehat{z} \to \widehat{H} & \textit{(unit)} \\ \widehat{s}: \widehat{H} \to \widehat{H} & \textit{(inverse)} \end{array}$$

such that the diagrams

$$\begin{array}{ccc} \widehat{H} \times \widehat{H} \times \widehat{H} & \xrightarrow{\mathrm{Id} \times \widehat{m}} & \widehat{H} \times \widehat{H} \\ {\scriptstyle \widehat{m} \times \mathrm{Id}} \downarrow & & \downarrow {\scriptstyle \widehat{m}} \\ \widehat{H} \times \widehat{H} & \xrightarrow{\widehat{m}} & \widehat{H} \end{array} \quad , \qquad \begin{array}{ccccc} \widehat{z} \times \widehat{H} & \xrightarrow{\widehat{e} \times \mathrm{Id}} & \widehat{H} \times \widehat{H} & \xleftarrow{\mathrm{Id} \times \widehat{e}} & \widehat{H} \times \widehat{z} \\ & \searrow & {\scriptstyle \widehat{m}} \downarrow & \swarrow & \\ & & \widehat{H} & & \end{array} \quad ,$$

and

$$\begin{array}{ccccc} \widehat{H} & \xrightarrow{(\widehat{s}, \mathrm{Id})} & \widehat{H} \times \widehat{H} & \xleftarrow{(\mathrm{Id}, \widehat{s})} & \widehat{H} \\ \downarrow & & {\scriptstyle \widehat{m}} \downarrow & & \downarrow \\ \widehat{z} & \xrightarrow{\widehat{e}} & \widehat{H} & \xleftarrow{\widehat{e}} & \widehat{z} \end{array}$$

commute.

A similar definition allows us to consider G^∞ Lie supergroups, and if one takes the 'unit' as $(z, B_{L'})$ instead of $\widehat{z}$, one can also define GH^∞ or H^∞ Lie supergroups. These different notions of Lie supergroups are closely related; indeed, if $(H, \mathcal{GH}^H)$ is a GH^∞ Lie supergroup, $(H, \mathcal{GH}^H \otimes_{B_{L'}} B_L)$ is a G-Lie supergroup, whilst if $\widehat{H} = (H, \mathcal{A})$ is an (m,n) dimensional G-Lie supergroup, $(H, \mathcal{A}^\infty)$ is a G^∞ Lie supergroup of dimension (m,n), and the underlying differentiable manifold H inherits a structure of ordinary Lie group of dimension $2^{L-1}(m+n)$.

We thus have the first and most important example of Lie supergroup.

EXAMPLE 1.1. Let us consider the general graded linear group $GL_L[p|q]$ over B_L (Section I.3) endowed with its natural structure of H^∞ supermanifold of dimension $(p^2+q^2, 2pq)$ as an open subset of the even sector of $\mathrm{Hom}_{B_L}(B_L^{p|q}, B_L^{p|q})$. Matrix multiplication gives a map

$$GL_L[p|q] \times GL_L[p|q] \to GL_L[p|q]$$

which is certainly H^∞, so that $GL_L[p|q]$ is an H^∞, and also a G-Lie supergroup. We shall denote it by $\widehat{GL}_L[p|q]$. ▲

Let $\widehat{H}$ be a G-Lie supergroup. Then points $g \in H$ in the underlying Lie group H define morphisms $\widehat{g}: \widehat{z} \to \widehat{H}$ of image g. More generally, as one usually does in algebraic geometry, morphisms $\widehat{g}: \widehat{T} \to \widehat{H}$, where $\widehat{T}$ is any G-supermanifold, can be regarded as '$\widehat{T}$-valued points'. Then, ordinary points of H correspond to 'points' with values in a graded single point $\widehat{z} = (z, B_L)$. For every 'point' $\widehat{g}: \widehat{T} \to \widehat{H}$, the point $\widehat{g}^{-1}: \widehat{T} \to \widehat{H}$ obtained by composition of $\widehat{g}$ with the inversion morphism, $\widehat{g}^{-1} = \widehat{s} \circ \widehat{g}$, is called the *inverse point of* $\widehat{g}$.

Let $\widehat{g}: \widehat{z} \to \widehat{H}$ be an ordinary 'point' in the previous sense.

Definition 1.2. *The left translation and the right translation by* $\widehat{g}$ *are the G-supermanifold morphisms* $\widehat{L}_{\widehat{g}}$, $\widehat{R}_{\widehat{g}}$ *given respectively by the diagrams*

$$\begin{array}{ccc} \widehat{H} & \xrightarrow{\widehat{L}_{\widehat{g}}} & \widehat{H} \\ \| & & \uparrow \widehat{m} \\ \widehat{z} \times \widehat{H} & \xrightarrow{\widehat{g} \times \mathrm{Id}} & \widehat{H} \times \widehat{H} \end{array} \quad \text{and} \quad \begin{array}{ccc} \widehat{H} & \xrightarrow{\widehat{R}_{\widehat{g}}} & \widehat{H} \\ \| & & \uparrow \widehat{m} \\ \widehat{H} \times \widehat{z} & \xrightarrow{\mathrm{Id} \times \widehat{g}} & \widehat{H} \times \widehat{H} \end{array} .$$

Clearly, the left translation $\widehat{L}_{\widehat{g}}: \widehat{H} \to \widehat{H}$ and the right translation $\widehat{R}_{\widehat{g}}: \widehat{H} \to \widehat{H}$ are G-supermanifold isomorphisms whose inverse morphisms are, respectively, the left and right translation by the inverse point $\widehat{g}^{-1}$.

REMARK 1.1. We would like to point out a rather odd phenomenon which arises in connection with the G-Lie supergroup structure of $GL_L[p|q]$ described in Example 1.1. If one of the two arguments in the multiplication morphism is fixed, and has entries in $B_L - B_{L'}$, then the ensuing map $GL_L[p|q] \to GL_L[p|q]$ is G^∞ but is neither H^∞ nor GH^∞. On the other hand, the related morphism $\widehat{GL}_L[p|q] \to \widehat{GL}_L[p|q]$ is a G-morphism, as follows from Definition 1.2. In this way we have obtained an example of a G-morphism which is not induced by a GH^∞ map (cf. Section IV.1). ▲

The actions on the underlying group H corresponding to the left and right translations by $\widehat{g}$ are, of course, the ordinary left and right translations by g,

$$L_g(g') = m(g, g') = gg' \qquad R_g(g') = m(g', g) = g'g\,.$$

It is now convenient to state the group axioms in terms of the sheaf $\mathcal{A}$. First, let us denote by $q\colon H \times H \times H \to H$ the map $q = m \circ (Id \times m) = m \circ (m \times \mathrm{Id})$. Then, one has sheaf morphisms

$$\begin{array}{ll} \widehat{m}^*\colon \mathcal{A} \to m_*(\mathcal{A}\hat{\otimes}_\pi\mathcal{A}) & \text{(comultiplication)} \\ \widehat{e}^*\colon \mathcal{A} \to e_*(B_L) & \text{(counit or augmentation)} \\ \widehat{s}^*\colon \mathcal{A} \to s_*\mathcal{A} & \text{(coinverse or antipode)}, \end{array}$$

and the group axioms are equivalent to the commutativity of the diagrams

$$\begin{array}{ccc} \mathcal{A} & \xrightarrow{\widehat{m}^*} & m_*(\mathcal{A}\hat{\otimes}_\pi\mathcal{A}) \\ {\scriptstyle \widehat{m}^*}\downarrow & & \downarrow{\scriptstyle \widehat{m}^*\otimes \mathrm{Id}} \\ m_*(\mathcal{A}\hat{\otimes}_\pi\mathcal{A}) & \xrightarrow{\mathrm{Id}\otimes\widehat{m}^*} & q_*(\mathcal{A}\hat{\otimes}_\pi\mathcal{A}\hat{\otimes}_\pi\mathcal{A}) \end{array}\,, \tag{1.1}$$

$$\begin{array}{ccccc} & \nearrow & \mathcal{A} & \nwarrow & \\ & & {\scriptstyle \widehat{m}^*}\uparrow & & \\ m_*(\mathcal{A}\hat{\otimes}_\pi e_*(B_L)) & \xrightarrow{\mathrm{Id}\otimes\widehat{e}^*} & m_*(\mathcal{A}\hat{\otimes}_\pi\mathcal{A}) & \xrightarrow{\widehat{e}^*\otimes\mathrm{Id}} & m_*(e_*(B_L)\hat{\otimes}_\pi\mathcal{A}) \end{array}\,, \tag{1.2}$$

$$\begin{array}{ccccc} e_*(B_L) & \xleftarrow{\widehat{e}^*} & \mathcal{A} & \xrightarrow{\widehat{e}^*} & e_*(B_L) \\ \downarrow & & {\scriptstyle \widehat{m}^*}\downarrow & & \downarrow \\ \mathcal{A} & \xleftarrow{\mathrm{Id}\cdot\widehat{s}^*} & m_*(\mathcal{A}\hat{\otimes}_\pi\mathcal{A}) & \xrightarrow{\widehat{s}^*\cdot\mathrm{Id}} & \mathcal{A} \end{array}\,, \tag{1.3}$$

which reflect the coassociativity, unit and antipode properties, respectively.

Accordingly, in a sense the structure sheaf $\mathcal{A}$ of a G-Lie supergroup $\widehat{H}$ can be considered as a sheaf of graded topological Hopf B_L-algebras; 'topological' means that the tensor product involved in the definition of a Hopf algebra must be completed in the Grothendieck π topology. The ring $\mathcal{A}(H)$ of global sections of $\mathcal{A}$ is a graded topological Hopf B_L-algebra, although unlike what happens for ordinary Lie groups, or even for graded Lie groups, it does not carry enough information about the Lie supergroup structure. Thus, we encounter once again the fact that the ring of global sections of the structure sheaf of a G-supermanifold does not convey complete information about the supermanifold structure.

The Lie superalgebra of a Lie supergroup. As is well known, the Lie algebra of a Lie group H is the algebra of left-invariant vector fields. The ordinary definition of left invariance, namely that a vector field D is left-invariant if $L_{g*}(D_e) = D_g$ for every point $g \in H$, can also be formulated for a graded vector field on a G-Lie supergroup $\widehat{H}$; however, in accordance with the previous discussion, the correct notion of left invariance must include the invariance under translations induced by $\widehat{T}$-valued points for any G-supermanifold $\widehat{T}$. This is achieved by means of the notion of invariant operators on a Hopf algebra and the corresponding notion of the Lie algebra of an affine group scheme ([**Wat**], page 92). This procedure has also been followed in [**Lop**] for Lie groups in the context of graded manifolds.

To do this, let us start by considering a graded vector field on a G-Lie supermanifold $\widehat{M}$ as an operator $D: \mathcal{A}_{\widehat{M}} \to \mathcal{A}_{\widehat{M}}$. Graded tangent vectors at a point $\widehat{y}: \widehat{z} \to \widehat{M}$ can be interpreted as sheaf morphisms $D_y: \mathcal{A}_{\widehat{M}} \to y_*(B_L)$, and, accordingly, the value at $\widehat{y}$ of a graded vector field $D: \mathcal{A}_{\widehat{M}} \to \mathcal{A}_{\widehat{M}}$ is the graded tangent vector

$$D_y = \widehat{y}^* \circ D: \mathcal{A}_{\widehat{M}} \to y_*(B_L).$$

Let $\widehat{H} = (H, \mathcal{A})$ be a G-Lie supergroup.

Definition 1.3. *A graded vector field $D: \mathcal{A} \to \mathcal{A}$ is left-invariant if the diagram*

$$\begin{array}{ccc} \mathcal{A} & \xrightarrow{\widehat{m}^*} & m_*(\mathcal{A}\hat{\otimes}_\pi\mathcal{A}) \\ {\scriptstyle D}\downarrow & & \downarrow{\scriptstyle \mathrm{Id}\otimes D} \\ \mathcal{A} & \xrightarrow{\widehat{m}^*} & m_*(\mathcal{A}\hat{\otimes}_\pi\mathcal{A}) \end{array}$$

commutes, that is, $\widehat{m}^* \circ D = (\mathrm{Id} \otimes D) \circ \widehat{m}^*$. Similarly, a graded vector field

$D: \mathcal{A} \to \mathcal{A}$ is right-invariant if the diagram

$$\begin{array}{ccc} \mathcal{A} & \xrightarrow{\widehat{m}^*} & m_*(\mathcal{A}\hat{\otimes}_\pi\mathcal{A}) \\ {\scriptstyle D}\downarrow & & \downarrow{\scriptstyle D\otimes \mathrm{Id}} \\ \mathcal{A} & \xrightarrow{\widehat{m}^*} & m_*(\mathcal{A}\hat{\otimes}_\pi\mathcal{A}) \end{array}$$

commutes, namely, $\widehat{m}^* \circ D = (D \otimes \mathrm{Id}) \circ \widehat{m}^*$.

This definition generalizes the classical one in the sense that if D is left-invariant (resp. right-invariant), one has

$$D_g = \widehat{L}_{\widehat{g}*}(D_e) \qquad (\text{resp. } D_g = \widehat{R}_{\widehat{g}*}(D_e))$$

for every point $g \in H$.

If D and D' are left-invariant graded vector fields, $[D, D']$ and $aD + bD'$ (for every $a, b \in B_L$) are left-invariant as well. If follows that all left-invariant graded vector fields on a G-Lie supergroup form a Lie superalgebra over B_L.

Definition 1.4. *The Lie superalgebra of a G-Lie supergroup $\widehat{H}$ is the Lie superalgebra $\mathfrak{h} = \mathrm{Lie}\,\widehat{H}$ over B_L formed by the left-invariant graded vector fields on $\widehat{H}$.*

The Lie superalgebra $\mathfrak{h}$ can be interpreted as the graded tangent space at the unit point, that is, the graded B_L-module $T_e\widehat{H} = \mathrm{Der}_{B_L}(\mathcal{A}_e, B_L)$. In this way, the graded tangent space $T_e\widehat{H}$ inherits the structure of a Lie superalgebra:

Proposition 1.1. *The morphism*

$$\begin{aligned} \mathfrak{h} = \mathrm{Lie}\,\widehat{H} &\to T_e\widehat{H} \\ D &\mapsto D_e \end{aligned}$$

is an isomorphism of graded B_L-modules. Therefore, the Lie superalgebra $\mathfrak{h}$ is a free rank (m, n) graded B_L-module.

Proof. Let $D_e: \mathcal{A} \to e_*(B_L)$ be a tangent graded vector at the unit point. D_e induces a graded derivation $(\mathrm{Id} \otimes D_e): m_*(\mathcal{A}\hat{\otimes}_\pi\mathcal{A}) \to \mathcal{A}\hat{\otimes}_\pi e_*(B_L) \simeq \mathcal{A}$. The composition

$$D = (\mathrm{Id} \otimes D_e) \circ \widehat{m}^*: \mathcal{A} \to \mathcal{A}$$

is a graded vector field whose value at the unit point is D_e. Moreover, D is left-invariant, because $(\mathrm{Id}\otimes D)\circ\widehat{m}^* = (\mathrm{Id}\otimes(\mathrm{Id}\otimes D_e)\circ\widehat{m}^*)\circ\widehat{m}^* = (\mathrm{Id}\otimes\mathrm{Id}\otimes D_e)\circ(\mathrm{Id}\otimes\widehat{m}^*)\circ\widehat{m}^*$, while $\widehat{m}^*\circ D = (\mathrm{Id}\otimes\mathrm{Id}\otimes D_e)\circ(\widehat{m}^*\otimes\mathrm{Id})\circ\widehat{m}^*$; the two quantities agree by coassociativity (cf. Eq. (1.1)). This proves surjectivity. One can show that the morphism is injective by proving that a left-invariant graded vector field D is determined by its value at the unit point $D_e = \widehat{e}^*\circ D$. But since $(\mathrm{Id}\otimes\widehat{e}^*)\circ\widehat{m}^* = \mathrm{Id}$, the invariance condition $\widehat{m}^*\circ D = (\mathrm{Id}\otimes D)\circ\widehat{m}^*$ implies that $D = (\mathrm{Id}\otimes\widehat{e}^*)\circ(\mathrm{Id}\otimes D)\circ\widehat{m}^* = (\mathrm{Id}\otimes D_e)\circ\widehat{m}^*$, thus determining D in terms of D_e. ■

Proposition 1.2. *The isomorphism $\mathfrak{h}\simeq T_e\widehat{H}$ induces a graded Lie bracket between elements in $T_e\widehat{H}$, according to the equation*

$$[X,Y] = (X\otimes Y - Y\otimes X)\circ\widehat{m}^*.$$

Proof. Let $X^* = (\mathrm{Id}\otimes X)\circ\widehat{m}^*$ and let $Y^* = (\mathrm{Id}\otimes Y)\circ\widehat{m}^*$ be the corresponding left-invariant graded vector fields so that $[X,Y] = \widehat{e}^*([X^*,Y^*])$. Writing $\widehat{m}^*(h) = \sum_k h^k\otimes h_k$ for every section $h\in\mathcal{A}_{\widehat{H}}(U)$, one has that $X^*(h) = \sum_k h^k X(h_k)$ and $Y^*(h) = \sum_k h^k Y(h_k)$. Then

$$\begin{aligned}[X^*,Y^*](h) &= \sum_{kj}(h^{kj}X(h^k{}_j)Y(h_k) - h^{hj}Y(h^k{}_j)X(h_k))\\ &= ((\mathrm{Id}\otimes X\otimes Y - \mathrm{Id}\otimes Y\otimes X)\circ(\widehat{m}^*\otimes\mathrm{Id})\circ\widehat{m}^*)(h)\end{aligned}$$

and

$$\begin{aligned}[X,Y] &= \widehat{e}^*\circ[X^*,Y^*]\\ &= (\mathrm{Id}\otimes X\otimes Y - \mathrm{Id}\otimes Y\otimes X)\circ(\widehat{e}^*\otimes\mathrm{Id}\otimes\mathrm{Id})\circ(\widehat{m}^*\otimes\mathrm{Id})\circ\widehat{m}^*\\ &= (X\otimes Y - Y\otimes X)\circ\widehat{m}^* \quad .\end{aligned}$$

■

A statement similar to Proposition 1.1 holds for right-invariant graded vector fields, namely, there is an isomorphism

$$\begin{aligned}\mathfrak{h}^R &\simeq T_e\widehat{H}\\ D &\mapsto D_e\end{aligned}$$

where $\mathfrak{h}^R$ stands for the graded B_L-module of right-invariant graded vector fields. However, as in the case of ordinary Lie groups, the structure of a Lie superalgebra inherited by $T_e\widehat{H}$ in this way is in general different from that considered before (the two structures are indeed *anti-isomorphic*). In the sequel, we shall always consider $T_e\widehat{H}$ as a Lie superalgebra through the isomorphism of Proposition 1.1; that is, by means of the graded Lie bracket calculated in Proposition 1.2.

Proposition 1.3. *There are isomorphisms of sheaves of graded $\mathcal{A}_{\widehat{H}}$-modules*

$$\mathcal{A}_{\widehat{H}} \otimes \mathfrak{h} \simeq \mathcal{D}er\mathcal{A}_{\widehat{H}}, \qquad \mathcal{A}_{\widehat{H}} \otimes \mathfrak{h}^R \simeq \mathcal{D}er\mathcal{A}_{\widehat{H}};$$

that is to say, the tangent sheaf $\mathcal{D}er\mathcal{A}_{\widehat{H}}$ is the globally free rank (m,n) sheaf of graded $\mathcal{A}_{\widehat{H}}$-modules generated by the left-invariant (resp. right-invariant) graded vector fields.

Proof. The proof of the first statement can be reduced to showing that if $(X^1,\dots,X^m,\Xi^1,\dots,\Xi^n)$ is a homogeneous basis of $\mathfrak{h}$ as a graded B_L-module, then it is also a basis of $\mathcal{D}er\mathcal{A}_{\widehat{H}}$ as a sheaf of $\mathcal{A}_{\widehat{H}}$-modules. For every point $g \in H$, the values $(X^1_g,\dots,X^m_g,\Xi^1_g,\dots,\Xi^n_g)$ form a basis of the tangent space $T_g\widehat{H}$, since the left translation $\widehat{L}_{\widehat{g}}$ induces an isomorphism of B_L-modules $\widehat{L}_{\widehat{g}*}: T_e\widehat{H} \simeq T_g\widehat{H}$ and $X^i_g = \widehat{L}_{\widehat{g}*}(X^i_e)$, $\Xi^\alpha_g = \widehat{L}_{\widehat{g}*}(\Xi^\alpha_e)$ by left-invariance. The graded Nakayama lemma (Proposition I.1.1) now implies that the germs of $(X^1,\dots,X^m,\Xi^1,\dots,\Xi^n)$ form a basis of $(\mathcal{D}er\mathcal{A}_{\widehat{H}})_g$ as an $(\mathcal{A}_{\widehat{H}})_g$-module for every $g \in H$, thus finishing the proof of the first claim. The second part is proved in the same way. ∎

2. Lie supergroup actions

We now wish to study the action of a G-Lie supergroup $\widehat{H}$ on a G-supermanifold.

Definition 2.1. *A right action of $\widehat{H}$ on a G-supermanifold $\widehat{P}$ is a G-morphism*

$$\widehat{\varsigma}: \widehat{P} \times \widehat{H} \to \widehat{P}$$

such that the diagram

$$\begin{array}{ccc} \widehat{P}\times\widehat{H}\times\widehat{H} & \xrightarrow{\widehat{\varsigma}\times \mathrm{Id}} & \widehat{P}\times\widehat{H} \\ {\scriptstyle \mathrm{Id}\times\widehat{m}}\downarrow & & \downarrow{\scriptstyle \widehat{\varsigma}} \\ \widehat{P}\times\widehat{H} & \xrightarrow{\widehat{\varsigma}} & \widehat{P} \end{array}$$

commutes, and the composition

$$\widehat{P}=\widehat{P}\times\widehat{z}\xrightarrow{\mathrm{Id}\times\widehat{e}}\widehat{P}\times\widehat{H}\xrightarrow{\widehat{\varsigma}}\widehat{P}$$

is the identity, $\widehat{\varsigma}\circ(\mathrm{Id}\times\widehat{e})=\mathrm{Id}$. *Similarly, a left action of* $\widehat{H}$ *on* $\widehat{P}$ *is a G-morphism*

$$\widehat{\varsigma}\colon\widehat{H}\times\widehat{P}\to\widehat{P}$$

such that the diagram

$$\begin{array}{ccc} \widehat{H}\times\widehat{H}\times\widehat{P} & \xrightarrow{\mathrm{Id}\times\widehat{\varsigma}} & \widehat{H}\times\widehat{P} \\ {\scriptstyle \widehat{m}\times\mathrm{Id}}\downarrow & & \downarrow{\scriptstyle \widehat{\varsigma}} \\ \widehat{H}\times\widehat{P} & \xrightarrow{\widehat{\varsigma}} & \widehat{P} \end{array}$$

commutes, and the composition

$$\widehat{P}=\widehat{z}\times\widehat{P}\xrightarrow{\widehat{e}\times\mathrm{Id}}\widehat{H}\times\widehat{P}\xrightarrow{\widehat{\varsigma}}\widehat{P}$$

is the identity, $\widehat{\varsigma}\circ(\widehat{e}\times\mathrm{Id})=\mathrm{Id}$.

EXAMPLE 2.1. Every G-Lie supergroup acts on itself both on the left and on the right by means of the multiplication morphism

$$\widehat{m}\colon\widehat{H}\times\widehat{H}\to\widehat{H}\,.$$

▲

EXAMPLE 2.2. There is a *trivial right action* of every G-Lie supergroup $\widehat{H}$ on a G-supermanifold $\widehat{M}$, given by

$$\widehat{\varsigma}=\widehat{p}_1\colon\widehat{M}\times\widehat{H}\to\widehat{M}$$

where $\widehat{p}_1$ is the projection onto the first factor. ▲

EXAMPLE 2.3. If $\widehat{M}$ is a G-supermanifold and $\widehat{H}$ is a G-Lie supergroup, there is a right action of $\widehat{H}$ on the product G-supermanifold $\widehat{M} \times \widehat{H}$, given by right multiplication of the second factor, that is:

$$\widehat{\varsigma} = \mathrm{Id} \times \widehat{m} \colon \widehat{M} \times \widehat{H} \times \widehat{H} \to \widehat{M} \times \widehat{H} .$$

In the same way, there is a left action on $\widehat{H} \times \widehat{M}$ by left multiplication on the first factor. ▲

EXAMPLE 2.4. The general linear supergroup $\widehat{H} = \widehat{GL_L}[p|q]$ (Example 1.1) acts linearly on $B_L^{p|q}$ endowed with its natural structure of G-supermanifold of dimension $(p+q, p+q)$, since, by (IV.3.4), the natural map

$$\widehat{GL_L}[p|q] \times B_L^{p|q} \to B_L^{p|q}$$

given by matrix multiplication is a G-morphism. Actually, this map is H^∞, and hence it defines an H^∞ left action of $GL_L[p|q]$ on $B_L^{p|q}$. ▲

EXAMPLE 2.5. The group morphism $\mathrm{Ber} \colon GL_L[m|n] \to GL_L[1|0]$ (cf. Proposition I.3.2) that maps a matrix $X \in GL_L[m|n]$ into its Berezinian $\mathrm{Ber}\, X$ (see Section I.3) is an H^∞ map, hence it induces a G-morphism $\mathrm{Ber} \colon \widehat{GL_L}[p|q] \to \widehat{GL_L}[1|0]$. The composition of $\mathrm{Ber} \times \mathrm{Id} \colon \widehat{GL_L}[p|q] \times B_L^{1|0} \to \widehat{GL_L}[1|0] \times B_L^{1|0}$ with the natural action $\widehat{GL_L}[1|0] \times B_L^{1|0} \to B_L^{1|0}$ provides a left action

$$\widehat{GL_L}[p|q] \times B_L^{1|0} \to B_L^{1|0} ;$$

that is, the action of the general linear supergroup $\widehat{GL_L}[p|q]$ on $B_L^{1|0}$ is such that a matrix acts by multiplication by its Berezinian. Analogously, since there is a natural isomorphism $GL_L[1|0] \simeq GL_L[0|1]$, one has an action

$$\widehat{GL_L}[p|q] \times B_L^{0|1} \to B_L^{0|1} .$$

▲

Let $\widehat{\varsigma} \colon \widehat{P} \times \widehat{H} \to \widehat{P}$ be a right action of a G-Lie supergroup $\widehat{H}$ on a G-supermanifold $\widehat{P}$. If $\widehat{Z}$ is a G-supermanifold, and $\widehat{f} \colon \widehat{Z} \to \widehat{P}$ and $\widehat{h} \colon \widehat{Z} \to \widehat{H}$ are G-morphisms, let us denote by $\widehat{f} \cdot \widehat{h} \colon \widehat{Z} \to \widehat{P}$ the G-morphism obtained by the composition:

$$\widehat{Z} \xrightarrow{(\widehat{f}, \widehat{h})} \widehat{P} \times \widehat{H} \xrightarrow{\widehat{\varsigma}} \widehat{P} . \tag{2.1}$$

In the same way, given a left action $\widehat{\varsigma}: \widehat{H} \times \widehat{P} \to \widehat{P}$, we shall denote by $\widehat{h}\cdot\widehat{f}: \widehat{Z} \to \widehat{P}$ the G-morphism obtained by composition

$$\widehat{Z} \xrightarrow{(\widehat{h},\widehat{f})} \widehat{H} \times \widehat{P} \xrightarrow{\widehat{\varsigma}} \widehat{P}.$$

In what follows, we shall focus our attention on right actions. This makes the exposition simpler without loosing generality, since the theories of right and left actions are completely symmetric. Let us then consider a right action $\widehat{\varsigma}: \widehat{P} \times \widehat{H} \to \widehat{P}$ of a G-Lie supergroup $\widehat{H}$ on a G-supermanifold $\widehat{P}$. Then, $\varsigma: P \times H \to P$ is a right action of the underlying ordinary Lie group H on the underlying differentiable manifold P. If $U \subset P$ is an open subset invariant under this action, $\varsigma(U \times H) \subset U$, then $\widehat{H}$ acts on the open G-submanifold $\widehat{U} = (U, \mathcal{A}_{\widehat{P}|U})$, and we shall say that $\widehat{U}$ is an invariant open submanifold of $\widehat{P}$.

We can also consider actions of a G-Lie supergroup on a 'relative' G-supermanifold, that is, on a G-morphism $\widehat{p}: \widehat{P} \to \widehat{M}$.

Definition 2.2. *A right action of a G-Lie supergroup $\widehat{H}$ on a relative G-supermanifold $\widehat{p}: \widehat{P} \to \widehat{M}$ is a right action $\widehat{\varsigma}: \widehat{P} \times \widehat{H} \to \widehat{P}$ such that the diagram*

$$\begin{array}{ccc} \widehat{P} \times \widehat{H} & \xrightarrow{\widehat{\varsigma}} & \widehat{P} \\ {\scriptstyle \widehat{p}_1}\downarrow & & \downarrow{\scriptstyle \widehat{p}} \\ \widehat{P} & \xrightarrow{\widehat{p}} & \widehat{M} \end{array}$$

is commutative.

Example 2.3 showed just such a situation: the action of $\widehat{H}$ on the product G-supermanifold $\widehat{M} \times \widehat{H}$, given by right multiplication of the second factor, is a right action on the relative G-supermanifold $\widehat{p}_1: \widehat{M} \times \widehat{H} \to \widehat{M}$.

Another very important example is given by the following construction.

EXAMPLE 2.6. Let $\widehat{q}: \widehat{\xi} \to \widehat{M}$ a rank (p,q) supervector bundle (SVB) over a G-supermanifold $\widehat{M}$ (Definition IV.3.3), and let $\widehat{\pi}: \mathrm{Iso}(\widehat{M} \times B_L^{p|q}, \widehat{\xi}) \to \widehat{M}$ be the superfibre bundle of isomorphisms of the trivial SVB $\widehat{M} \times B_L^{p|q}$ of rank (p,q) with $\widehat{\xi}$. By (IV.3.12), the G-morphism

$$\mathrm{Iso}(\widehat{M} \times B_L^{p|q}, \widehat{\xi}) \times \widehat{GL_L}[p|q] \to \mathrm{Iso}(\widehat{M} \times B_L^{p|q}, \widehat{\xi})$$

is a right action of the general linear supergroup $\widehat{GL_L}[p|q]$ on the relative G-supermanifold $\widehat{\pi}\colon \mathrm{Iso}(\widehat{M}\times B_L^{p|q},\widehat{\xi})\to\widehat{M}$. ▲

If $\widehat{\varsigma}\colon \widehat{P}\times\widehat{H}\to\widehat{P}$ is a right action of a G-Lie supergroup $\widehat{H}$ on a relative G-supermanifold $\widehat{p}\colon\widehat{P}\to\widehat{M}$, for every open $U\subset M$ the morphism $\widehat{\varsigma}$ induces a right action, denoted with the same symbol, on the relative G-supermanifold $\widehat{p}\colon \widehat{P}_{|\widehat{U}}\to\widehat{U}$ obtained by restricting $\widehat{p}$ to the pre-image $\widehat{P}_{|\widehat{U}}=(p^{-1}(U),\mathcal{A}_{\widehat{P}|p^{-1}(U)})$.

Let $\widehat{P}$ and $\widehat{N}$ be G-supermanifolds that are acted on by a G-Lie supergroup $\widehat{H}$.

Definition 2.3. *A morphism of G-supermanifolds $\widehat{f}\colon\widehat{P}\to\widehat{N}$ is said to be $\widehat{H}$-invariant if it is compatible with the action of $\widehat{H}$; that is, if the diagram*

$$\begin{array}{ccc} \widehat{P}\times\widehat{H} & \xrightarrow{\widehat{f}\times \mathrm{Id}} & \widehat{N}\times\widehat{H} \\ \downarrow{\scriptstyle\widehat{\varsigma}} & & \downarrow{\scriptstyle\widehat{\varsigma}} \\ \widehat{P} & \xrightarrow{\widehat{f}} & \widehat{N} \end{array}$$

is commutative.

In the same way, if $\widehat{p}\colon\widehat{P}\to\widehat{M}$ and $\widehat{q}\colon\widehat{N}\to\widehat{M}$ are G-morphisms (that is, relative G-supermanifolds) acted on by $\widehat{H}$ (Definition 2.2), one can define:.

Definition 2.4. *A morphism of G-supermanifolds $\widehat{f}\colon\widehat{P}\to\widehat{N}$ is said to be an $\widehat{H}$-invariant morphism of relative G-supermanifolds over $\widehat{M}$ if $\widehat{p}=\widehat{q}\circ\widehat{f}$ and $\widehat{f}$ is $\widehat{H}$-invariant.*

There is an important class of $\widehat{H}$-invariant morphisms; since the sections of the structure sheaf on an open subset $V\subset P$ are exactly the G-morphisms of $\widehat{V}$ into $B_L\equiv B_L^{1,1}$ (Proposition IV.1.2), for every $\widehat{H}$-invariant open G-submanifold $\widehat{V}$ of $\widehat{P}$ one has:

Definition 2.5. *The invariant subring $\mathcal{A}_{\widehat{P}}(V)^{\widehat{H}}$ of $\mathcal{A}_{\widehat{P}}(V)$ is the subring of the sections that are $\widehat{H}$-invariant when considered as G-morphisms $\widehat{V}\to B_L$, where one takes the trivial action of $\widehat{H}$ on B_L.*

In this way we have introduced the notion of G-invariant 'functions' on a G-Lie supergroup.

The notion of quotient by the action of a G-Lie supergroup is again taken from the theory of algebraic groups (see [**Pen**], [**Lop**]). Let $\widehat{\varsigma}: \widehat{P} \times \widehat{H} \to \widehat{P}$ be an action of a G-Lie supergroup on a G-supermanifold.

Definition 2.6. *A quotient of the action of $\widehat{H}$ on $\widehat{P}$ is a pair $(\widehat{M}, \widehat{p})$, where $\widehat{M}$ is a G-supermanifold and $\widehat{p}: \widehat{P} \to \widehat{M}$ is morphism of G-supermanifolds such that:*

(1) *$\widehat{\varsigma}$ acts on the relative G-supermanifold $\widehat{p}: \widehat{P} \to \widehat{M}$ (Definition 2.2);*
(2) *for every morphism $\widehat{f}: \widehat{P} \to \widehat{N}$ such that $\widehat{f} \circ \widehat{\varsigma} = \widehat{f} \circ \widehat{p}_1$, there is a unique morphism $\widehat{g}: \widehat{M} \to \widehat{N}$ with $\widehat{f} = \widehat{g} \circ \widehat{p}$.*

In general, given an action of a G-Lie supergroup on a G-supermanifold, the quotient may fail to exist. Later on we shall see an important class of morphisms that are quotients, namely, principal superfibre bundles.

If a quotient $\widehat{p}: \widehat{P} \to \widehat{M}$ of an action of $\widehat{H}$ on $\widehat{P}$ exists, the structure sheaf $\mathcal{A}_{\widehat{M}}$ can be described in terms of $\mathcal{A}_{\widehat{P}}$ as its invariant subsheaf (cf. Definition 2.5):

Proposition 2.1. *Let $\widehat{p}: \widehat{P} \to \widehat{M}$ be a quotient of an action $\widehat{\varsigma}: \widehat{P} \times \widehat{H} \to \widehat{P}$ of a G-Lie supergroup $\widehat{H}$ on a G-supermanifold $\widehat{P}$. For every open subset $U \subset M$, there is a graded B_L-algebra isomorphism*

$$\mathcal{A}_{\widehat{M}}(U) \simeq \mathcal{A}_{\widehat{P}}(p^{-1}(U))^{\widehat{H}}$$

between the sections on U of the structure sheaf of the quotient G-supermanifold and the $\widehat{H}$-invariant sections of the structure sheaf of $\widehat{P}$ on $p^{-1}(U)$. In sheaf notation:

$$\mathcal{A}_{\widehat{M}} \simeq (\widehat{p}_* \mathcal{A}_{\widehat{P}})^{\widehat{H}} .$$

Proof. This follows from (2) of the definition of quotient, taking $\widehat{N} = B_L$, and from the definition of $\widehat{H}$-invariant sections of the structure sheaf $\mathcal{A}_{\widehat{P}}$ on an invariant open submanifold. ∎

The invariant sections of $\mathcal{A}_{\widehat{P}}(p^{-1}(U))$ are precisely the elements that have the same image under the morphisms

$$\mathcal{A}_{\widehat{P}}(p^{-1}(U)) \xrightarrow{\widehat{\varsigma}^*} (\mathcal{A}_{\widehat{H}} \otimes_\pi \mathcal{A}_{\widehat{P}})(\varsigma^{-1}(p^{-1}(U))$$
$$\mathcal{A}_{\widehat{P}}(p^{-1}(U)) \xrightarrow{\widehat{p_1}^*} (\mathcal{A}_{\widehat{H}} \otimes_\pi \mathcal{A}_{\widehat{P}})(p_1^{-1}(p^{-1}(U))$$

Since $\widehat{p} \circ \widehat{\varsigma} = \widehat{p} \circ \widehat{p}_1$, if one writes $\widehat{\pi} = \widehat{p} \circ \widehat{\varsigma} = \widehat{p} \circ \widehat{p}_1 : \widehat{P} \times \widehat{H} \to \widehat{M}$, one has that $\mathcal{A}_{\widehat{M}}(U) \simeq (\mathcal{A}_{\widehat{P}}(p^{-1}(U)))^{\widehat{H}}$ is the kernel of the morphism of graded B_L-modules $(\widehat{\varsigma}^* - \widehat{p}_1^*) : \mathcal{A}_{\widehat{P}}(p^{-1}(U)) \to (\mathcal{A}_{\widehat{P}} \otimes_\pi \mathcal{A}_{\widehat{H}})(\pi^{-1}(U))$. This can be summarized by the following:

Proposition 2.2. *Let $\widehat{p}: \widehat{P} \to \widehat{M}$ be the quotient of an action $\widehat{\varsigma}: \widehat{P} \times \widehat{H} \to \widehat{P}$ of a G-Lie supergroup $\widehat{H}$ on a G-supermanifold $\widehat{P}$. The sequence of sheaves of B_L-modules on M*

$$0 \to \mathcal{A}_{\widehat{M}} \xrightarrow{\widehat{p}^*} p_* \mathcal{A}_{\widehat{P}} \xrightarrow{\widehat{\varsigma}^* - \widehat{p}_1^*} \pi_*(\mathcal{A}_{\widehat{P}} \otimes_\pi \mathcal{A}_{\widehat{H}})$$

is exact. ∎

G-Lie supergroup actions and graded vector fields. In this section, we study the effect of the action of a G-Lie supergroup on graded vector fields. Let $\widehat{\varsigma}: \widehat{P} \times \widehat{H} \to \widehat{P}$ be an action of a G-Lie supergroup $\widehat{H}$ on a G-supermanifold $\widehat{P}$. As in Definition 1.2, for every point $\widehat{g}: \widehat{z} \to \widehat{H}$ one can consider the right translation by $\widehat{g}$, which is the morphism of G-supermanifolds given by the diagram

$$\begin{array}{ccc} \widehat{P} & \xrightarrow{\widehat{R}_{\widehat{g}}} & \widehat{P} \\ \Vert & & \uparrow{\scriptstyle \widehat{\varsigma}} \\ \widehat{P} \times \widehat{z} & \xrightarrow{\mathrm{Id} \times \widehat{g}} & \widehat{P} \times \widehat{H} \end{array} . \tag{2.2}$$

The effect of $\widehat{R}_{\widehat{g}}$ on the manifold P is given by $R_g(z) = \varsigma(z, g) = zg$. In a similar way, if $\widehat{y}: \widehat{z} \to \widehat{P}$ is a point of $\widehat{P}$, there is a morphism $\widehat{L}_{\widehat{y}}: \widehat{H} \to \widehat{P}$ defined by the diagram

$$\begin{array}{ccc} \widehat{H} & \xrightarrow{\widehat{L}_{\widehat{y}}} & \widehat{P} \\ \Vert & & \uparrow{\scriptstyle \widehat{\varsigma}} \\ \widehat{z} \times \widehat{H} & \xrightarrow{\widehat{y} \times \mathrm{Id}} & \widehat{P} \times \widehat{H} \end{array} . \tag{2.3}$$

The effect of $\widehat{L}_{\widehat{y}}$ on points is the G^∞ morphism

$$\begin{aligned} L_y \colon H &\to P \\ g &\mapsto \varsigma(y, g) = yg \end{aligned} .$$

The morphisms $\widehat{L}_{\widehat{g}}$ may fail to be one-to-one; however, if $\widehat{P} = \widehat{H}$ with the action given by the multiplication morphism (Example 2.1), $\widehat{L}_{\widehat{g}}$ is, for every point $\widehat{g}$, an isomorphism with inverse $\widehat{L}_{\widehat{g}^{-1}}$. In this case, the composition

$$\widehat{H} \xrightarrow{\widehat{L}_{\widehat{g}}} \widehat{H} \xrightarrow{\widehat{R}_{\widehat{g}^{-1}}} \widehat{H} \tag{2.4}$$

is an isomorphism of G-Lie supergroups that preserves the unit point, so that it induces an isomorphism of Lie superalgebras

$$\begin{aligned} \operatorname{ad}(\widehat{g}) = (\widehat{R}_{\widehat{g}^{-1}} \circ \widehat{L}_{\widehat{g}})_*\colon \mathfrak{h} &\to \mathfrak{h} \\ X &\mapsto \operatorname{ad}(\widehat{g}) \cdot X \end{aligned}\ , \tag{2.5}$$

called the *adjoint morphism* corresponding to $\widehat{g}$.

There is a notion of $\widehat{H}$-invariant graded vector field on $\widehat{P}$, that generalizes the notion of an invariant graded vector field on $\widehat{H}$.

Definition 2.7. *A graded vector field $D\colon \mathcal{A}_{\widehat{P}} \to \mathcal{A}_{\widehat{P}}$ on $\widehat{P}$ is $\widehat{H}$-invariant if the diagram*

$$\begin{array}{ccc} \mathcal{A}_{\widehat{P}} & \xrightarrow{\widehat{\varsigma}^*} & \varsigma_*(\mathcal{A}_{\widehat{P}}\hat{\otimes}_\pi\mathcal{A}_{\widehat{H}}) \\ {\scriptstyle D}\downarrow & & \downarrow{\scriptstyle D\otimes \mathrm{Id}} \\ \mathcal{A}_{\widehat{P}} & \xrightarrow{\widehat{\varsigma}^*} & \varsigma_*(\mathcal{A}_{\widehat{P}}\hat{\otimes}_\pi\mathcal{A}_{\widehat{H}}) \end{array}$$

is commutative, that is, $\widehat{\varsigma}^* \circ D = (D \otimes \mathrm{Id}) \circ \widehat{\varsigma}^*$.

As for invariant graded vector fields on a G-Lie supergroups, if D is a $\widehat{H}$-invariant graded vector field on $\widehat{P}$, one has that

$$\widehat{R}_{\widehat{g}*}(D_z) = D_{gz}$$

for every $z \in P$, $g \in H$.

The elements $X \in T_e\widehat{H} = \mathfrak{h}$ in the Lie superalgebra of $\widehat{H}$ induce graded vector fields on $\widehat{P}$.

Definition 2.8. *The fundamental graded vector field on $\widehat{P}$ associated with an* *element $X \in \mathfrak{h}$ is the graded vector field X^* defined by the diagram*

$$\begin{array}{ccc} \mathcal{A}_{\widehat{P}} & \xrightarrow{\widehat{\varsigma}^*} & \varsigma_*(\mathcal{A}_{\widehat{P}}\hat{\otimes}_\pi\mathcal{A}_{\widehat{H}}) \\ {\scriptstyle X^*}\downarrow & & \downarrow{\scriptstyle \mathrm{Id}\otimes X} \\ \mathcal{A}_{\widehat{P}} & \xrightarrow{\sim} & \mathcal{A}_{\widehat{P}}\hat{\otimes}_\pi e_*(B_L) \end{array} .$$

One should note that fundamental graded vector fields are not $\widehat{H}$-invariant. The action of a right translation on a fundamental graded vector field is given by the following statement:

Proposition 2.3. *Let $X \in \mathfrak{h}$ be an element of the Lie superalgebra of $\widehat{H}$ and X^* the associated fundamental graded vector field. For every pair of points $g \in H$ and $y \in P$, one has:*

$$\widehat{R}_{\widehat{g}*}((X^*)_y) = ((\mathrm{ad}(\widehat{g}^{-1})\cdot X)^*)_{gy} .$$

Proof. The proof is a straightforward adaptation of the calculation that proves the analogous ordinary result. One has that $\widehat{R}_{\widehat{g}*}((X^*)_y) = \widehat{R}_{\widehat{g}*}\widehat{L}_{\widehat{y}*}(X) = \widehat{L}_{\widehat{y}*}\widehat{R}_{\widehat{g}*}(X) = \widehat{L}_{\widehat{yg}*}\widehat{L}_{\widehat{g}^{-1}*}\widehat{R}_{\widehat{g}*}(X) = \widehat{L}_{\widehat{yg}*}(\mathrm{ad}(\widehat{g}^{-1})\cdot X) = ((\mathrm{ad}(\widehat{g}^{-1})\cdot X)^*)_{yg}$. ■

Proposition 2.4. *The map*

$$\begin{aligned} \mathfrak{h} &\to \mathrm{Der}_{B_L}\mathcal{A}_{\widehat{P}}(P) \\ X &\mapsto X^* \end{aligned}$$

is a morphism of Lie superalgebras over B_L. That is, the graded Lie bracket of two fundamental graded vector fields is the fundamental graded vector field associated with the corresponding graded Lie bracket, $[X^*, Y^*] = [X, Y]^*$.

Proof. Proceeding as in Proposition 1.2, one can prove that:

$$[X^*, Y^*] = (\mathrm{Id}\otimes X\otimes Y - \mathrm{Id}\otimes Y\otimes X)\circ(\widehat{\varsigma}^*\otimes\mathrm{Id})\circ\widehat{\varsigma}^* .$$

However, Proposition 1.2 implies that

$$\begin{aligned} [X, Y]^* &= (\mathrm{Id}\otimes[X, Y])\circ\widehat{\varsigma}^* \\ &= (\mathrm{Id}\otimes X\otimes Y - \mathrm{Id}\otimes Y\otimes X)\circ(\mathrm{Id}\otimes\widehat{m}^*)\circ\widehat{\varsigma}^* \\ &= (\mathrm{Id}\otimes X\otimes Y - \mathrm{Id}\otimes Y\otimes X)\circ(\widehat{\varsigma}^*\otimes\mathrm{Id})\circ\widehat{\varsigma}^* \quad , \end{aligned}$$

thus finishing the proof. ■

EXAMPLE 2.7. (Fundamental graded vector fields for the right action $\widehat{m}\colon \widehat{H} \times \widehat{H} \to \widehat{H}$, cf. Example 2.1). In this case, fundamental graded vector fields are exactly left-invariant graded vector fields in the sense of Definition 1.3, and, by Proposition 1.3, there is an isomorphism of sheaves of $\mathcal{A}_{\widehat{H}}$-modules

$$\begin{aligned} \mathcal{A}_{\widehat{H}} \otimes_{B_L} \mathfrak{h} &\simeq \mathcal{D}er\,\mathcal{A}_{\widehat{H}} \\ f \otimes X &\mapsto f \cdot X^* \end{aligned} . \tag{2.6}$$

▲

EXAMPLE 2.8. (Fundamental graded vector fields for the right action $\widehat{H}$ on $\widehat{M} \times \widehat{H}$ given by multiplication of the second factor, $\widehat{M}$ being an arbitrary G-supermanifold, cf. Example 2.3). The fundamental graded vector field X^* determined by an element $X \in \mathfrak{h}$ is the operator $X^*\colon \mathcal{A}_{\widehat{M}} \hat{\otimes}_\pi \mathcal{A}_{\widehat{H}} \to \mathcal{A}_{\widehat{M}} \hat{\otimes}_\pi \mathcal{A}_{\widehat{H}}$ defined by $\mathrm{Id} \otimes (\mathrm{Id} \otimes X) \circ \widehat{m}^*$. ▲

The adjoint representation. Let $\widehat{H}$ be a G-Lie supergroup.

Definition 2.9. *The adjoint representation of $\widehat{H}$ is the left action of $\widehat{H}$ on itself, $\widehat{\mathrm{Ad}}\colon \widehat{H} \times \widehat{H} \to \widehat{H}$, obtained by composition of the G-morphisms*

$$\widehat{H} \times \widehat{H} \xrightarrow{\widehat{\Delta} \times \mathrm{Id}} \widehat{H} \times \widehat{H} \times \widehat{H} \xrightarrow{\mathrm{Id} \times \widehat{\varrho}} \widehat{H} \times \widehat{H} \times \widehat{H}$$
$$\xrightarrow{\mathrm{Id} \times \mathrm{Id} \times \widehat{s}} \widehat{H} \times \widehat{H} \times \widehat{H} \xrightarrow{\widehat{m} \times \mathrm{Id}} \widehat{H} \times \widehat{H} \xrightarrow{\widehat{m}} \widehat{H}\,,$$

where $\widehat{\Delta}\colon \widehat{H} \to \widehat{H} \times \widehat{H}$ is the diagonal morphism and $\widehat{\varrho}\colon \widehat{H} \times \widehat{H} \to \widehat{H} \times \widehat{H}$ is the morphism that exchanges factors.

The map $\mathrm{Ad}\colon H \times H \to H$ is the usual adjoint representation of H on itself, $\mathrm{Ad}(g,h) = ghg^{-1}$. For every point $\widehat{g}\colon \widehat{g} \to \widehat{H}$, the composition

$$\widehat{H} = \widehat{g} \times \widehat{H} \xrightarrow{\widehat{g} \times \mathrm{Id}} \widehat{H} \times \widehat{H} \xrightarrow{\widehat{\mathrm{Ad}}} \widehat{H}$$

is no more than the morphism (2.4) that induces the adjoint morphism $\mathrm{ad}(\widehat{g}) = (\widehat{R}_{\widehat{g}^{-1}} \circ \widehat{L}_{\widehat{g}})_*\colon \mathfrak{h} \to \mathfrak{h}$ (cf. Eq. (2.5)).

The action of $\widehat{\mathrm{Ad}}^*$ is easy to compute; writing $\widehat{m}^*(h) = \sum_k h^k \otimes h_k$ for every section $h \in \mathcal{A}_{\widehat{H}}$, one easily observes that

$$\widehat{\mathrm{Ad}}^*(h) = \sum_{kj} (-1)^{|h_k{}^j||h_{kj}|} h^k \widehat{s}^*(h_{kj}) \otimes h_k{}^j \tag{2.7}$$

(here we have also set $\widehat{m}^*(h_k) = \sum_j h_k{}^j \otimes h_{kj}$). Our next step is to prove that the map

$$\begin{aligned} H \times \mathfrak{h} &\to \mathfrak{h} \\ (g, X) &\mapsto \mathrm{ad}(g) \cdot X \end{aligned}$$

is actually a G-morphism. The definition of the adjoint representation as a morphism of supermanifolds, and not merely as a map, needs a more complicated construction; the corresponding theory in the framework of graded manifolds is dealt with in [Kos]. Let us denote as before by $\mathcal{A}_e$ the graded local ring of germs of $\mathcal{A}_{\widehat{H}}$ at the unit point $e \in H$, and by $\mathfrak{L}_e$ the ideal of the germs that vanish when evaluated at e, so that there is a natural isomorphism of rank (m, n) free B_L-modules $\mathfrak{L}_e/\mathfrak{L}_e^2 \simeq \mathfrak{h}^*$. Then, the sheaf morphism $(\widehat{\mathrm{Ad}})^*: \mathcal{A}_{\widehat{H}} \to \mathrm{Ad}_*(\mathcal{A}_{\widehat{H}} \hat{\otimes}_\pi \mathcal{A}_{\widehat{H}})$ induces a morphism

$$(\widehat{\mathrm{Ad}})^*: \mathfrak{L}_e/\mathfrak{L}_e^2 = \mathfrak{h}^* \to \mathrm{Ad}_*(\mathcal{A}_{\widehat{H}} \otimes (\mathfrak{L}_e/\mathfrak{L}_e^2)) = \mathrm{Ad}_*(\mathcal{A}_{\widehat{H}} \otimes \mathfrak{h}^*),$$

which in turn induces a sheaf morphism

$$\widehat{ST}(\mathfrak{h}^*) \to \mathrm{Ad}_*(\mathcal{A}_{\widehat{H}} \hat{\otimes}_\pi \widehat{ST}(\mathfrak{h}^*)).$$

Here $\widehat{ST}$ has the same meaning as in Section IV.2. Now, $\widehat{ST}(\mathfrak{h}^*)$ is the structure sheaf of the Lie superalgebra $\mathfrak{h}$, when considered as a G-supermanifold $\widehat{\mathfrak{h}}$ of dimension $(m + n, m + n)$ by means of a B_L-module isomorphism $\mathfrak{h} \simeq B_L^{m|n}$, that is, as an SVB over a single point (Definition IV.3.2). One can thus give the following result.

Proposition 2.5. *The adjoint representation of $\widehat{H}$ on its Lie superalgebra $\mathfrak{h}$ is the left action of $\widehat{H}$ on $\widehat{\mathfrak{h}}$ given by the G-morphism*

$$\widehat{\mathrm{ad}}: \widehat{H} \times \widehat{\mathfrak{h}} \to \widehat{\mathfrak{h}}$$

induced by the above sheaf morphism.

As one would expect, the effect of the adjoint representation on points is exactly the map

$$H \times \mathfrak{h} \to \mathfrak{h}$$
$$(g, X) \mapsto \mathrm{ad}(g) \cdot X$$

Furthermore, for every point $\widehat{g}: \widehat{z} \to \widehat{H}$, the composition

$$\widehat{\mathfrak{h}} = \widehat{z} \times \widehat{\mathfrak{h}} \xrightarrow{\widehat{g} \times \mathrm{Id}} \widehat{H} \times \widehat{\mathfrak{h}} \xrightarrow{\widehat{\mathrm{ad}}} \widehat{\mathfrak{h}}$$

is no more than the morphism (2.5), which is thus proved to be a G-morphism.

3. Principal superfibre bundles

Our wish is now to devise a suitable notion of a principal bundle within the category of G-supermanifolds. Let $\widehat{p}: \widehat{P} \to \widehat{M}$ be a G-morphism, and let $\widehat{\varsigma}: \widehat{P} \times \widehat{H} \to \widehat{P}$ be a right action of a G-Lie supergroup $\widehat{H}$ on the relative G-supermanifold $\widehat{p}: \widehat{P} \to \widehat{M}$ (Definition 2.2). We know that this action induces an action $\widehat{\varsigma}$ of $\widehat{H}$ on the pre-image $\widehat{p}: \widehat{P}_{|\widehat{U}} = (p^{-1}(U), \mathcal{A}_{\widehat{P}|_{p^{-1}(U)}}) \to \widehat{U}$ of every open subset $U \subset M$.

Definition 3.1. *A principal superfibre bundle of supergroup $\widehat{H}$ (for brevity, an $\widehat{H}$-PSFB) is a G-morphism $\widehat{p}: \widehat{P} \to \widehat{M}$ endowed with a $\widehat{H}$ action $\widehat{\varsigma}: \widehat{P} \times \widehat{H} \to \widehat{P}$, such that:*

(1) *$\widehat{H}$ acts on the relative G-supermanifold $\widehat{p}: \widehat{P} \to \widehat{M}$ (Definition 2.2).*
(2) *$\widehat{p}: \widehat{P} \to \widehat{M}$ is locally trivial; that is, there exist an open cover $\{U_i\}$ of M and G-invariant isomorphisms of relative G-supermanifolds (Definition 2.4)*

$$\widehat{\phi}_i: \widehat{P}_{|\widehat{U}_i} \simeq \widehat{U}_i \times \widehat{H}, \tag{3.1}$$

where $\widehat{H}$ acts on the relative G-supermanifold $\widehat{p}_1: \widehat{U}_i \times \widehat{H} \to \widehat{U}_i$ by right multiplication.

Condition (2) implies that an $\widehat{H}$-PSFB is a locally trivial G-superbundle (Definition IV.3.1) with standard fibre $\widehat{H}$.

EXAMPLE 3.1. The natural projection

$$\widehat{p}: \widehat{M} \times \widehat{H} \to \widehat{M},$$

where $\widehat{H}$ acts on $\widehat{M} \times \widehat{H}$ by right multiplication, is an $\widehat{H}$-PSFB, that will be called the *standard trivial $\widehat{H}$-PSFB* over $\widehat{M}$. ▲

A morphism of $\widehat{H}$-PSFB's over the same G-supermanifold is defined as an $\widehat{H}$-invariant G-morphism (Definition 2.4). Now, if $\widehat{p}: \widehat{P} \to \widehat{M}$ is an $\widehat{H}$-PSFB, the morphisms (3.1) are in fact $\widehat{H}$-PSFB isomorphisms of the restrictions $\widehat{p}: \widehat{P}_{|\widehat{U}_i} \to \widehat{U}_i$ with the standard trivial $\widehat{H}$-PSFB's over $\widehat{U}_i$. In other words, any $\widehat{H}$-PSFB is locally isomorphic with the standard trivial $\widehat{H}$-PSFB.

EXAMPLE 3.2. If $\widehat{p}: \widehat{P} \to \widehat{M}$ is an $\widehat{H}$-PSFB and $U \subset M$ is an open subset, the restriction $\widehat{p}: \widehat{P}_{|\widehat{U}} \to \widehat{U}$ is again an $\widehat{H}$-PSFB. ▲

EXAMPLE 3.3. Let $\widehat{q}: \widehat{\xi} \to \widehat{M}$ a rank (p, q) supervector bundle (SVB) over a G-supermanifold $\widehat{M}$. Then, the superfibre bundle $\widehat{\pi}: \text{Iso}(\widehat{M} \times B_L^{p|q}, \widehat{\xi}) \to \widehat{M}$ of isomorphisms of the trivial SVB $\widehat{M} \times B_L^{p|q}$ of rank (p, q) with $\widehat{\xi}$ (Definition IV.3.5), endowed with the right action

$$\text{Iso}(\widehat{M} \times B_L^{p|q}, \widehat{\xi}) \times \widehat{GL_L[p|q]} \to \text{Iso}(\widehat{M} \times B_L^{p|q}, \widehat{\xi}),$$

given by Example 2.4, is a $\widehat{GL_L[p|q]}$-PSFB. ▲

EXAMPLE 3.4. Taking in the previous example $\widehat{\xi}$ as the graded tangent bundle $T(M, \mathcal{A}_{\widehat{M}})$ to the G-supermanifold $\widehat{M}$, one has that the *superbundle of graded frames* $Fr(M, \mathcal{A}_{\widehat{M}}) \to \widehat{M}$ (cf. Example IV.3.1) is a $\widehat{GL_L[p|q]}$-PSFB. ▲

The following proposition is an analogue of the Galois theorem, in that it states that the base of an $\widehat{H}$-PSFB is the orbit space of the total space and that its structure sheaf is the invariant sheaf under the action of the supergroup.

Proposition 3.1. *Let $\widehat{p}: \widehat{P} \to \widehat{M}$ be an $\widehat{H}$-PSFB; the pair $(\widehat{M}, \widehat{p})$ is then a quotient of the action of $\widehat{H}$ on $\widehat{P}$.*

Proof. One can easily see that the question is local on $\widehat{M}$ and can thus assume that $\widehat{p}: \widehat{P} \to \widehat{M}$ is the standard $\widehat{H}$-PSFB $\widehat{p}: \widehat{M} \times \widehat{H} \to \widehat{M}$. Since the first condition in the definition of quotient, namely that $\widehat{H}$ acts on the relative G-supermanifold $\widehat{p}: \widehat{P} \to \widehat{M}$, is obviously fulfilled, one only has to prove that if $\widehat{f}: \widehat{M} \times \widehat{H} \to \widehat{N}$ is a G-morphism such that $\widehat{f} \circ \widehat{\varsigma} = \widehat{f} \circ \widehat{p}_1$, then there is a unique morphism $\widehat{g}: \widehat{M} \to \widehat{N}$ with $\widehat{f} = \widehat{g} \circ \widehat{p}$. Now, the unit point $\widehat{e}: \widehat{z} = (z, \mathcal{B}_L) \to \widehat{H}$ induces a section $\widehat{s}_e: \widehat{M} = \widehat{M} \times \widehat{z} \to \widehat{M} \times \widehat{H}$ of $\widehat{p}: \widehat{M} \times \widehat{H} \to \widehat{M}$ such that $\widehat{s}_e \circ \widehat{p}_1 = \widehat{p}_1 \circ (\widehat{s}_e \times \mathrm{Id})$ as morphisms from $\widehat{M} \times \widehat{H}$ into itself. Let us define $\widehat{g}: \widehat{M} \to \widehat{N}$ by $\widehat{g} = \widehat{f} \circ \widehat{s}_e$. The condition $\widehat{f} \circ \widehat{\varsigma} = \widehat{f} \circ \widehat{p}_1$ implies that $\widehat{f} \circ \widehat{\varsigma} \circ (\widehat{s}_e \times \mathrm{Id}) = \widehat{f} \circ \widehat{p}_1 \circ (\widehat{s}_e \times \mathrm{Id}) = \widehat{f} \circ \widehat{s}_e \circ \widehat{p}_1$. Since $\widehat{\varsigma} \circ (\widehat{s}_e \times \mathrm{Id}) = \mathrm{Id}$ by the first condition in the definition of an action of a G-Lie supergroup, one obtains $\widehat{f} = \widehat{g} \circ \widehat{p}$ as expected. The uniqueness of $\widehat{g}$ is proved straightforwardly. ∎

Corollary 3.1. *Let $\widehat{p}: \widehat{P} \to \widehat{M}$ be an $\widehat{H}$-PSFB.*

(1) *One has an isomorphism of sheaves of graded B_L-algebras*

$$\mathcal{A}_{\widehat{M}} \simeq (\widehat{p}_* \mathcal{A}_{\widehat{P}})^{\widehat{H}} .$$

(2) *There is an exact sequence of sheaves of graded B_L-modules on M*

$$0 \to \mathcal{A}_{\widehat{M}} \xrightarrow{\widehat{p}^*} p_* \mathcal{A}_{\widehat{P}} \xrightarrow{\widehat{\varsigma}^* - \widehat{p}_1^*} \pi_*(\mathcal{A}_{\widehat{P}} \otimes_\pi \mathcal{A}_{\widehat{H}}) ,$$

where $\widehat{\pi} = \widehat{p} \circ \widehat{\varsigma} = \widehat{p} \circ \widehat{p}_1 : \widehat{P} \times \widehat{H} \to \widehat{M}$.

Proof. This follows from Propositions 2.1 and 2.2. ∎

Transition morphisms. We now describe how trivial $\widehat{H}$-PSFB's can be glued to yield another $\widehat{H}$-PSFB, and, conversely, that any $\widehat{H}$-PSFB can be obtained in this way. The first question is to determine the automorphisms of the trivial standard $\widehat{H}$-PSFB $\widehat{p}_1: \widehat{M} \times \widehat{H} \to \widehat{M}$. Let $\widehat{\phi}: \widehat{M} \times \widehat{H} \simeq \widehat{M} \times \widehat{H}$ be an isomorphism of $\widehat{H}$-PSFB's; let us consider the G-morphism $\widehat{\psi}: \widehat{M} \to \widehat{H}$ from the base G-supermanifold to the G-Lie supergroup $\widehat{H}$, defined by the diagram

$$\begin{array}{ccc} \widehat{M} \times \widehat{H} & \xrightarrow{\widehat{\phi}} & \widehat{M} \times \widehat{H} \\ {\scriptstyle \widehat{s}_e}\uparrow & & \downarrow{\scriptstyle \widehat{p}_2} \\ \widehat{M} & \xrightarrow{\widehat{\psi}} & \widehat{H} \end{array} ,$$

where $\widehat{s}_e: \widehat{M} = \widehat{M} \times \widehat{z} \to \widehat{M} \times \widehat{H}$ is the section of $\widehat{p}_1$ induced by the unit point $\widehat{e}: \widehat{z} = (z, \mathcal{B}_L) \to \widehat{H}$. The original isomorphism $\widehat{\phi}$ can easily be described in terms of $\widehat{\psi}$; in fact, the two components of $\widehat{\phi}: \widehat{M} \times \widehat{H} \simeq \widehat{M} \times \widehat{H}$ are $\widehat{p}_1$ and $(\widehat{\psi} \circ \widehat{p}_1) \cdot \widehat{p}_2$, namely,

$$\widehat{\phi} = (\widehat{p}_1, (\widehat{\psi} \circ \widehat{p}_1) \cdot \widehat{p}_2) .$$

Let us now consider an arbitrary $\widehat{H}$-PSFB $\widehat{p}: \widehat{P} \to \widehat{M}$, and let $\{U_i\}$ be a trivializing open cover for $\widehat{p}: \widehat{P} \to \widehat{M}$ (Definition 3.1), so that there are isomorphisms of $\widehat{H}$-PSFB's

$$\widehat{\phi}_i: \widehat{P}_{|\widehat{U}_i} \simeq \widehat{U}_i \times \widehat{H}\,.$$

The family of such isomorphisms is called a *trivialization* of $\widehat{p}: \widehat{P} \to \widehat{M}$. If we write $U_{ij} = U_i \cap U_j$ for every i, j, there are isomorphisms of $\widehat{H}$-PSFB's

$$\widehat{\phi}_{ij} = \widehat{\phi}_{i|U_{ij}} \circ (\widehat{\phi}_{j|U_{ij}})^{-1}: \widehat{U}_{ij} \times \widehat{H} \simeq \widehat{U}_{ij} \times \widehat{H}\,,$$

that fulfill the glueing condition (II.4.5) (cocycle condition)

$$\widehat{\phi}'_{ik} = \widehat{\phi}'_{ij} \circ \widehat{\phi}'_{jk} \tag{3.2}$$

for every i, j, k, where primes denote restriction to $U_{ijk} = U_i \cap U_j \cap U_k$.

Definition 3.2. *The G-morphisms $\widehat{\psi}_{ij}: \widehat{U}_{ij} \to \widehat{H}$ constructed as above from the isomorphisms $\widehat{\phi}_{ij}$ are called the transition morphisms for the $\widehat{H}$-PSFB $\widehat{p}: \widehat{P} \to \widehat{M}$ relative to the fixed trivialization.*

One can easily show that the transition morphisms enjoy the property

$$\widehat{\psi}_{ik} = \widehat{\psi}_{ij} \cdot \widehat{\psi}_{jk} \tag{3.3}$$

for every triple (i, j, k). The dot here has the same meaning as in Eq. (2.1).

Let us consider, conversely, an open cover $\{U_i\}$ of a G-supermanifold $\widehat{M}$ and a family $\{\widehat{\psi}_{ij}\}$ of G-morphisms

$$\widehat{\psi}_{ij}: \widehat{U}_{ij} \to \widehat{H}$$

fulfilling the condition (3.3).

Proposition 3.2. *There exists an $\widehat{H}$-PSFB $\widehat{p}: \widehat{P} \to \widehat{M}$ and a trivialization of it on the open cover $\{U_i\}$ whose transition morphisms are the G-morphisms $\widehat{\psi}_{ij}$.*

Proof. The G-morphisms $\widehat{\psi}_{ij}: \widehat{U}_{ij} \to \widehat{H}$ determine isomorphisms of trivial $\widehat{H}$-PSFB's

$$\widehat{\phi}_{ij}: \widehat{U}_{ij} \times \widehat{H} \simeq \widehat{U}_{ij} \times \widehat{H}$$

defined by $\widehat{\phi}_{ij} = (\widehat{p}_1, (\widehat{\psi}_{ij} \circ \widehat{p}_1) \cdot \widehat{p}_2)$. Now, (3.3) implies that the isomorphisms $\widehat{\phi}_{ij}$ fulfill the cocycle condition (3.2). By glueing (Lemma IV.1.2), we can construct a G-supermanifold $\widehat{P}$ and G-isomorphisms $\widehat{\phi}_i: \widehat{P}_{|\widehat{U}_i} \simeq \widehat{U}_i \times \widehat{H}$ such that $\widehat{\phi}_{i|U_{ij}} = \widehat{\phi}_{ij} \circ \widehat{\phi}_{j|U_{ij}}$. The rest of the proof is straightforward. ■

4. Connections

Connections on supervector bundles where introduced in Section VI.1. Here we wish to reformulate that notion in the case of principal superfibre bundles; as we shall see in the next Section, any SVB can be regarded as a bundle associated with a PSFB, so that the two notions of connection can be related as in the ordinary case.

Let $\widehat{p}: \widehat{P} \to \widehat{M}$ be $\widehat{H}$ be an $\widehat{H}$-PSFB. Then, $\widehat{P}$ is acted on by $\widehat{H}$ so that $\widehat{H}$-invariant graded vector fields (Definition 2.7) and fundamental graded vector fields (Definition 2.8) can be considered on $\widehat{P}$.

Proposition 4.1. *Let $\widehat{p}: \widehat{P} \to \widehat{M}$ be an $\widehat{H}$-PSFB. Every $\widehat{H}$-invariant graded vector field on $\widehat{P}$ is $\widehat{p}$-projectable to $\widehat{M}$.*

Proof. Let $D: \mathcal{A}_{\widehat{P}} \to \mathcal{A}_{\widehat{P}}$ be an $\widehat{H}$-invariant graded vector field, so that $\widehat{\varsigma}^* \circ D = (D \otimes \mathrm{Id}) \circ \widehat{\varsigma}^*$. Then, from the exact sequence in Corollary 3.1, one obtains a commutative diagram

$$\begin{array}{ccccccc}
0 & \longrightarrow & \mathcal{A}_{\widehat{M}} & \xrightarrow{\widehat{p}^*} & p_*\mathcal{A}_{\widehat{P}} & \xrightarrow{\widehat{\varsigma}^* - \widehat{p}_1^*} & \pi_*(\mathcal{A}_{\widehat{P}} \otimes_\pi \mathcal{A}_{\widehat{H}}) \\
 & & & & \Big\downarrow D & & \Big\downarrow D\otimes\mathrm{Id} \\
0 & \longrightarrow & \mathcal{A}_{\widehat{M}} & \xrightarrow{\widehat{p}^*} & p_*\mathcal{A}_{\widehat{P}} & \xrightarrow{\widehat{\varsigma}^* - \widehat{p}_1^*} & \pi_*(\mathcal{A}_{\widehat{P}} \otimes_\pi \mathcal{A}_{\widehat{H}})
\end{array} .$$

It follows that there exists a graded vector field $\widehat{p}(D): \mathcal{A}_{\widehat{M}} \to \mathcal{A}_{\widehat{M}}$ that fits into the diagram. ■

We can then associate with every open subset $V \subset M$ the $\mathcal{A}_{\widehat{M}}(V)$-module $\mathcal{D}er(p_*\mathcal{A}_{\widehat{P}})^{\widehat{H}}(V)$ of all $\widehat{H}$-invariant graded vector fields on $p^{-1}(V)$, thus defining a sheaf $\mathcal{D}er(p_*\mathcal{A}_{\widehat{P}})^{\widehat{H}}$ of $\mathcal{A}_{\widehat{M}}$-modules.

Vertical graded vector fields are in turn generated by fundamental graded vector fields.

Proposition 4.2. *There is an isomorphism of sheaves of $\mathcal{A}_{\widehat{P}}$-modules*

$$\begin{aligned} v\colon \mathcal{A}_{\widehat{P}} \otimes_{B_L} \mathfrak{h} &\simeq \mathcal{V}er\mathcal{A}_{\widehat{P}} \\ f \otimes X &\mapsto f \cdot X^* \end{aligned}$$

Proof. The morphism is globally defined, so that one can check that it is an isomorphism only locally, that is to say, assuming that $\widehat{p}\colon \widehat{P} \to \widehat{M}$ is the trivial $\widehat{H}$-PSFB $\widehat{p}\colon \widehat{M} \times \widehat{H} \to \widehat{M}$. Now, $\mathcal{V}er\mathcal{A}_{\widehat{P}} \simeq \widehat{p}_2^*(\mathcal{D}er\mathcal{A}_{\widehat{H}})$ by Proposition 2.3, and one concludes by (2.6). ■

REMARK 4.1. If we endow $\mathcal{A}_{\widehat{P}} \otimes_{B_L} \mathfrak{h}$ with the Lie superalgebra structure induced by that of $\mathfrak{h}$; i.e.

$$[f \otimes X, g \otimes Y] = (-1)^{|X||g|} fg[X, Y],$$

the isomorphism $v\colon \mathcal{A}_{\widehat{P}} \otimes_{B_L} \mathfrak{h} \simeq \mathcal{V}er\mathcal{A}_{\widehat{P}}$ *is not* a morphism of Lie superalgebras, because the graded Lie bracket of the corresponding vertical graded vector fields is given by

$$[fX^*, gY^*] = fX^*(g)Y^* - (-1)^{|fX^*||gY^*|} gY^*(f)X^* + (-1)^{|fX^*||g|} fg[X^*, Y^*].$$

Nevertheless, the restriction $\mathfrak{h} \to \mathcal{V}er\mathcal{A}_{\widehat{P}}$ is a Lie superalgebra morphism, that is, $[X, Y]^* = [X^*, Y^*]$ (Proposition 2.4). ▲

Let us consider the sheaf $(p_*\mathcal{V}er\mathcal{A}_{\widehat{P}})^{\widehat{H}} = p_*(\mathcal{V}er\mathcal{A}_{\widehat{P}}) \cap \mathcal{D}er(p_*\mathcal{A}_{\widehat{P}})^{\widehat{H}}$ on M whose sections are the vertical $\widehat{H}$-invariant graded vector fields; the local structure of this sheaf is quite simple. Actually, if $\widehat{p}\colon \widehat{M} \times \widehat{H} \to \widehat{M}$ is a trivial $\widehat{H}$-PSFB, the same techniques of Proposition 1.1 and Definition 2.7 allow us to prove that there is an isomorphism of $\mathcal{A}_{\widehat{M}}$-modules

$$\begin{aligned} v\colon \mathcal{A}_{\widehat{M}} \otimes \mathfrak{h} &\to (p_*\mathcal{V}er\mathcal{A}_{\widehat{M}\times\widehat{H}})^{\widehat{H}} \\ f \otimes X &\mapsto f \otimes (X \otimes \mathrm{Id}) \circ \widehat{m}^* \end{aligned}, \tag{4.1}$$

where the elements $X \in \mathfrak{h}$ are interpreted as graded tangent vectors $X\colon (\mathcal{A}_{\widehat{H}})_e \to B_L$ at the unit point. The global structure of vertical $\widehat{H}$-invariant graded vector fields is given by the following result.

Proposition 4.3. *Let $\widehat{p}\colon \widehat{P} \to \widehat{M}$ be an $\widehat{H}$-PSFB. There is an exact sequence of sheaves of $\mathcal{A}_{\widehat{M}}$-modules*

$$0 \to (p_*\mathcal{V}er\mathcal{A}_{\widehat{P}})^{\widehat{H}} \to \mathcal{D}er(p_*\mathcal{A}_{\widehat{P}})^{\widehat{H}} \xrightarrow{p} \mathcal{D}er\mathcal{A}_{\widehat{M}} \to 0,$$

which is called the Atiyah sequence of $\widehat{p}\colon \widehat{P} \to \widehat{M}$.

Proof. As in the proof of Proposition IV.5.1, one has only to prove that if V is a trivializing open subset of M, so that $\widehat{p}\colon \widehat{P}_{|p^{-1}(V)} \to \widehat{V}$ is the trivial $\widehat{H}$-PSFB $\widehat{p}\colon \widehat{V} \times \widehat{H} \to \widehat{V}$, every graded vector field D' on $\widehat{V}$ is the projection of an $\widehat{H}$-invariant graded vector field on $\widehat{V} \times \widehat{H}$. But $D = D' \otimes \mathrm{Id}$ defines a graded vector field on $\widehat{V} \times \widehat{H}$ that is $\widehat{H}$-invariant and projects onto D'. ■

Fundamental graded vector fields and $\widehat{H}$-invariant vertical graded fields are related as follows.

Lemma 4.1. *If X^* is a fundamental graded vector field and D is an $\widehat{H}$-invariant vertical graded field on an $\widehat{H}$-PSFB $\widehat{p}\colon \widehat{P} \to \widehat{M}$, then $[X^*, D] = 0$.*

Proof. This question is local on $\widehat{M}$, and so we can assume again that $\widehat{p}\colon \widehat{P} \to \widehat{M}$ is the trivial bundle $\widehat{p}\colon \widehat{M} \times \widehat{H} \to \widehat{M}$. Then $X^* = \mathrm{Id} \otimes (\mathrm{Id} \otimes X) \circ \widehat{m}^*$ (as in Example 2.7), whilst $D = fD'$ for some section f of $\mathcal{A}_{\widehat{M}}$; here $D' = \mathrm{Id} \otimes (Y \otimes \mathrm{Id}) \circ \widehat{m}^*$ for some $Y \in \mathfrak{h}$, according to (4.1). Since $X^*(f) = 0$, one has to prove that $[X^*, D'] = 0$, or equivalently, that the graded vector fields $X^* = (\mathrm{Id} \otimes X) \circ \widehat{m}^*$ and $Y^\vee = (Y \otimes \mathrm{Id}) \circ \widehat{m}^*$ on $\widehat{H}$ have a vanishing graded Lie bracket. An easy computation shows that

$$\begin{aligned} Y^\vee \circ X^* &= (Y \otimes \mathrm{Id} \otimes X) \circ (\widehat{m}^* \otimes \mathrm{Id}) \circ \widehat{m}^* \\ X^* \circ Y^\vee &= (-1)^{|X||Y|} (Y \otimes \mathrm{Id} \otimes X) \circ (\mathrm{Id} \otimes \widehat{m}^*) \circ \widehat{m}^* \end{aligned}$$

However, $(\widehat{m}^* \otimes \mathrm{Id}) \circ \widehat{m}^* = (\mathrm{Id} \otimes \widehat{m}^*) \circ \widehat{m}^*$ by coassociativity (1.1), thus finishing the proof. ■

Definition 4.1. *A connection on an $\widehat{H}$-PSFB $\widehat{p}\colon \widehat{P} \to \widehat{M}$ is a splitting (cf. Proposition VI.1.1) of the Atiyah sequence, that is, an even morphism of $\mathcal{A}_{\widehat{M}}$-modules*

$$\nabla\colon \mathcal{D}er\mathcal{A}_{\widehat{M}} \to \mathcal{D}er(p_*\mathcal{A}_{\widehat{P}})^{\widehat{H}}$$

such that $p \circ \nabla = \mathrm{Id}$.

The image of ∇ is called the *horizontal $\widehat{H}$-invariant distribution* associated with the connection, and one has a decomposition

$$\mathcal{D}er(p_*\mathcal{A}_{\widehat{P}})^{\widehat{H}} \simeq (p_*\mathcal{V}er\mathcal{A}_{\widehat{P}})^{\widehat{H}} \oplus \nabla(\mathcal{D}er\mathcal{A}_{\widehat{M}}), \tag{4.2}$$

that is, a split exact sequence of sheaves of $\mathcal{A}_{\widehat{M}}$-modules

$$0 \to \mathcal{D}er\mathcal{A}_{\widehat{M}} \xrightarrow{\nabla} \mathcal{D}er(p_*\mathcal{A}_{\widehat{P}})^{\widehat{H}} \xrightarrow{\varpi} (p_*\mathcal{V}er\mathcal{A}_{\widehat{P}})^{\widehat{H}} \to 0 .$$

The horizontal $\widehat{H}$-invariant distribution of ∇ is now given by

$$\nabla(\mathcal{D}er\mathcal{A}_{\widehat{M}}) = \mathrm{Ker}(\varpi) .$$

If D' is a graded vector field on an open subset $V \subset M$, the graded vector field $\nabla(D')$ is called the *horizontal lift* of D' with respect to the connection ∇.

Isomorphisms of connections. Let $\widehat{p}: \widehat{P} \to \widehat{M}$ and $\widehat{p}': \widehat{P}' \to \widehat{M}$ be $\widehat{H}$-PSFB's, and $\widehat{\phi}: \widehat{P} \to \widehat{P}'$ an isomorphism of $\widehat{H}$-SPFB's. Then, $\widehat{\phi}$ induces a sheaf isomorphism

$$\begin{aligned} \breve{\phi}: \mathcal{D}er\mathcal{A}_{\widehat{P}} &\to \mathcal{D}er\mathcal{A}_{\widehat{P}'} \\ D &\mapsto \breve{\phi}\cdot D = (\widehat{\phi}^{-1})^* \circ D \circ \widehat{\phi}^* \end{aligned} \tag{4.3}$$

which in turn yield isomorphisms

$$\begin{aligned} &\breve{\phi}: \mathcal{D}er(p_*\mathcal{A}_{\widehat{P}})^{\widehat{H}} \simeq \mathcal{D}er(p_*\mathcal{A}_{\widehat{P}'})^{\widehat{H}} \\ &\breve{\phi}: (p_*\mathcal{V}er\mathcal{A}_{\widehat{P}})^{\widehat{H}} \simeq (p_*\mathcal{V}er\mathcal{A}_{\widehat{P}'})^{\widehat{H}} \end{aligned} \tag{4.4}$$

and then an automorphism

$$\begin{array}{ccccccccc} 0 & \longrightarrow & (p_*\mathcal{V}er\mathcal{A}_{\widehat{P}})^{\widehat{H}} & \longrightarrow & \mathcal{D}er(p_*\mathcal{A}_{\widehat{P}})^{\widehat{H}} & \xrightarrow{p} & \mathcal{D}er\mathcal{A}_{\widehat{M}} & \longrightarrow & 0 \\ & & \breve{\phi}\downarrow & & \breve{\phi}\downarrow & & \| & & \\ 0 & \longrightarrow & (p_*\mathcal{V}er\mathcal{A}_{\widehat{P}'})^{\widehat{H}} & \longrightarrow & \mathcal{D}er(p_*\mathcal{A}_{\widehat{P}'})^{\widehat{H}} & \xrightarrow{p} & \mathcal{D}er\mathcal{A}_{\widehat{M}} & \longrightarrow & 0 \end{array}$$

of the Atiyah sequence. It follows that if ∇ is a connection on $\widehat{p}: \widehat{P} \to \widehat{M}$, then $\breve{\phi} \circ \nabla$ is a connection on $\widehat{p}': \widehat{P}' \to \widehat{M}$, called the connection obtained from ∇ through the $\widehat{H}$-SPFB automorphism $\widehat{\phi}$.

Existence of connections. The proof of Proposition 4.3 shows that the trivial $\widehat{H}$-PSFB $\widehat{p}: \widehat{M} \times \widehat{H} \to \widehat{M}$ carries a connection, given by

$$\begin{aligned} \nabla^0: \mathcal{D}er\mathcal{A}_{\widehat{M}} &\to \mathcal{D}er(p_*\mathcal{A}_{\widehat{M}\times\widehat{H}})^{\widehat{H}} \\ D' &\mapsto D' \otimes \mathrm{Id} , \end{aligned}$$

called the *canonical flat connection* of the trivial PSFB. Since $\widehat{H}$-PSFB's are locally trivial, any $\widehat{H}$-PSFB always admits connections locally; the existence of a globally defined connection is then a cohomological question. As long as sheaves of $\mathcal{A}_{\widehat{M}}$-modules may have non-vanishing cohomology, an arbitrary $\widehat{H}$-PSFB need not carry connections.

Actually, proceeding as in Section VI.1, we can attach to every $\widehat{H}$-PSFB $\widehat{p}: \widehat{P} \to \widehat{M}$ a cohomology class

$$b(\widehat{P}) \in H^1(M, \mathcal{H}om_{\mathcal{A}_{\widehat{M}}}(\mathcal{D}er\mathcal{A}_{\widehat{M}}, (p_*\mathcal{V}er\mathcal{A}_{\widehat{P}})^{\widehat{H}}))$$

of the sheaf

$$\mathcal{H}om_{\mathcal{A}_{\widehat{M}}}(\mathcal{D}er\mathcal{A}_{\widehat{M}}, (p_*\mathcal{V}er\mathcal{A}_{\widehat{P}})^{\widehat{H}}) = (p_*\mathcal{V}er\mathcal{A}_{\widehat{P}})^{\widehat{H}} \otimes_{\mathcal{A}_{\widehat{M}}} \Omega^1_{\mathcal{A}_{\widehat{M}}}$$

of $(p_*\mathcal{V}er\mathcal{A}_{\widehat{P}})^{\widehat{H}}$-valued graded differential 1-forms on $\widehat{M}$, that vanishes if and only if there exists a connection on $\widehat{p}: \widehat{P} \to \widehat{M}$. This class is called the *Atiyah class* of the $\widehat{H}$-PSFB. Given a trivialization

$$\widehat{\phi}_i: \widehat{P}_{|\widehat{U}_i} \simeq \widehat{U}_i \times \widehat{H}$$

of the $\widehat{H}$-PSFB $\widehat{p}: \widehat{P} \to \widehat{M}$ on an open cover $\{U_i\}$ of M, if $\nabla_i = \breve{\phi}_i^{-1} \circ \nabla_i^0$ denotes the connection on $\widehat{p}: \widehat{P}_{|\widehat{U}_i} \to \widehat{U}_i$ obtained from the canonical flat connection ∇_i^0 on $\widehat{U}_i \times \widehat{H} \to \widehat{U}_i$ through the $\widehat{H}$-PSFB isomorphism $\widehat{\phi}_i^{-1}$, then the 1-cocycle

$$\{\omega_{ij} \equiv \nabla_{i|\widehat{U}_{ij}} - \nabla_{j|\widehat{U}_{ij}}\} \tag{4.5}$$

is a representative of $b(\widehat{P})$.

Connection forms. A connection ∇ on an $\widehat{H}$-PSFB $\widehat{p}: \widehat{P} \to \widehat{M}$ can be described in terms of an $\mathfrak{h}$-valued graded differential 1-form on $\widehat{P}$ (the *connection form* of ∇); we recall that such a form can be regarded as a morphism of sheaves of $\mathcal{A}_{\widehat{P}}$-modules

$$\omega: \mathcal{D}er\mathcal{A}_{\widehat{P}} \to \mathcal{A}_{\widehat{P}} \otimes_{B_L} \mathfrak{h}$$

(cf. Section VI.4). It follows from Proposition 4.2 that a $\mathfrak{h}$-valued graded differential 1-form on $\widehat{P}$ can be considered as a morphism of sheaves of $\mathcal{A}_{\widehat{P}}$-modules

$$\nu: \mathcal{D}er\mathcal{A}_{\widehat{P}} \to \mathcal{V}er\mathcal{A}_{\widehat{P}}.$$

Now let $\nabla\colon \mathcal{D}er\mathcal{A}_{\widehat{M}} \to \mathcal{D}er(p_*\mathcal{A}_{\widehat{P}})^{\widehat{H}}$ be a connection; it induces a morphism of $\mathcal{A}_{\widehat{P}}$-modules

$$\gamma\colon \widehat{p}^*(\mathcal{D}er\mathcal{A}_{\widehat{M}}) \to \widehat{p}^*(\mathcal{D}er(p_*\mathcal{A}_{\widehat{P}})^{\widehat{H}}) \simeq \mathcal{D}er\mathcal{A}_{\widehat{P}}$$

which is a splitting of the exact sequence

$$0 \to \mathcal{V}er\mathcal{A}_P \to \mathcal{D}er\mathcal{A}_P \xrightarrow{p} \widehat{p}^*(\mathcal{D}er\mathcal{A}_{\widehat{M}}) \to 0\,.$$

Therefore there is an exact sequence of $\mathcal{A}_{\widehat{P}}$-modules

$$0 \to \widehat{p}^*(\mathcal{D}er\mathcal{A}_{\widehat{M}}) \xrightarrow{\gamma} \mathcal{D}er\mathcal{A}_{\widehat{P}} \xrightarrow{\nu=\widehat{p}^*(\varpi)} \mathcal{V}er\mathcal{A}_{\widehat{P}} \to 0\,,$$

so that the connection ∇ induces a $\mathfrak{h}$-valued graded differential 1-form

$$\omega = \upsilon \circ \nu\colon \mathcal{D}er\mathcal{A}_{\widehat{P}} \to \mathcal{A}_{\widehat{P}} \otimes_{B_L} \mathfrak{h}$$

on $\widehat{P}$, called the *connection form* of ∇. By its very definition, $\nu\colon \mathcal{D}er\mathcal{A}_{\widehat{P}} \to \mathcal{V}er\mathcal{A}_{\widehat{P}}$ is the identity on vertical graded vector fields; this means that on fundamental graded vector fields the connection form ω acts as the inverse of υ:

$$\omega(X^*) = X \tag{4.6}$$

for every $X \in \mathfrak{h}$.

The distribution $\mathcal{H}or\mathcal{A}_{\widehat{P}} = \gamma(\widehat{p}^*(\mathcal{D}er\mathcal{A}_{\widehat{M}})) = \operatorname{Ker}\nu = \operatorname{Ker}\omega$ is called the *horizontal distribution* of ∇. There is a decomposition

$$\mathcal{D}er\mathcal{A}_{\widehat{P}} \simeq \mathcal{V}er\mathcal{A}_{\widehat{P}} \oplus \mathcal{H}or\mathcal{A}_{\widehat{P}}\,, \tag{4.7}$$

and one has the corresponding vertical and horizontal projections

$$\begin{aligned} \nu\colon \mathcal{D}er\mathcal{A}_{\widehat{P}} &\to \mathcal{V}er\mathcal{A}_{\widehat{P}} \\ h = \gamma \circ p\colon \mathcal{D}er\mathcal{A}_{\widehat{P}} &\to \mathcal{H}or\mathcal{A}_{\widehat{P}} \end{aligned}\,,$$

that enable us to write a graded vector field on $\widehat{P}$ as the sum of its vertical and horizontal components with respect to the connection ∇:

$$D = \nu(D) + h(D)\,.$$

Let us notice that a horizontal graded vector field $D = h(D)$ on $\widehat{P}$ need not be the horizontal lift of a graded vector field on $\widehat{M}$, because it may fail to be $\widehat{H}$-invariant. If one considers $\widehat{H}$-invariant graded vector fields, the decomposition (4.7) may be identified with the decomposition (4.2), so that an $\widehat{H}$-invariant horizontal graded vector field D is the horizontal lift of its projection $D' = p(D)$, that is, $D = \nabla(D')$.

Equation (4.7) also induces a decomposition of the graded tangent space at a point $\widehat{y}: \widehat{z} \to \widehat{P}$ into the sum of the vertical and horizontal graded tangent spaces:

$$T_y\widehat{P} \simeq V_y\widehat{P} \oplus H_y\widehat{P},$$

where

$$V_y\widehat{P} = \widehat{y}^*(\mathcal{V}er\mathcal{A}_{\widehat{P}}), \qquad H_y\widehat{P} = \widehat{y}^*(\mathcal{H}or\mathcal{A}_{\widehat{P}}).$$

Conversely, we can give the following result.

Proposition 4.4. *Let ω be a $\mathfrak{h}$-valued graded differential 1-form on $\widehat{P}$ satisfying the following properties:*

(1) *$\nu = \upsilon^{-1} \circ \omega: \mathcal{D}er\mathcal{A}_{\widehat{P}} \to \mathcal{V}er\mathcal{A}_{\widehat{P}}$ is the identity on vertical graded vector fields; that is, the composition*

$$\mathcal{V}er\mathcal{A}_{\widehat{P}} \hookrightarrow \mathcal{D}er\mathcal{A}_{\widehat{P}} \xrightarrow{\nu} \mathcal{V}er\mathcal{A}_{\widehat{P}}$$

is the identity morphism;

(2) *ν transforms $\widehat{H}$-invariant graded vector fields into $\widehat{H}$-invariant vertical graded vector fields; that is, it induces a morphism*

$$\varpi: \mathcal{D}er(p_*\mathcal{A}_{\widehat{P}})^{\widehat{H}} \to (p_*\mathcal{V}er\mathcal{A}_{\widehat{P}})^{\widehat{H}}.$$

Then ω is the connection form of a unique connection on $\widehat{P}$.

Proof. The form ω induces a splitting of the Atiyah sequence, namely, a connection ∇ such that $\nabla(\mathcal{D}er\mathcal{A}_{\widehat{M}}) = \operatorname{Ker}\varpi$. It follows immediately that ω is the connection form of ∇. ■

Curvature form. Let us consider a connection ∇ on an $\widehat{H}$-PSFB $\widehat{p}: \widehat{P} \to \widehat{M}$, and let $\omega: \mathcal{D}er\mathcal{A}_{\widehat{P}} \to \mathcal{A}_{\widehat{P}} \otimes_{B_L} \mathfrak{h}$ be the corresponding connection form. ω is a $\mathfrak{h}$-valued even graded differential 1-form, and we can apply to it the differential calculus of graded differential forms with values in a free module as developed in Section VI.4. In particular, the exterior differential of ω is given by

$$2d\omega(D_1, D_2) = (-1)^{|\omega||D|_1} D_1(\omega(D_2)) - \\ (-1)^{(|D_1|+|\omega|)|D_2|} D_2(\omega(D_1)) - \omega([D_1, D_2]) . \quad (4.7)$$

Definition 4.2. *The curvature form of the connection ∇ is the $\mathfrak{h}$-valued graded differential 2-form R described by*

$$R(D_1, D_2) = (d\omega)(h(D_1), h(D_2)) .$$

Proposition 4.5. (Structure equation) *One has*

$$d\omega(D_1, D_2) = -\tfrac{1}{2}[\omega(D_1), \omega(D_2)] + R(D_1, D_2)$$

for any graded vector fields D_1, D_2 on $\widehat{P}$, where the graded Lie bracket is induced by that of $\mathfrak{h}$ (see Remark 4.1).

Proof. Both members of the equation are $\mathfrak{h}$-valued graded differential 2-forms. Then, as long as fundamental graded vector fields and $\widehat{H}$-invariant horizontal graded vector fields generate all graded vector fields on $\widehat{P}$, it is enough to prove the claim in three cases:

(1) D_1 and D_2 are $\widehat{H}$-invariant horizontal graded vector fields.
Then, $\omega(D_1) = \omega(D_2) = 0$, and the formula is the definition of R.

(2) D_1 and D_2 are fundamental graded vector fields.
Then, $D_1 = X_1^*$, $D_2 = X_2^*$ for certain X_1, X_2 in $\mathfrak{h}$ and $\omega(D_1) = X_1$, $\omega(D_2) = X_2$ (cf. (4.6)); that is, they are 'constant' $\mathfrak{h}$-valued sections of $\mathcal{A}_{\widehat{P}} \otimes_{B_L} \mathfrak{h}$. As a consequence, $D_1(\omega(D_2)) = D_2(\omega(D_1)) = 0$ by Eq. (4.7). Moreover, $\omega([X_1^*, X_2^*]) = \omega([X_1, X_2]^*) = [X_1, X_2]$ by Proposition 2.4 and (4.6), so that $2d\omega(D_1, D_2) = -[X_1, X_2]$ by (4.9), thus proving the equation since $R(D_1, D_2) = 0$.

(3) D_1 is fundamental and D_2 is horizontal and $\widehat{H}$-invariant.
Then, $\omega(D_2) = 0$ and $D_1 = X^*$ for some $X \in \mathfrak{h}$, so that $D_2(\omega(X^*)) = D_2(X) = 0$ as above. Moreover, $R(D_1, D_2) = 0$, and by (4.9), the proof is reduced to showing that $\omega([D_1, D_2]) = 0$. We shall prove that in fact $[D_1, D_2] = 0$. The question being local on $\widehat{M}$, we can assume that $\widehat{p}\colon \widehat{P} \to \widehat{M}$ is the trivial $\widehat{H}$-PSFB $\widehat{p}\colon \widehat{M} \times \widehat{H} \to \widehat{M}$. Then, if D' is the projection of D_2 to $\widehat{M}$, one has $D_2 = D' \otimes \mathrm{Id} + \tilde{D}$ for some vertical $\widehat{H}$-invariant graded

vector field $\tilde{D}$. Now $[D_1, D' \otimes \mathrm{Id}] = 0$ trivially, and $[D_2, \tilde{D}] = 0$ by Lemma 4.1. ∎

5. Associated superfibre bundles

It is possible to introduce the notion of *associated superfibre bundles* with a certain PSFB; in particular, supervector bundles can be regarded as associated superfibre bundles.

Let $\widehat{p}: \widehat{P} \to \widehat{M}$ be an $\widehat{H}$-SPFB; as usual, we denote by $\widehat{\varsigma}: \widehat{P} \times \widehat{H} \to \widehat{P}$ the right action of $\widehat{H}$ on $\widehat{P}$. Let $\widehat{\rho}: \widehat{H} \times \widehat{F} \to \widehat{F}$ be a left action of $\widehat{H}$ on a G-supermanifold $\widehat{F}$; then, $\widehat{H}$ acts on the product supermanifold $\widehat{P} \times \widehat{F}$ on the right as follows. Let us denote by $\widehat{\rho}^{-1}: \widehat{H} \times \widehat{F} \to \widehat{F}$ the composition $\widehat{\rho}^{-1} = \widehat{\rho} \circ (\widehat{s} \times \mathrm{Id})$, where $\widehat{s}: \widehat{H} \to \widehat{H}$ is the inversion morphism; the following commutative diagram defines a G-morphism $\widehat{\tau}: \widehat{P} \times \widehat{F} \times \widehat{H} \to \widehat{P} \times \widehat{F}$ which yields a right action of $\widehat{H}$ on $\widehat{P} \times \widehat{F}$:

$$
\begin{array}{ccc}
\widehat{P} \times \widehat{F} \times \widehat{H} & \xrightarrow{\ \widehat{\tau}\ } & \widehat{P} \times \widehat{F} \\
{\scriptstyle \mathrm{Id} \times \widehat{\varrho}} \downarrow & & \uparrow {\scriptstyle \widehat{\varsigma} \times \widehat{\rho}^{-1}} \\
\widehat{P} \times \widehat{H} \times \widehat{F} & \xrightarrow{\ \mathrm{Id} \times \widehat{\Delta} \times \mathrm{Id}\ } & \widehat{P} \times \widehat{H} \times \widehat{H} \times \widehat{F}
\end{array}
\quad ;
$$

here $\widehat{\Delta}: \widehat{H} \to \widehat{H} \times \widehat{H}$ is the diagonal morphism, and $\widehat{\varrho}: \widehat{F} \times \widehat{H} \to \widehat{H} \times \widehat{F}$ is the morphism that exchanges the factors.

As one would expect, the action $\widehat{\tau}$ induces a right action of the ordinary Lie group H on the ordinary underlying differentiable manifold $P \times F$ defined by

$$(z, f)g = (zg, g^{-1}f),$$

where $z \in P$, $f \in F$, $g \in H$.

We now prove that, as in the ordinary case, the right action of $\widehat{H}$ on $\widehat{P} \times \widehat{F}$ gives rise to a quotient G-supermanifold $\widehat{\Theta}$ (Definition 2.2), which is a superfibre bundle over the base G-supermanifold $\widehat{M}$. Indeed, by the theory of (ordinary) associated bundles, the quotient space $\Theta = P \times F/H$ has the structure of a differentiable manifold. If $\pi: P \times F \to \Theta = P \times F/H$ is the natural projection, the map $p_\Theta: \Theta \to M$ described by $p_\Theta(\pi(z, f)) = p(z)$ endows Θ with a structure of differentiable bundle of fibre F that trivializes on the open subsets where

$P \to M$ is trivial. On the other hand, Proposition 2.1 tells us that the structure sheaf of a quotient G-supermanifold is the subsheaf invariant under the action of $\widehat{H}$.

Let us consider the graded ringed space

$$\widehat{\Theta} = (\Theta, \mathcal{A}_{\widehat{\Theta}}) = (\Theta, \pi_*(\mathcal{A}_{\widehat{P}} \hat{\otimes}_\pi \mathcal{A}_{\widehat{F}})^{\widehat{H}})$$

together with the natural graded ringed space morphism $\widehat{\pi}\colon \widehat{P} \times \widehat{F} \to \widehat{\Theta}$.

Proposition 5.1. *$\widehat{\Theta}$ is a G-supermanifold and $\widehat{\pi}\colon \widehat{P} \times \widehat{F} \to \widehat{\Theta}$ is the quotient of the action of $\widehat{H}$ on $\widehat{P} \times \widehat{F}$.*

Proof. The question is local on $\widehat{M}$, so that we can assume that $\widehat{p}\colon \widehat{P} \to \widehat{M}$ is the trivial $\widehat{H}$-PSFB $\widehat{p}\colon \widehat{M} \times \widehat{H} \to \widehat{M}$. Now, $\Theta \simeq M \times F$ and $\pi\colon M \times H \times F \to M \times F$ is described by $\pi(m, g, f) = (m, gf)$. Thus, it suffices to prove that the G-morphism $\widehat{\phi} = (\mathrm{Id}, \widehat{\rho})\colon \widehat{M} \times \widehat{H} \times \widehat{F} \to \widehat{M} \times \widehat{F}$ is the quotient of the action of $\widehat{H}$ on $\widehat{M} \times \widehat{H} \times \widehat{F}$. The first condition of Definition 2.6 is the commutativity of the diagram

$$\begin{array}{ccc} \widehat{M} \times \widehat{H} \times \widehat{F} \times \widehat{H} & \xrightarrow{\widehat{\tau}} & \widehat{M} \times \widehat{H} \times \widehat{F} \\ {\scriptstyle \widehat{p}_1}\downarrow & & \downarrow{\scriptstyle \widehat{\pi}} \\ \widehat{M} \times \widehat{H} \times \widehat{F} & \xrightarrow{\widehat{\pi}} & \widehat{M} \times \widehat{F} \end{array},$$

which is verified trivially. Concerning the second condition, we notice that the G-morphism $\widehat{s}_e = \mathrm{Id} \times \widehat{e} \times \mathrm{Id}\colon \widehat{M} \times \widehat{F} = \widehat{M} \times \widehat{z} \times \widehat{F} \to \widehat{M} \times \widehat{H} \times \widehat{F}$ is a section of $\widehat{\pi}$, i.e. $\widehat{\pi} \circ \widehat{s}_e = \mathrm{Id}$; then, if $\widehat{f}\colon \widehat{M} \times \widehat{H} \times \widehat{F} \to \widehat{N}$ is a G-morphism such that $\widehat{f} \circ \widehat{\tau} = \widehat{f} \circ \widehat{p}_1$, the morphism $\widehat{g} = \widehat{f} \circ \widehat{s}_e\colon \widehat{N} \times \widehat{F} \to \widehat{N}$ fulfills $\widehat{f} = \widehat{g} \circ \widehat{\pi}$. ∎

The G-morphism $\widehat{p} \circ \widehat{p}_1\colon \widehat{P} \times \widehat{F} \to \widehat{M}$ satisfies the condition $(\widehat{p} \circ \widehat{p}_1) \circ \widehat{\tau} = (\widehat{p} \circ \widehat{p}_1) \circ \widehat{p}_1$, so that there exist a G-morphism $\widehat{p}_{\widehat{\Theta}}\colon \widehat{\Theta} \to \widehat{M}$ such that $\widehat{p} \circ \widehat{p}_1 = \widehat{p}_{\widehat{\Theta}} \circ \widehat{\pi}$. The above proof in fact shows that $\widehat{p}_{\widehat{\Theta}}\colon \widehat{\Theta} \to \widehat{M}$ is a locally trivial superbundle with fibre $\widehat{F}$.

Definition 5.1. *The superbundle*

$$\widehat{p}_{\widehat{\Theta}}\colon \widehat{\Theta} \to \widehat{M}$$

is called the associated superfibre bundle (ASFB) with $\widehat{p}\colon \widehat{P} \to \widehat{M}$ with typical fibre $\widehat{F}$ with respect to the given left action of $\widehat{H}$ on $\widehat{F}$.

Given a trivialization $\widehat{\phi}_i: \widehat{P}_{|\widehat{U}_i} \simeq \widehat{U}_i \times \widehat{H}$, of the $\widehat{H}$-PSFB $\widehat{p}: \widehat{P} \to \widehat{M}$ on an open cover $\{U_i\}$ of M, there is an induced trivialization

$$\widehat{\eta}_i: \Theta_{|\widehat{U}_i} \simeq \widehat{U}_i \times \widehat{F}$$

of the ASFB $\widehat{p}_{\widehat{\Theta}}: \widehat{\Theta} \to \widehat{M}$. Moreover, if $\widehat{\psi}_{ij}: \widehat{U}_{ij} \to \widehat{H}$ are the transition morphisms corresponding to the trivialization of $\widehat{p}: \widehat{P} \to \widehat{M}$ (Definition 3.2), the isomorphisms

$$\widehat{\eta}_{ij} = \widehat{\eta}_{i|U_{ij}} \circ (\widehat{\eta}_{j|U_{ij}})^{-1}: \widehat{U}_{ij} \times \widehat{F} \simeq \widehat{U}_{ij} \times \widehat{F} \tag{5.1}$$

are given by

$$\widehat{\eta}_{ij} = (\widehat{p}_1, (\widehat{\psi}_{ij} \circ \widehat{p}_1) \cdot \widehat{p}_2), \tag{5.2}$$

where, as usual, $(\widehat{\psi}_{ij} \circ \widehat{p}_1) \cdot \widehat{p}_2$ denotes the composition $\widehat{\rho} \circ ((\widehat{\psi}_{ij} \circ \widehat{p}_1), \widehat{p}_2)$. In this sense, it can be said that an ASFB has the same transition morphisms as the corresponding $\widehat{H}$-PSFB.

Supervector bundles as associated superbundles. Let us take $\widehat{H} = \widehat{GL_L}[p|q]$, the general linear supergroup over B_L (Example 1.1), and $\widehat{F}$ as the free rank (p,q) B_L-module $B_L^{p|q}$, endowed with its natural structure of a G-supermanifold of dimension $(p+q, p+q)$. If $\widehat{p}: \widehat{P} \to \widehat{M}$ is a $\widehat{GL_L}[p|q]$-PSFB, the ASFB $\widehat{p}_{\widehat{\Theta}}: \widehat{\Theta} \to \widehat{M}$ corresponding to the left action of $\widehat{GL_L}[p|q]$ on $\widehat{F}$ (Example 2.4) is a supervector bundle (Definition IV.3.3) since, by (5.2), the isomorphisms (5.1) are B_L-linear when restricted to the fibres, because they are given by the left action of $GL_L[p|q]$.

This example is typical in the sense that all SVB's are associated superfibre bundles: let us take a rank (p,q) SVB $\widehat{q}: \widehat{\Theta} \to \widehat{M}$ over a G-supermanifold $\widehat{M}$. Then, the superbundle of isomorphisms $\widehat{\pi}: \mathrm{Iso}(\widehat{M} \times B_L^{p|q}, \widehat{\xi}) \to \widehat{M}$ of the trivial SVB with $\widehat{\xi}$ is a principal superfibre bundle with respect to the natural right action of $\widehat{GL_L}[p|q]$ (3.12). One can then consider the superfibre bundle $\widehat{p}_{\widehat{\Theta}}: \widehat{\Theta} \to \widehat{M}$ with typical fibre $B_L^{p|q}$ associated with $\widehat{\pi}: \mathrm{Iso}(\widehat{M} \times B_L^{p|q}, \widehat{\xi}) \to \widehat{M}$ with respect to the left action of $\widehat{GL_L}[p|q]$ defined in Example 2.4.

Proposition 5.2. *There is an isomorphism of SVB's over* $\widehat{M}$

$$\widehat{\Theta} \simeq \widehat{\xi};$$

that is, every SVB $\widehat{q}\colon \widehat{\xi} \to \widehat{M}$ is the ASFB with the $\widehat{GL_L}[p|q]$-PSFB $\widehat{\pi}\colon \mathrm{Iso}(\widehat{M} \times B_L^{p|q}, \widehat{\xi}) \to \widehat{M}$ of typical fibre $B_L^{p|q}$, with respect to the natural left action of $\widehat{GL_L}[p|q]$.

Proof. The G-morphism $\mathrm{Iso}(\widehat{M} \times B_L^{p|q}, \widehat{\xi}) \times B_L^{p|q} \to \widehat{\xi}$ given by (IV.3.10) fits into a commutative diagram

$$\begin{array}{ccc} \mathrm{Iso}(\widehat{M} \times B_L^{p|q}, \widehat{\xi}) \times B_L^{p|q} \times \widehat{GL_L}[p|q] & \xrightarrow{\widehat{\tau}} & \mathrm{Iso}(\widehat{M} \times B_L^{p|q}, \widehat{\xi}) \times B_L^{p|q} \\ \widehat{p}_1 \downarrow & & \downarrow \\ \mathrm{Iso}(\widehat{M} \times B_L^{p|q}, \widehat{\xi}) \times B_L^{p|q} & \longrightarrow & \widehat{\xi} \end{array}.$$

By definition of a quotient, there exists a G-morphism $\widehat{\Theta} \to \xi$ that commutes with the natural projections onto $\widehat{M}$. It remains only to prove that this morphism is an isomorphism of SVB's; we can assume that $\widehat{\xi}$ is the trivial SVB $\widehat{M} \times B_L^{p|q}$. In this case, $\mathrm{Iso}(\widehat{M} \times B_L^{p|q}, \widehat{\xi})$ is the trivial $\widehat{GL_L}[p|q]$-PSFB $\widehat{M} \times \widehat{GL_L}[p|q]$, and one easily concludes. ■

EXAMPLE 5.1. Let $\widehat{M}$ be a G-supermanifold of dimension (m, n), and let $Fr(M, \mathcal{A}_{\widehat{M}}) \to \widehat{M}$ be the superbundle of graded frames (cf. Example 3.4), that is a $\widehat{GL_L}[m|n]$-PSFB. When n is even, we consider the left action $\widehat{GL_L}[p|q] \times B_L^{1|0} \to B_L^{1|0}$ defined as the multiplication by the Berezinian (cf. Example 2.5), while when n is odd we consider the analogous action $\widehat{GL_L}[p|q] \times B_L^{0|1} \to B_L^{0|1}$. The corresponding ASFB, denoted by $\mathrm{Ber}\,\widehat{M} \to \widehat{M}$, is a superline bundle either of rank $(1, 0)$ or $(0, 1)$, depending on the parity of n, and is called the *Berezinian bundle* of the G-supermanifold $\widehat{M}$ (cf. [**LeY, HeM2**]). ▲

The adjoint superbundle. A remarkable example of ASFB is the *adjoint superbundle* associated with a given $\widehat{H}$-PSFB; in this case the fibre is the Lie superalgebra $\mathfrak{h}$ of $\widehat{H}$ and the action of $\widehat{H}$ on it is the adjoint representation of $\widehat{H}$ over $\mathfrak{h}$.

Definition 5.2. *The adjoint superbundle of an $\widehat{H}$-PSFB $\widehat{p}\colon \widehat{P} \to \widehat{M}$ is the ASFB*

$$\widehat{q}\colon \widehat{\mathrm{Ad}}(\widehat{P}) \to \widehat{M}$$

with typical fibre $\widehat{\mathfrak{h}}$, taken with respect to the adjoint representation $\widehat{\mathrm{Ad}}\colon \widehat{H} \times \widehat{\mathfrak{h}} \to \widehat{\mathfrak{h}}$ (Definition 2.8).

If a trivialization $\widehat{\phi}_i: \widehat{P}_{|\widehat{U}_i} \simeq \widehat{U}_i \times \widehat{H}$ of $\widehat{p}: \widehat{P} \to \widehat{M}$ on an open cover $\{U_i\}$ of M with transition morphisms $\widehat{\psi}_{ij}: \widehat{U}_{ij} \to \widehat{H}$ is given, the corresponding trivialization of $\widehat{q}: \widehat{\mathrm{Ad}}(\widehat{P}) \to \widehat{M}$ is described by $\widehat{\eta}_i: \widehat{\mathrm{Ad}}(\widehat{P})_{|\widehat{U}_i} \simeq \widehat{U}_i \times \widehat{\mathfrak{h}}$, where, according to (5.2), the isomorphisms

$$\widehat{\eta}_{ij} = \widehat{\eta}_{i|U_{ij}} \circ (\widehat{\eta}_{j|U_{ij}})^{-1}: \widehat{U}_{ij} \times \widehat{\mathfrak{h}} \simeq \widehat{U}_{ij} \times \widehat{\mathfrak{h}}$$

are given by

$$\widehat{\eta}_{ij} = (\widehat{p}_1, \widehat{\mathrm{ad}} \circ (\widehat{\psi}_{ij} \circ \widehat{p}_1, \widehat{p}_2)) .$$

These morphisms are linear when restricted to the fibres, so that $\widehat{q}: \widehat{\mathrm{Ad}}(\widehat{P}) \to \widehat{M}$ is an SVB.

Let us describe the isomorphisms $\widehat{\eta}_{ij}: \widehat{U}_{ij} \times \widehat{\mathfrak{h}} \simeq \widehat{U}_{ij} \times \widehat{\mathfrak{h}}$ for this SVB, or equivalently, the corresponding isomorphisms of free $\mathcal{A}_{\widehat{U}_{ij}}$-modules, $\widehat{\Lambda}_{ij}: \mathcal{A}_{\widehat{U}_{ij}} \otimes \mathfrak{h} \to \mathcal{A}_{\widehat{U}_{ij}} \otimes \mathfrak{h}$. If we consider the isomorphism $\widehat{\zeta}_{ij}: \widehat{U}_{ij} \times \widehat{H} \simeq \widehat{U}_{ij} \times \widehat{H}$ of relative G-supermanifolds defined by $\widehat{\zeta}_{ij} = (\widehat{p}_1, \mathrm{Ad} \circ (\widehat{\psi}_{ij} \circ \widehat{p}_1, \widehat{p}_2))$ then, by the very definition of the adjoint representation, one has that

$$\widehat{\Lambda}_{ij}(f \otimes X) = (f \otimes X) \circ \widehat{\zeta}^*_{ij} ,$$

where the elements $X \in \mathfrak{h}$ are considered as graded tangent vectors $X: \mathcal{A}_{\widehat{H}} \to B_L$ at the unit point.

Our next aim is to give an alternative description of the adjoint superfibre bundle. To do that, let us recall the relationship between the Lie superalgebra $\mathfrak{h} = T_e\widehat{H}$ and the vertical $\widehat{H}$-invariant vector fields on a trivial PSFB, which is given, according to (4.1), by the isomorphism

$$\begin{aligned} \gamma: \mathcal{A}_{\widehat{U}_{ij}} \otimes \mathfrak{h} &\to (p_* \mathcal{V}er \mathcal{A}_{\widehat{U}_{ij} \times \widehat{H}})^{\widehat{H}} \\ f \otimes X &\mapsto f \otimes (X \otimes \mathrm{Id}) \circ \widehat{m}^* \end{aligned} \quad . \tag{5.3}$$

Then one has:

Lemma 5.1. *There is a commutative diagram of isomorphisms of $\mathcal{A}_{\widehat{U}_{ij}}$-mod-*

ules

$$\begin{array}{ccc} \mathcal{A}_{\widehat{U}_{ij}} \otimes \mathfrak{h} & \xrightarrow{\widehat{\Lambda}_{ij}} & \mathcal{A}_{\widehat{U}_{ij}} \otimes \mathfrak{h} \\ \gamma \downarrow & & \gamma \downarrow \\ (p_* \mathcal{V}er \mathcal{A}_{\widehat{U}_{ij} \times \widehat{H}})^{\widehat{H}} & \xrightarrow{\check{\phi}_{ij}} & (p_* \mathcal{V}er \mathcal{A}_{\widehat{U}_{ij} \times \widehat{H}})^{\widehat{H}} \end{array} , \tag{5.4}$$

where $\check{\phi}_{ij} \cdot (D) = (\widehat{\phi}_{ij}^{-1})^* \circ D \circ \widehat{\phi}_{ij}^*$ *is the isomorphism (4.4) induced by the* $\widehat{H}$*-PSFB isomorphism* $\widehat{\phi}_{ij}$.

Proof. Let us start by describing the G-morphisms $\widehat{\phi}_{ij}$ and $\widehat{\zeta}_{ij}$ as compositions of morphisms by means of the commutative diagrams

$$\begin{array}{ccc} \widehat{U}_{ij} \times \widehat{H} & \xrightarrow{\widehat{\phi}_{ij}} & \widehat{U}_{ij} \times \widehat{H} \\ \widehat{\Delta} \times \mathrm{Id} \downarrow & & \uparrow \mathrm{Id} \times \widehat{m} \\ \widehat{U}_{ij} \times \widehat{U}_{ij} \times \widehat{H} & \xrightarrow{\mathrm{Id} \times \psi_{ij} \times \mathrm{Id}} & \widehat{U}_{ij} \times \widehat{H} \times \widehat{H} \end{array}$$

and

$$\begin{array}{ccc} \widehat{U}_{ij} \times \widehat{H} & \xrightarrow{\widehat{\zeta}_{ij}} & \widehat{U}_{ij} \times \widehat{H} \\ \widehat{\Delta} \times \mathrm{Id} \downarrow & & \uparrow \mathrm{Id} \times \widehat{\mathrm{Ad}} \\ \widehat{U}_{ij} \times \widehat{U}_{ij} \times \widehat{H} & \xrightarrow{\mathrm{Id} \times \psi_{ij} \times \mathrm{Id}} & \widehat{U}_{ij} \times \widehat{H} \times \widehat{H} \end{array} ;$$

in this way, the morphisms $\widehat{\phi}_{ij}^*$ and $\widehat{\zeta}_{ij}^*$ are easily computed. Since, by (2.7),

$$\widehat{\zeta}_{ij}^*(f' \otimes h) = \sum_{kj} (-1)^{|h_k{}^j||h_{kj}|} f' \widehat{\psi}_{ij}^*(h^k) \widehat{\psi}_{ij}^*(\widehat{s}^* h_{kj}) h_k{}^j ,$$

one obtains the equation

$$\widehat{\Lambda}_{ij}(f \otimes X)(f' \otimes h) = \sum_{kj} \epsilon_{kj}(X, f', h) f f' \widehat{\psi}_{ij}^*(h^k) \widehat{\psi}_{ij}^*(\widehat{s}^* h_{kj}) X(h_k{}^j) ,$$

where $\epsilon_{kj}(X, f', h) = (-1)^{|X|(|f'|+|h^k|+|h_{kj}|)+|h_k{}^j||h_{kj}|}$. Furthermore, $\check{\phi}_{ij} \cdot \gamma(f \otimes$

X) is given by

$$[(\widehat{\phi}_{ij}^{-1})^* \circ (f \otimes (X \otimes \mathrm{Id}) \circ \widehat{m}^*)) \circ \widehat{\phi}_{ij}^*](f' \otimes h)$$
$$= \sum_{kjl} (-1)^{|X|(|f'|+|h^k|)} f f' \widehat{\psi}_{ij}(h^k) X(h_k{}^j) \widehat{\psi}_{ij}(s^* h_{kj}{}^l) \otimes h_{kjl} \quad .$$

The inverse γ^{-1} of γ is $\mathrm{Id} \otimes \widehat{e}^*$, and so,

$$(\gamma^{-1}[\breve{\phi}_{ij} \cdot \gamma(f \otimes X)])(f' \otimes h)$$
$$= \sum_{kjl} (-1)^{|X|(|f'|+|h^k|)} f f' \widehat{\psi}_{ij}(h^k) X(h_k{}^j) \widehat{\psi}_{ij}(s^* h_{kj}{}^l) \widehat{e}^*(h_{kjl}) \quad .$$

However, from $(\widehat{s} \times \widehat{e}) \circ \widehat{m} = \mathrm{Id}$, we have $\widehat{s}^*(h_{kj}) = \sum_l \widehat{s}^*(h_{kj}{}^l) \widehat{e}^*(h_{kjl})$, which completes the proof. ■

Proposition 5.3. *The adjoint superbundle $\widehat{q}\colon \widehat{\mathrm{Ad}}(\widehat{P}) \to \widehat{M}$ is the SVB associated with the rank (m,n) locally free $\mathcal{A}_{\widehat{M}}$-module $(p_* \mathcal{V}er \mathcal{A}_{\widehat{P}})^{\widehat{H}}$ of $\widehat{H}$-invariant vertical graded vector fields.*

Proof. Since the ASFB $\widehat{q}\colon \widehat{\mathrm{Ad}}(\widehat{P}) \to \widehat{M}$ is the graded locally ringed space obtained by glueing the trivial SVB's $\widehat{U}_{ij} \times \mathfrak{h}$ by means of the isomorphisms $\widehat{\eta}_{ij}$, the $\mathcal{A}_{\widehat{M}}$-module of sections of the adjoint superbundle is the sheaf $\mathcal{F}$ obtained by glueing the corresponding sheaves of sections $\mathcal{A}_{\widehat{U}_{ij}} \otimes \mathfrak{h}$ by means of the sheaf isomorphisms $\widehat{\Lambda}_{ij}$. By the previous lemma, $\mathcal{F}$ is isomorphic with the $\mathcal{A}_{\widehat{M}}$-module $\mathcal{F}'$ obtained by glueing the sheaves $(p_* \mathcal{V}er \mathcal{A}_{\widehat{U}_i \times \widehat{H}})^{\widehat{H}}$ through the sheaf isomorphisms $\breve{\phi}_{ij}$. Then, the sheaf isomorphisms

$$\breve{\phi}_i\colon (p_* \mathcal{V}er \mathcal{A}_{\widehat{P}})^{\widehat{H}}{}_{|\widehat{U}_i} \simeq (p_* \mathcal{V}er \mathcal{A}_{\widehat{U}_i \times \widehat{H}})^{\widehat{H}}$$
$$D \mapsto \breve{\phi}_i \cdot D = (\widehat{\phi}_i^{-1})^* \circ D \circ \widehat{\phi}_i^*$$

obtained from $\widehat{\phi}_i\colon \widehat{P}_{|\widehat{U}_i} \simeq \widehat{U}_i \times \widehat{H}$ fulfill the condition $\breve{\phi}_{ij} = \breve{\phi}_{i|U_{ij}} \circ (\breve{\phi}_{j|U_{ij}})^{-1}$, thus defining an isomorphism of $\mathcal{A}_{\widehat{M}}$-modules

$$\mathcal{F}' \simeq (p_* \mathcal{V}er \mathcal{A}_{\widehat{P}})^{\widehat{H}}$$

as claimed. ■

Bibliography

Arnowitt, R., Nath, P. & Zumino, B.

[**ANZ**] Superfield densities and action principle in superspace, *Phys. Lett.* **56B** (1975) 81–84.

Atiyah, M.F.

[**Ati**] Complex analytic connections in fibre bundles, *Trans. Amer. Math. Soc.* **85** (1957), 181–207.

Atiyah, M.F. & Macdonald, I.G.

[**AtM**] *Introduction to commutative algebra,* Addison-Wesley, Reading, MA, 1969.

Bartocci, C.

[**Ba**] *Elementi di geometria globale delle supervarietà,* Ph.D. Thesis, University of Milano, 1990.

Bartocci, C. & Bruzzo, U.

[**BB1**] Some remarks on the differential-geometric approach to supermanifolds, *J. Geom. Phys.* **4** (1987) 391–404.

[**BB2**] Cohomology of supermanifolds, *J. Math. Phys.* **28** (1987) 2363–2368.

[**BB3**] Cohomology of the structure sheaf of real and complex supermanifolds, *J. Math. Phys.* **29** (1988) 1789–1795.

[**BB4**] Existence of connections on superbundles, *Lett. Math. Phys.* **17** (1989) 61–68.

[**BB5**] Super line bundles, *Lett. Math. Phys.* **17** (1989) 263–274.

[**BB6**] On DeWitt supermanifolds and their Picard variety, *C. R. Acad. Sci. Paris Sér. I Math.* **309** (1989) 75–80.

Bartocci, C., Bruzzo, U. & Hernández Ruipérez, D.

[**BBH**] A remark on a new category of supermanifolds, *J. Geom. Phys.* **6** (1989) 509–516.

Bartocci, C., Bruzzo, U., Hernández Ruipérez, D. & Pestov, V.G.

[**BBHP1**] Ob aksiomaticheskom podhode k supermnogoobraziyam, submitted to *Dokl. Acad. Nauk* (in Russian); English translation: On an axiomatic approach to supermanifolds, Preprint 180/1991, Dipartimento di Matematica, Università di Genova

[BBHP2] On an axiomatics for supermanifolds, Preprint 190/1991, Dipartimento di Matematica, Università di Genova

BARTOCCI, C., BRUZZO, U. & LANDI, G.

[BBL] Geometry of standard constraints and Weil triviality in supersymmetric gauge theories, *Lett. Math. Phys.* **18** (1989) 235–245.

BATCHELOR, M.

[Bch1] The structure of supermanifolds, *Trans. Amer. Math. Soc.* **253** (1979) 329–338.

[Bch2] Two approaches to supermanifolds, *Trans. Amer. Math. Soc.* **258** (1980) 257–270.

BECCHI, C., ROUET, A. & STORA, R.

[BcRS1] Renormalization of gauge theories, *Ann. Phys. (N.Y.)* **98** (1976) 287–321.

[BcRS2] Gauge field models and renormalized models with broken symmetries, in: *Renormalization of gauge theories*, Erice Lectures 1975, G. Velo and A.S. Wightman eds., *NATO ASI Series C* **132**, Reidel, Dordrecht, 1976.

BEREZIN, F.A.

[Be] *The method of second quantization,* Academic Press, New York, 1966.

BEREZIN, F.A. & KATS, G.I.

[BK] Lie groups with commuting and anticommuting parameters, *Math. USSR Sb.* **11** (1970) 311–325.

BEREZIN, F.A. & LEĬTES, D.A.

[BL] Supermanifolds, *Soviet Math. Dokl.* **16** (1975) 1218–1222.

BLATTNER, R.J. & RAWNSLEY, J.H.

[BlR] Remarks on Batchelor's theorem, in: *Mathematical aspects of superspace*, H.J. Seifert, C.J.S. Clarke and A. Rosenblum eds., *NATO ASI Ser. C* **132**, Reidel, Dordrecht, 1984, pp. 161–171.

BLYTH, T.S.

[Bly] *Module theory*, Clarendon Press, Oxford, 1977.

BONORA, L., PASTI, P. & TONIN, M.

[BoPT1] Supermanifolds and BRS transformations, *J. Math. Phys.* **23** (1982) 839–845.

[BoPT2] Chiral anomalies in higher dimensional supersymmetric theories, *Nucl. Phys. B* **286** (1987) 150–174.

BOURBAKI, N.

[Bou] *Eléments de mathématique. Algèbre I (Chapitres 1 à 3)*, Hermann, Paris, 1970.

BOYER, C.P. & GITLER, S.

[BoyG] The theory of G^∞ supermanifolds, *Trans. Amer. Math. Soc.* **285** (1984) 241–261.

BREDON, G.E.

[Bre] *Sheaf theory*, McGraw-Hill, New York, 1967.

BRUZZO, U.

[Bru] Supermanifolds, supermanifold cohomology, and super vector bundles, in: *Differential geometric methods in theoretical physics*, K. Bleuler and M. Werner eds., *NATO ASI Ser. C* **250**, Kluwer, Dordrecht, 1988, pp. 417–440.

BRUZZO, U. & HERNÁNDEZ RUIPÉREZ, D.

[BruH] Characteristic classes of super vector bundles, *J. Math. Phys.* **30**(1989) 1233–1237.

BRUZZO, U. & LANDI, G.

[BruL] Simple proof of Weil triviality in supersymmetric gauge theories, *Phys. Rev. D* **39** (1989) 1174–1178.

BRYANT, P.

[Bry1] GH^∞ supermanifolds, *J. Math. Phys.* **29** (1988) 1575–1579.

[Bry2] The structure of de Witt supermanifolds, *Rivista Mat. Pura Appl.* N. 3 (1988) 57–79.

BUCKINGHAM, S.

[Buc] Weil triviality and anomalies in two-dimensional supergravity, *Class. Quantum Grav.* **5** (1988) 1615–1625.

CATENACCI, R., REINA, C. & TEOFILATTO, P.

[CaRT] On the body of supermanifolds, *J. Math. Phys.* **26** (1985) 671–674.

CORWIN, L., NE'EMAN, Y. & STERNBERG, S.

[CNS] Graded Lie algebras in mathematics and physics (Bose-Fermi symmetry), *Rev. Modern Phys.* **47** (1975) 573–603.

DELIGNE, P.

[Del] *Equations différentielles à points singuliers réguliers*, Lecture Notes Math. **163**, Springer-Verlag, Berlin, 1970.

DELL, J. & SMOLIN, L.

[DeS] Graded manifold theory as the geometry of supersymmetry, *Commun.*

Math. Phys. **66** (1979) 197–221.

DE RHAM, G.

[DR] *Variétés différentiables*, Hermann, Paris, 1955.

DEWITT, B.

[DW] *Supermanifolds*, Cambridge University Press, London, 1984.

FAYET, P. FERRARA, S

[FF] Supersymmetry, *Physics Report* **32** (1977) 249–334.

FREUND, P.G.O.

[Fre] *Introduction to supersymmetry*, Cambrige University Press, Cambridge, 1986.

FRIEDAN, D.

[Fri] Notes on string theory and two conformal field theories, in: *Unified string theories*, M. Green and D. Gross eds., World Scientific, Singapore, 1986.

FRÖLICHER, A. & NIJENHUIS, A.

[FröN] Theory of vector-valued differential forms. Parts I–II, *Indag. Math.* **29** (1956) 338–359.

GATES, S.J., GRISARU, M.T., ROČEK, M. & SIEGEL, W.

[GGRS] *Superspace, or one thousand and one lectures in supersymmetry*, Benjamin/Cummings, Reading (Mass.), 1983.

GELFAND, I.M., RAIKOV, A.D. & CHILOV, G.E.

[GlRC] *Les anneaux normés commutatifs*, Gauthier-Villars, Paris, 1964

GIDDINGS, S. & NELSON, P

[GN] Line bundles on super Riemann surfaces, *Commun. Math. Phys.* **118** (1988) 289–302.

GODEMENT, R.

[Go] *Théorie des faisceaux*, Hermann, Paris, 1964.

GREEN, P.

[Gre] On holomorphic graded manifolds, *Proc. Amer. Math. Soc.* **85** (1982) 587–590.

GRIFFITHS, P. & HARRIS, J.

[GrH] *Principles of algebraic geometry*, Wiley-Interscience, New York, 1978.

GROTHENDIECK, A.

[Gro1] Produits tensoriels topologiques et espaces nucléaires, *Mem. Amer. Math. Soc.* **16** (1955).

[Gro2] Sur quelques points d'algèbre homologique, *Tôhoku Math. J.* **9** (1957) 119–221.

[Gro3] La théorie des classes de Chern, *Bull. Soc. Math. de France* **86** (1958) 137–154.

GROTHENDIECK, A. & DIEUDONNÉ, J.

[GroD] *Eléments de géometrie algébrique. I,* Grundlehren Math. Wiss. **166**, Springer-Verlag, Berlin, 1971.

HARTSHORNE, R.

[Har] *Algebraic geometry,* Springer-Verlag, New York, 1977.

HERNÁNDEZ RUIPÉREZ, D. & MUÑOZ MASQUÉ, J.

[HeM1] Global variational calculus on graded manifolds, I: graded jet bundles, structure 1-form and graded infinitesimal contact transformations, *J. Math. Pures Appl.* **63** (1984) 283–309.

[HeM2] Construction intrinsèque du faisceau de Berezin d'une variété graduée *C. R. Acad. Sc. Paris* **301** (1985) 915–918.

HILTON, P.J. & STAMMBACH, U.

[HiS] *A course in homological algebra,* Springer-Verlag, New York, 1971.

HIRZEBRUCH, F.

[Hirz] *Topological methods in algebraic topology,* Grundlehren Math. Wiss. **131**, Springer-Verlag, Berlin, 1966.

HODGKIN, L.

[Hod] Problems of fields on super Riemann surfaces, *J. Geom. Phys.* **6** (1989) 334–348.

HOYOS, J., QUIRÓS, M., RAMÍREZ MITTELBRUNN, J. & DE URRÍES, F.J.

[HQ1] Generalized supermanifolds. III. ρ-supermanifolds, *J. Math. Phys.* **25** (1984) 847–854.

[HQ2] Superfield formalism for gauge theories with ghosts and gauge-fixing terms from dynamics on superspace, *Nucl. Phys. B* **218** (1983) 159–172, and references therein.

HUMPHREYS, J.E.

[Hum] *Linear algebraic groups,* Graduate Texts in Mathematics **21**, Springer-Verlag, New York, 1975.

HUSEMOLLER, D.

[Hus] *Fibre bundles,* McGraw-Hill, New York, 1966.

JADCZYK, A. & PILCH, K.

[JP] Superspace and supersymmetries, *Commun. Math. Phys.* **78** (1981) 373–390.

KAC, V.G.

[Ka1] Lie superalgebras, *Adv. in Math.* **26** (1977) 8–96.

[Ka2] A sketch of Lie superalgebra theory, *Commun. Math. Phys.* **53** (1977) 31–64.

KASHIWARA, M. & SCHAPIRA, P.

[KaS] *Sheaves on manifolds,* Grundlehren Math. Wiss. **292**, Springer-Verlag, Berlin, 1990.

KOBAYASHI, S. & NOMIZU, K.

[KN] *Foundations of differential geometry. Vol I,* Interscience Publ., New York, 1963.

KOSTANT, B.

[Kos] Graded manifolds, graded Lie theory, and prequantization, in: *Differential geometric methods in mathematical physics,* K. Bleuler and A. Reetz eds., Lecture Notes Math. **570**, Springer-Verlag, Berlin, 1977, pp. 177–306.

KOSZUL, J.

[Ksz] *Lectures on fibre bundles and differential geometry,* Tata Institute, Bombay, 1960.

LEĬTES, D.A.

[Leĭ] Introduction to the theory of supermanifolds, *Russian Math. Surveys,* **35** (1980) 1–64.

LÓPEZ ALMOROX, A.

[Lop] Supergauge theories in graded manifolds, in *Differential geometric methods in mathematical physics,* P. L. García and A. Pérez-Rendón eds., Lecture Notes Math. **1251**, Springer-Verlag, Berlin, 1987, pp. 114–136.

MANIN, YU.I.

[Ma1] Quantum strings and algebraic curves, in: *Proceedings of the International Congress of Mathematicians, August 3–11, 1986,* Amer. Math. Soc., Providence (RI), 1987, pp. 1286-1295.

[Ma2] *Gauge field theory and complex geometry,* Grundlehren Math. Wiss. **289**, Springer-Verlag, Berlin, 1988 (original Russian edition 1984).

MATSUMOTO, S. & KAKAZU, K.

[MK] A note on topology of supermanifolds, *J. Math. Phys.* **27** (1986) 2960–2962.

MILNOR, J.W. & STASHEFF, J.D.

[MiS] *Characteristic classes,* Annals of Mathematical Studies **76**, Princeton University Press, Princeton, 1974.

MOLOTKOV, V.

[Mol] Infinite-dimensional $\mathbb{Z}_2^k$-supermanifolds, Preprint ICTP/84/183, Trieste,

1984.

NAIMARK, M.A.

[Nai] *Normed algebras*, Wolters-Noordhoff Publ., Groningen, 1972.

NĂSTĂSESCU, C. & VAN OYSTAEYEN F.

[NVO] *Graded ring theory*, North-Holland, Amsterdam, 1982.

PENKOV, I.B.

[Pen] *Classical Lie supergroups and Lie superalgebras and their representations*, Prépublication de l'Institut Fourier **117**, Grenoble, 1988.

PIETSCH, A.

[Pie] *Nuclear locally convex spaces, Ergebnisse der Mathematik und ihrer Grenzgebiete* **66**, Springer-Verlag, Berlin, 1972.

RABIN, J.M.

[Ra] Supermanifold cohomology and the Wess-Zumino term of the covariant superstring action, *Commun. Math. Phys.* **108** (1987) 375–389.

RABIN, J. & CRANE, L.

[RC1] How different are the supermanifolds of Rogers and DeWitt?, *Commun. Math. Phys.* **102** (1985) 123–137.

[RC2] Global properties of supermanifolds, *Commun. Math. Phys.* **100**(1985) 141–160.

RITTENBERG, V. & SCHEUNERT, M.

[RiS] Elementary construction of graded Lie groups, *J. Math. Phys.* **19** (1978) 709–713.

ROBERTSON, A.P. & ROBERTSON, W.

[RR] *Topological vector spaces*, Cambridge Univ. Press, Cambridge, 1973.

ROGERS, A.

[Rs1] A global theory of supermanifolds, *J. Math. Phys.* **21** (1980) 1352–1365.

[Rs2] Graded manifolds, supermanifolds and infinite-dimensional Grassmann algebras, *Commun. Math. Phys.* **105** (1986) 375–384.

[Rs3] Some examples of compact supermanifolds with non-Abelian fundamental group, *J. Math. Phys.* **22** (1981) 443–444.

[Rs4] On the existence of global integral superforms on supermanifolds, *J. Math. Phys.* **26** (1985) 2749–2753.

ROSLY, A.A., SCHWARZ, A.S. & VORONOV, A.A.

[RSV1] Geometry of superconformal manifolds, *Commun. Math. Phys.* **119** (1988) 129–152.

[RSV2] Superconformal geometry and string theory, *Commun. Math. Phys.* **120** (1989) 437–458.

ROTHSTEIN, M.

[Rt1] Deformations of complex supermanifolds, *Trans. Amer. Math. Soc.* **95** (1985) 255–260.

[Rt2] The axioms of supermanifolds and a new structure arising from them, *Trans. Amer. Math. Soc.* **297** (1986) 159–180.

SALAM, A. & STRATHDEE, J.

[SaS] Super-gauge transformations, *Nucl. Phys. B* **76** (1974) 477–482.

SCHEUNERT, M.

[Sch] *The theory of Lie superalgebras*, *Lecture Notes Math.* **716**, Springer-Verlag Berlin, 1979.

SCHMITT, T.

[Scm] Infinite-dimensional supermanifolds. I, Report R-Math-08/88, Akademie der Wissenschaften der DDR, Berlin, 1988.

SERRE, J.-P.

[Ser] Un théorème de dualité, *Comment. Math. Helv.* **29** (1955) 9–26.

SINGER, I.M.

[Sin] *String theory for mathematicians*, unpublished lecture notes, Ecole Normale Supérieure, Paris, 1987.

SPANIER, E.H.

[Spa] *Algebraic topology,* McGraw-Hill, New York, 1966.

TENNISON, B.R.

[Ten] *Sheaf theory, London Math. Soc. Lecture Note Ser.* **20**, Cambridge University Press, Cambridge, 1975.

VAISMAN, I.

[Vai] *Cohomology and differential forms*, M. Dekker, New York, 1973.

VLADIMIROV, V.S. & VOLOVICH, I.V.

[VlV] Superanalysis – I. Differential calculus, *Theor. Math. Phys.* **60** (1984) 317–335.

VOLKOV, D.V. & AKULOV, V.P.

[VoA] Is the neutrino a Goldstone particle?, *Phys. Lett. B* **46** (1973) 109–110.

WARNER, F.W.

[War] *Foundations of differentiable manifolds and Lie groups,* *Graduate Text in Mathematics* **95**, Springer-Verlag, New York, 1987.

WATERHOUSE, W.C.

[Wat] *Introduction to affine group schemes, Graduate Texts in Mathematics* **66**, Springer-Verlag, New York, 1979.

WELLS, R.O. JR.

[Wel] *Differential analysis on complex manifolds, Graduate Texts in Mathematics* **65**, Springer-Verlag, New York, 1980.

WESS, J. & BAGGER, J.

[WsB] *Supersymmetry and supergravity,* Princeton University Press, Princeton (NJ), 1984.

WESS, J. & ZUMINO, B.

[WZ] Supergauge transformations in four dimensions, *Nucl. Phys. B* **70** (1974) 39–50.

Index